Atlas of Minerals and Igneous and Metamorphic Rocks in Thin-Section

The *Atlas of Minerals and Igneous and Metamorphic Rocks in Thin-Section* provides the geology student and geoscientist with a stunning new color atlas of the main rock-forming minerals and igneous and metamorphic rocks in thin-section. It showcases minerals in various settings and degrees of alteration and preservation to allow users to best identify their own specimens in practice. Chapter 1 highlights the distinctive characteristics used to identify different minerals. Building on this base, following chapters describe rock textures and types, summarizing their petrogenesis within a plate tectonic framework. This book also includes insights into how additional information from petrographic thin-sections can be obtained using modern analytical methods to increase our understanding of geological processes. The *Atlas* is an indispensable reference textbook for all facilities that use a petrographic microscope, for professional geoscientists, and as an aid for any student studying minerals and rocks.

Alessandro Da Mommio completed his PhD in geology at the University of Milan in 2015. With a long-standing expertise in petrography since his undergraduate studies, he has developed a distinguished website dedicated to his specialty: the petrography of minerals and rocks. In 2014, one of his images secured second place in the Nikon Small World Photomicrography Competition. He's currently teaching science and mathematics.

Victoria Pease is Professor of Tectonics and Magmatism at Stockholm University. Her research focuses on Arctic tectonics. She has been teaching the next generation of geologists for more than 20 years. She sits on a number of international scientific committees and is Chief Editor of *Precambrian Research*. Victoria is co-author (with Hugh Rollinson) of the reference textbook *Using Geochemical Data: To Understand Geological Processes, Second Edition* (2021, Cambridge University Press).

"*Atlas of Minerals and Igneous and Metamorphic Rocks in Thin-Section* by A. Da Mommio and V. Pease is a well-researched and easy-to-use reference. Fundamental characteristics of minerals and rocks in thin-section are combined with high-quality images that will be useful to beginners and experts alike. Thin-sections are the windows into igneous and metamorphic processes, and this atlas will become an essential component of the petrologist's toolbox."

Professor Wendy Bohrson, *Colorado School of Mines*

"The atlas contains a wealth of spectacular color photomicrographs that beautifully illustrate the essential information about minerals and rocks and provide a visual basis for identifying their textures and fabrics. Information boxes explain key aspects of mineralogy and petrology in greater detail, and application boxes provide examples of advanced techniques and broader interpretations of rock and mineral data. An indispensable reference for all geoscientists."

Professor Carol D. Frost, *University of Wyoming*

"Observing rocks and minerals through a petrological microscope offers a captivating experience that continues to engage both students and researchers. While each thin-section reveals a unique glimpse into the origin and formation of Earth's building blocks, a systematic approach is essential to unravel this information. *Atlas of Minerals and Igneous and Metamorphic Rocks in Thin-Section* by Alessandro Da Mommio and Victoria Pease introduces the remarkable optical shifts that occur when light rays of different wavelengths pass through crystal structures. The book explores the optical properties of common minerals found in igneous and metamorphic rocks, along with their microscopic textural relationships, enhanced by numerous beautiful color images. It will serve as a valuable resource for those beginning their studies in petrology."

Professor Sally A. Gibson, *University of Cambridge*

"This book is the unmistakable descendent of the classic mineralogy text of Deer, Howie, and Zussman, but is richly illustrated with beautiful photomicrographs of the minerals themselves. Collectors, students, and professionals will all enjoy and benefit from it."

Professor Robert J. Stern, *University of Texas at Dallas*

Atlas of Minerals and Igneous and Metamorphic Rocks in Thin-Section

Alessandro Da Mommio
University of Milan

Victoria Pease
Stockholm University

Shaftesbury Road, Cambridge CB2 8EA, United Kingdom

One Liberty Plaza, 20th Floor, New York, NY 10006, USA

477 Williamstown Road, Port Melbourne, VIC 3207, Australia

314–321, 3rd Floor, Plot 3, Splendor Forum, Jasola District Centre, New Delhi – 110025, India

103 Penang Road, #05-06/07, Visioncrest Commercial, Singapore 238467

Cambridge University Press is part of Cambridge University Press & Assessment, a department of the University of Cambridge.

We share the University's mission to contribute to society through the pursuit of education, learning and research at the highest international levels of excellence.

www.cambridge.org
Information on this title: www.cambridge.org/9781009100229

DOI: 10.1017/9781009110020

When citing this work, please include a reference to the DOI 10.1017/9781009110020

First published 2025

Printed in the United Kingdom by CPI Group Ltd, Croydon CR0 4YY

A catalogue record for this publication is available from the British Library.

Library of Congress Cataloging-in-Publication Data
Names: Da Mommio, Alessandro, author. | Pease, Victoria, author.
Title: Atlas of minerals and igneous and metamorphic rocks in thin-section / Alessandro Da Mommio, Victoria Pease.
Description: Cambridge ; New York, NY : Cambridge University Press, 2024. | Includes bibliographical references and index.
Identifiers: LCCN 2023054266 (print) | LCCN 2023054267 (ebook) | ISBN 9781009100229 (hardback) | ISBN 9781009112055 (paperback) | ISBN 9781009110020 (ebook)
Subjects: LCSH: Thin sections (Geology) | Rocks–Identification. | Rock-forming minerals–Identification. | Mineralogy, Determinative.
Classification: LCC QE434 .D36 2024 (print) | LCC QE434 (ebook) | DDC 552/.8–dc23/eng/20240105
LC record available at https://lccn.loc.gov/2023054266
LC ebook record available at https://lccn.loc.gov/2023054267

ISBN 978-1-009-10022-9 Hardback
ISBN 978-1-009-11205-5 Paperback

Additional resources for this publication at www.cambridge.org/atlas.

Contents

List of Information Boxes ix
List of Applications x
Preface xi
Acknowledgments xii

1 Rock-Forming Minerals 1

1.1 Introduction **2**

1.2 Framework Silicates **10**

The Feldspars

Plagioclase Feldspar 10

Alkali Feldspar

Anorthoclase 12
Microcline 13
Orthoclase 14
Sanidine 16

The Feldspathoids

Kalsilite 18
Leucite 19
Nepheline 21
Scapolite 23

The Sodalites

Sodalite 24
Nosean 24
Haüyne 24

The Silica Minerals

Coesite 26
Cristobalite 27
Quartz 28
Tridymite 31

1.3 Sheet Silicates **36**

The Micas

Biotite 36
Muscovite 38
Sericite 40

Chlorite Group 41
Prehnite 43
Pyrophyllite 45

The Serpentine Minerals

Antigorite 46
Chrysotile 47
Lizardite 49

Stilpnomelane 50

Talc 51

1.4 Chain Silicates **53**

The Amphiboles

Anthophyllite 53
Gedrite 53
Cummingtonite–Grunerite 55
Eckermannite–Arfvedsonite 57

Glaucophane–Riebeckite 59
Hornblende 61
Kaersutite–Ferrokaersutite 63
Richterite–Ferrorichterite 65
Katophorite 65
Tremolite–Ferroactinolite 66

The Pyroxenes
Aegirine 68
Aegirine–Augite 68
Augite–Ferroaugite 69
Diopside–Hedenbergite 71
Jadeite 73
Omphacite 74
Pigeonite 76
Enstatite–Ferrosilite 77

Sapphirine 79
Wollastonite 81

1.5 Disilicates, Orthosilicates, and Ring Silicates 84

The Aluminosilicates
Andalusite 84
Kyanite 86
Sillimanite 88

Beryl 90
Chloritoid 91
Cordierite 93

The Epidotes
Allanite 95
Clinozoisite 96
Epidote 97
Piemontite 99
Zoisite 100

The Garnets
Almandine 101
Andradite 103
Grossular–Hydrogrossular 104
Pyrope 105
Spessartine 106
Uvarovite 107

Lawsonite 108

The Melilites
Gehlenite 109
Åkermanite 109

The Olivines
Forsterite–Fayalite 111
Monticellite 113

Pumpellyite 115
Staurolite 117
Titanite 119
Topaz 120
Tourmaline 121
Zircon 122

1.6 Nonsilicates 125

The Carbonates
Aragonite 125
Calcite 126
Dolomite 128
Magnesite 129
Siderite 130

The Halides
Fluorite 131

The Oxides and Hydroxides
Brucite 133
Cassiterite 134
Corundum 137
Perovskite 139
Rutile 141
Spinel Group 143

The Phosphates
Apatite 145
Monazite 147
Xenotime 149

The Sulfates
Barite 150

Bibliography 154

2 Igneous Rocks 157

2.1 Introduction 158

Textures of Igneous Rocks 158

Naming Igneous Rocks 158

2.2 Crystallinity 163

2.3 Granularity 164

Phaneritic 164

Equigranular 164

Aphanitic 168

2.4 Crystal Form 171

Crystal Faces 171

Crystal Shapes 171

2.5 Mutual Relations of Crystals 175

Granular Textures 175

Inequigranular Textures 176

Oriented Textures 182

Banded Textures 187

Intergrowth Textures 190

Radiate Textures 193

Overgrowth Textures 196

Cavity Textures 199

2.6 Plutonic Rocks 201

Ultramafic Rocks

Peridotites

Lherzolite 202

Wehrlite 204

Harzburgite 206

Dunite 208

Pyroxenites

Clinopyroxenite 210

Olivine Clinopyroxenite 212

Websterite 213

Olivine Websterite 214

Orthopyroxenite 215

Olivine Orthopyroxenite 217

Gabbroic Rocks

Diorite 220

Gabbro 222

Norite 224

Troctolite 226

Anorthosite 228

Syenitic Rocks

Monzogabbro–Monzodiorite 230

Monzonite 232

Syenite 234

Alkali Feldspar Syenite 236

Granitic Rocks

Tonalite 239

Granodiorite 240

Granite 244

Alkali Feldspar Granite 246

Feldspathoid Rocks

Foid-Bearing Rocks 248

Foid Rocks 250

Foidolite 252

2.7 Volcanic Rock Types 254

Silica: (Over)saturated

Basalt 254

Basaltic Andesite 256

Andesite 258

Trachyandesite 260

Trachyte 262

Dacite 264

Rhyolite 266

Silica: Undersaturated

Basanite/Tephrite 270

Phonotephrite–Tephriphonolite 272

Phonolite 274

Foidite 276

Glassy Rocks

Obsidian 278

Pitchstone 280

Tuffaceous Rocks, Welded and Unwelded 282

Bibliography 286

3 Metamorphic Rocks 293

3.1 Introduction 294

3.2 Metamorphic Textures 297

Grain Size and Shape 297
- Acicular 297
- Augen 297
- Granoblastic Textures: Decussate 298
- Granoblastic Textures: Polygonal 299
- Porphyroblasts 300
- Porphyroclasts 302
- Strain/Pressure Shadows 303

Deformation Fabrics 305
- Deformation Twins 305
- Foliation and Schistosity 307
- Crenulation Cleavage 309
- Kink Bands 312
- Poikiloblasts and Poikiloclasts 313
- Ribbons 316
- Stylolites 319
- Veins and Strain Fringes 322

Reaction and Disequilibrium Textures 324
- Atoll 324
- Zoning 327

3.3 Metamorphic Facies 330
- Zeolite Facies 333
- Prehnite–Pumpellyite Facies 335
- Greenschist Facies 337
- Amphibolite Facies 339
- Granulite Facies 343
- Blueschist Facies 345
- Eclogite Facies 347
- Hornfels Facies 349

3.4 Metamorphic Rocks 354

Root
- Schist 354
- Gneiss 356
- Granofels 358

Protolith
- Slate 362
- Phyllite 364
- Spilite 366
- Serpentinite 368
- Hornfels 370
- Quartzite 374
- Greenschist 376
- Amphibolite 378
- Granulite 380
- Blueschist 382
- Eclogite 384
- Migmatite 386

Bibliography 389

Index 391

Information Boxes

1.1 Feldspar Solid Solution 17
1.2 Polymorphism 30
1.3 The Zeolites 34
1.4 Olivine Alteration 114
1.5 The Fe–Ti Oxides 140
1.6 The Sulfides 142

2.1 Nucleation and Crystal Growth 167
2.2 Textures Related to Rapid Cooling 170
2.3 Embayment and Resorption 174
2.4 Symplectites and Myrmekites 189
2.5 Exsolution in Igneous Rocks 192
2.6 Magmatic Zoning 195
2.7 Opacite Rims 198
2.8 Describing Igneous Rocks 238
2.9 Pegmatites 249
2.10 Crystal Cargo 257

3.1 Pseudomorphs 304
3.2 Boudinage and Microboudinage 306
3.3 Foliation Terminology 311
3.4 Lineations 315
3.5 Reaction Rims and Epitaxy 325
3.6 Symplectites 326
3.7 Exsolution in Metamorphic Rocks 329
3.8 Metamorphic Zones 332
3.9 Transport and Diffusion 341
3.10 Fluids, Temperature, and Metamorphism 342
3.11 Cataclasite, Mylonite, and Pseudotachylite 372
3.12 Marble 373
3.13 Skarn Formation 388

Applications

1.1	Common Polymorphs	32
1.2	Imaging Minerals in Thin-Section	82
1.3	Thermobarometry	123
1.4	Temperature and Oxygen Fugacity Using Fe–Ti Oxides	135
1.5	Geochronology	151
2.1	Melt Generation from Mantle Rocks	218
2.2	Composition, Zoning, and Magmatic Processes	242
2.3	Crystal Size Distribution Analysis	268
2.4	Crustal Anatexis	284
3.1	Shear Sense Indicators	317
3.2	Stylolites as Strain Markers and Conduits for Diagenetic Fluids	320
3.3	Pressure and Temperature Paths	351
3.4	Time and P–T Paths	360

Preface

Determining what is a "common" versus "uncommon" mineral is somewhat subjective. We tend to be inclusive; for example, insofar as a mineral such as kalsilite may be relatively rare, it is not uncommon in its normal paragenesis – that is, unusual ultra-high-K rocks – and so we include it here. Given that our intended audience is not just undergraduates but also advanced students and beyond, we feel this is warranted.

We have depended on the foundational work of Deer, Howie, and Zussman (DHZ) throughout our careers and have continued to do so for much of the optical information included in Chapter 1. We cannot stress enough the value of their work. For the student who anticipates utilizing petrography beyond their undergraduate education, DHZ provide a wealth of information on recognition, characterization, experimentation, and paragenesis of the rock-forming minerals well beyond that conveyed here, and we highly recommend it as a life-long reference.

Chapter 1 focuses on the minerals found in igneous and metamorphic rocks. Chapter 2 combines these minerals with the textures of igneous rocks and introduces the rocks themselves, following the naming conventions of the International Union of Geological Sciences (IUGS). Chapter 3 addresses the minerals, textures, and fabrics of metamorphic rocks, as well as the rocks themselves, and broadly follows IUGS naming conventions, although we acknowledge that this is not necessarily the most widely adopted nomenclature at present.

Each chapter provides more detailed information on minerals, textures, and rocks in the form of *information boxes*, while *applications* explore how the information gleaned from the petrographic evaluation of thin-sections can be used to further our understanding of mineral- and rock-forming processes – that is, how what you see in a thin-section can be applied. Relevant textbooks and references are given at the end of each chapter, allowing those interested to delve more deeply into a given topic.

The observations and textural descriptions of minerals and their relationships in rock thin-sections is a fundamental part of geology and a necessary foundation for most of today's more advanced applications, from geochronology to pressure–temperature–time paths. It is amazing to us how a simple petrographic microscope provides such a wealth of information – its scientific added value is completely disproportionate to this relatively inexpensive investigative tool!

We hope you enjoy looking at minerals and rocks under the microscope as much as we have enjoyed sharing them with you.

Acknowledgments

ADM wishes to acknowledge help and encouragement from Aku P. Heinonen, Alberto Zanetti, Alessandra Montanini, Alessandro Zara, Andy Tindle, Anton Chakhmouradian, Axel Sjoqvist, Bernardo Cesare, Cherfi Youcef, Chiara Groppo, Chiara Montomoli, Christian Biagioni, Dalila Grilli, David Alderton, David P. West, David Sherrod, Enrique Gómez Rivas, Fabrizio Innocenti, Flavio Milazzo, Francesco Stoppa, Gianluca Sessa, Giovanni Grieco, Haakon Fossen, Hans-Peter Schertl, Hermes García, Hildegard Wilske, James Connelly, James St. John, John Booth, John Bowles, John W. Goodge, Kevin Walsh, Leone Melluso, Marco Filippi, Marco Merlini, Marco Pistolesi, Marilena Moroni, Massimo D'Orazio, Mattia Bonazzi, Michael C. Lesher, Michele Zucali, Micol Bussolesi, Mikhail Sidorov, Miłosz Huber, Nicoletta Marinoni, Niels Jöns, Olga Ageeva, Patrizia Fumagalli, Pietro Vignola, Rodolfo Carosi, Salvatore Iaccarino, Samuele Papeschi, Sarah J. Barnes, Sergio Rocchi, Simone Tumiati, Simone Vezzoni, Stefan Marincea, Stefano Poli, Teresa Trua, Thair Al-Ani, Valentine Troll, Valeria Caironi, Victor Cardenes, William S. Cordua.

VP thanks everyone who provided images for this project and her undergraduate microscopy instructors, Tom, Dan, and Patrick, whose enthusiasm inspired her own love of minerals and rocks under the microscope. Thanks also to Martin for his support – he seldom complained about the seemingly endless commentary during this project!

1

ROCK-FORMING MINERALS

1.1 Introduction

This book is intended to be a reference for the advanced undergraduate and beyond. Depending on your skill level, it may be a primary guide or used to refresh your knowledge. This book does not replace a proper mineralogy course or textbook; it does not replace an optical petrography course or textbook. It is intended to be used in conjunction with such courses and to be a long-term reference after the completion of such courses. You should be learning or already know how to use an optical microscope. You may already know, or be learning, how to recognize and determine the distinctive properties of the most common igneous and metamorphic rock-forming minerals. This includes:

- descriptive matter: crystal shape/habit, twinning, alteration, etc.
- color
- pleochroism
- anisotropy
- crystal systems and uniaxial (tetragonal, hexagonal) versus biaxial (orthorhombic, monoclinic, triclinic) character
- optic signs as positive (+) or negative (–)
- birefringence ($1° < 0.018$, $2° < 0.036$, $3° < 0.055$) and use of the Michel-Levy color chart
- relief (high, moderate, low)
- cleavage (1, 2 or >2, and how to determine the angle between them)
- extinction (parallel, inclined, symmetrical, undulose) and the angle of extinction.

The recognition and characterization of minerals and textures of igneous and metamorphic rocks is the primary purpose of this book. A person with little experienced can follow the steps outlined in Fig. 1.1 to identify unknown minerals in thin-section. These steps define a systematic method for mineral identification. The information obtained is then used in conjunction with the overview of mineral characteristics given in Table 1.1. We use a simplified approach to mineral classification based primarily on the Michel-Levy color chart (Fig. 1.2) and birefringence. The Michel-Levy color chart provides a quantified visual reference of mineral birefringence in which lower-order colors are more vibrant, and higher-order colors become increasingly pastel (Fig. 1.2). The minerals in Table 1.1 are listed in order of increasing birefringence. Once the birefringence of a mineral is determined, Table 1.1 is used to narrow the range of possible choices. Other characteristics – such as relief, cleavage, pleochroism, and optic sign – may be needed for final identification of the mineral.

The use of Table 1.1 assumes that thin-sections have the "correct" thickness (30 μm) for the accurate determination of birefringence. This is not always the case: Thinner sections will produce lower interference colors and thicker sections will result in higher interference colors than those of the ideal 30 μm thickness. The more experienced person will recognize whether a thin-section is the correct thickness by assessing, for example, the interference colors of common minerals such as quartz or plagioclase. If you are more experienced, you can proceed directly to the overview of mineral characteristics (Table 1.1). Mineral abbreviations follow the recommendations of Whitney and Evans (2010) (Table 1.2).

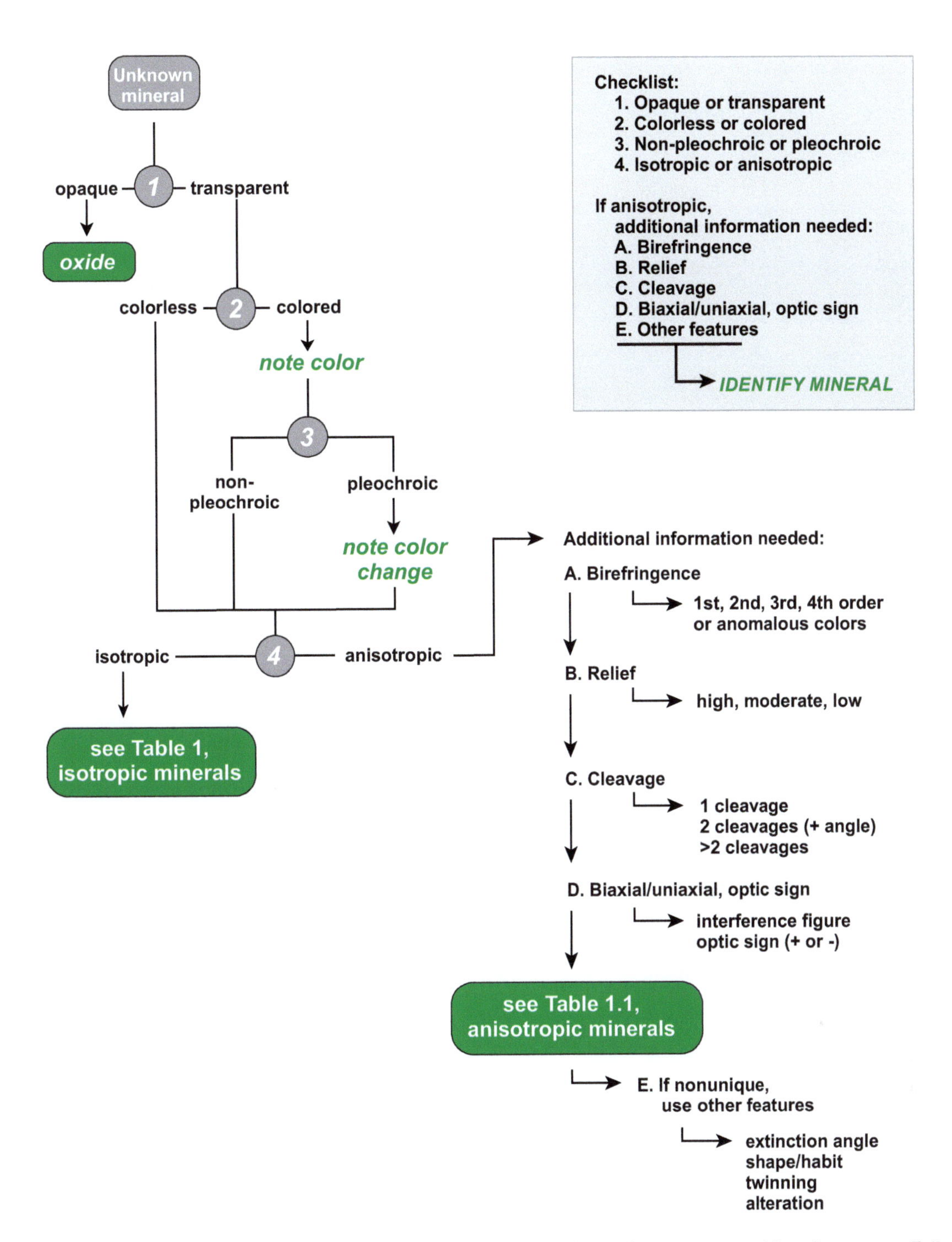

Figure 1.1 Flow chart for the identification of a mineral in thin-section using a petrographic microscope. Follow the numbered steps from the upper-left to the lower-right. At each number, a decision is made regarding whether to stop or continue. At step 4, additional information may be needed, as specified by letters A–E. Final identification can be made with cross-reference to Table 1.1.

Table 1.1 Overview of common rock-forming minerals and their distinguishing characteristics

Birefringence	Mineral	Relief	Cleavage	Pleochroic	U	B	+	–
Isotropic minerals								
1.433	Fluorite	H–	*					
1.479–1.493	Analcime	M–						
1.483–1.490	Sodalite	M–						
1.508–1.511	Leucite	L–						
1.71–1.89	Garnet	H						
1.719–2.74	Spinel	H						
2.30–2.38	Perovskite	H						
Anisotropic minerals								
0.00–0.01 (avg.)	Chlorite	M	*	*		*		*
0.001–0.005	Chabazite	M–				*	*	*
0.001–0.007	Apatite	M		*	*			*
0.001–0.013	Melilite	M	*	*		*	*	*
0.002–0.004	Tridymite (α)	M–				*	*	
0.003 (approx.)	Cristobalite (α)	M–				*	*	
0.003–0.005	Nepheline	L–				*	*	*
0.003–0.008	Zoisite	H	*			*	*	
0.003–0.009	Heulandite	M–	*			*	*	*
0.003–0.010	Phillipsite	M–				*	*	
0.004–0.005	Na-scapolite (marialite)	L	**		*			*
0.004–0.007	Antigorite	L	*			*		*
0.004–0.009	Beryl	L			*			*
0.004–0.015	Clinozoisite	H	*			*	*	
0.005–0.006	Kalsilite	L–				*		*
0.005–0.009	Sapphirine	H	*	*		*	*	*
0.005–0.022	Chloritoid	L	*	*	*		*	
0.006	Coesite	M				*		*
0.006–0.008	Lizardite	L	*			*		*
0.006–0.010	K-spar, Na-spar	L	**			*		*
0.006–0.016	Riebeckite	H	**	*		*	*	*
0.006–0.021	Jadeite	H	**			*	*	
0.007–0.016	Enstatite	H	**	*		*		*
0.007–0.021	Katophorite	H	**	*		*		*
0.008 (approx.)	Microcline	L	**			*		*
0.008 (approx.)	Åkermanite	M			*		*	
0.008–0.009	Corundum	H		*	*			*

Table 1.1 (*cont.*)

Birefringence	Mineral	Relief	Cleavage	Pleochroic	U	B	+	–
0.008–0.011	Topaz	M	*			*	*	
0.008–0.018	Cordierite	L	*	*		*		*
0.009	Quartz (α)	L				*		*
0.009–0.010	Stilbite	L–	*			*		*
0.009–0.012	Andalusite	M	*			*		*
0.01 (approx.)	Albite (Na-Pl)	L	**			*	*	*
0.010–0.012	Arfvedsonite	H	**	*		*		*
0.010–0.020	Pumpellyite	H	**	*		*	*	*
0.011 (approx.)	Gehlenite	H			*			*
0.011–0.014	Staurolite	H	*	*		*	*	
0.012 (approx.)	Barite	M	**			*		*
0.012 (approx.)	Natrolite	M	*			*	*	
0.012–0.016	Kyanite	H	*	*		*		*
0.012–0.028	Omphacite	H	**	*		*	*	
0.013 (approx.)	Anorthite (Ca-Pl)	M	**			*		*
0.013–0.014	Wollastonite	M	**			*		*
0.013–0.017	Chrysotile	L				*		*
0.013–0.020	Eckermannite	H	**	*		*		*
0.013–0.021	Anthophyllite	M	**			*	*	*
0.013–0.036	Allanite	H		*		*	*	*
0.014–0.017	Ca,Mg-olivine (monticellite)	M		*		*		*
0.014–0.020	Brucite	M	*		*		*	
0.015–0.022	Mg-richterite	H	**	*		*		*
0.015–0.051	Epidote	H	*	*		*		*
0.016–0.022	Ferrosilite	H	**	*		*		*
0.017–0.021	Li-tourmaline (elbaite)	M		*	*			*
0.017–0.032	Ferroactinolite	H	**	*		*		*
0.018–0.022	Sillimanite	H	*	*		*	*	
0.018–0.025	Augite	H	**	*		*	*	
0.019–0.021	Lawsonite	H	**			*	*	
0.02 (approx.)	Hornblende	H	**	*		*	*	*
0.020–0.032	Cummingtonite	H	**			*	*	
0.021–0.029	Gedrite	H	**	*		*	*	*
0.021–0.029	Mg-tourmaline (dravite)	M		*	*			*
0.022–0.027	Tremolite	M	**			*		*
0.022–0.029	Fe-richterite	H	**	*		*		*

Table 1.1 (*cont.*)

Birefringence	Mineral	Relief	Cleavage	Pleochroic	U	B	+	–
0.022–0.051	Prehnite	M	*			*	*	
0.018–0.020	Glaucophane	M	**	*		*		*
0.023–0.029	Pigeonite	H	**	*		*	*	
0.024–0.034	Ca-scapolite (meionite)	L			*			*
0.025–0.033	Ferroaugite	H	**			*	*	
0.025–0.035	Fe-tourmaline (schorl)	H		*	*			*
0.025–0.073	Piemontite	H	*	*		*	*	
0.024–0.028	Hedenbergite	H	**	*		*	*	
0.028–0.047	Kaersutite	H	**	*		*	*	
0.028–0.07	Biotite	M	*	*		*		*
0.03–0.11	Stilpnomelane	H	*	*		*		*
0.030–0.045	Aegirine–augite	H	**	*		*	*	*
0.032–0.045	Grunerite	H	**	*		*		*
0.028–0.034	Diopside	H	**			*	*	
0.035	Mg-olivine (forsterite)	H				*	*	
0.035–0.042	Muscovite	M	*		*			*
0.036–0.054	Sericite	L,M				*		*
0.040–0.060	Aegirine	H	**	*		*		*
0.042–0.065	Zircon	H			*		*	
0.045–0.075	Monazite	H	*			*	*	
0.050 (approx.)	Pyrophyllite	M	*			*		*
0.050 (approx.)	Talc	M	*			*		*
0.052	Fe-olivine (fayalite)	H		*		*		*
0.080	Perovskite	VH	*			*	*	
0.096	Xenotime	VH	*	*	*		*	
0.096–0.098	Cassiterite	H			*		*	
0.100–0.192	Titanite	H	*	*		*	*	
0.155–0.156	Aragonite	L,H	*			*		*
0.172–0.190	Calcite	L,H	*		*			*
0.179–0.185	Dolomite	L,H	*		*			*
0.0190–0.218	Magnesite	L,H	*		*			*
0.207–0.242	Siderite	H	*		*			*
0.286–0.296	Rutile	VH	*		*		*	

L, low; M, moderate; H, high; VH, very high.

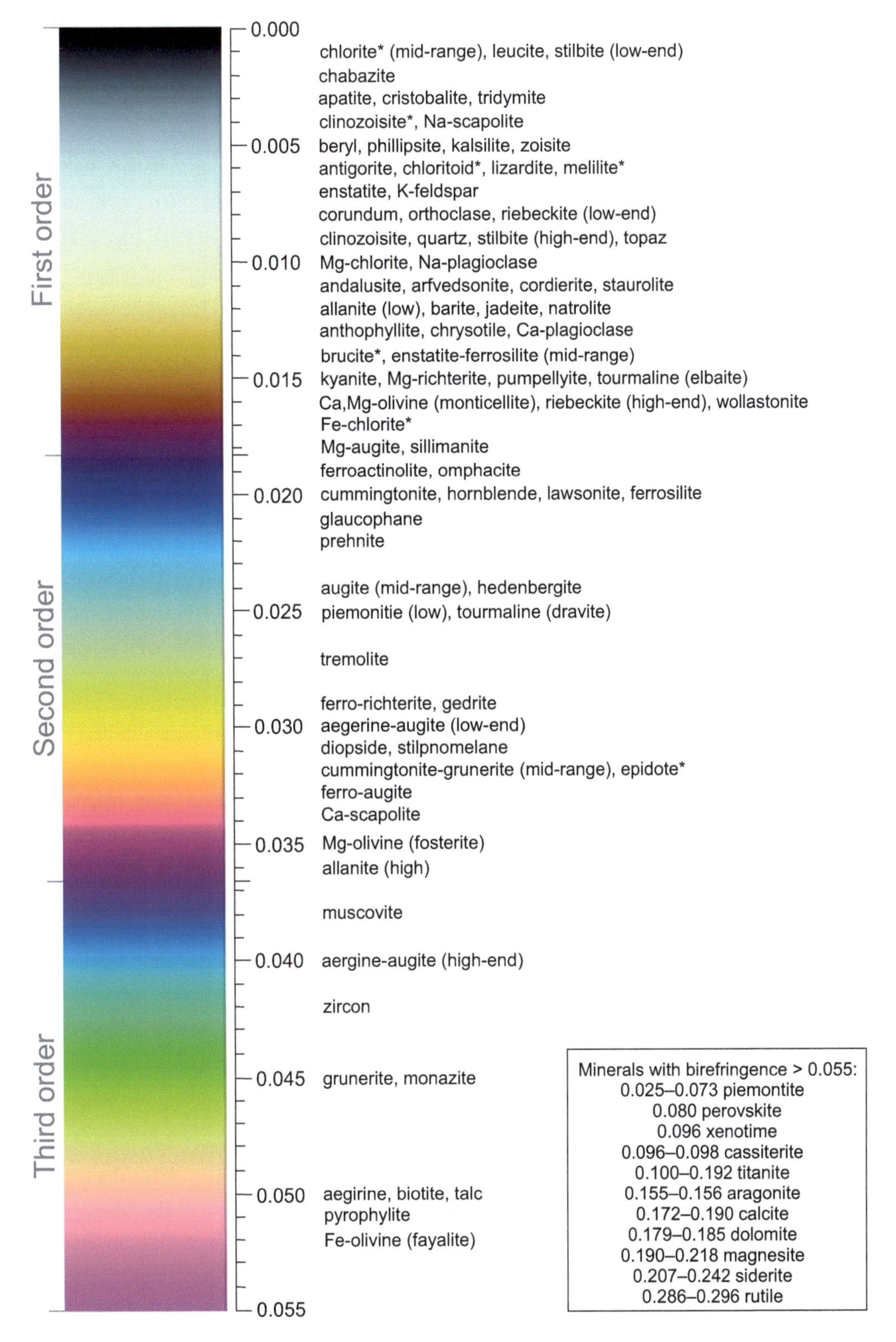

Figure 1.2 Michel-Levy birefringence chart for mineral identification (after Kato, 2001). Uses a 30 μm thin-section to determine birefringence. Cross-reference the birefringence to Table 1.1 and compare the properties of minerals with similar birefringence in order to make the final mineral identification. Note that minerals with an asterisk (*) can show anomalous colors.

Table 1.2 Mineral abbreviations (after Whitney & Evans, 2010)

Ab, Albite	Cst, Cassiterite	Hul, Heulandite series
Act, Actinolite	Ctl, Chrysotile	
Adr, Andradite	Cum, Cummingtonite	Ilt, Illite series
Aeg, Aegirine	Czo, Clinozoisite	Ilm, Ilmenite
Afs, Alkali feldspar		
Agt, Aegirine–augite	Di, Diopside	Jd, Jadeite
Ak, Åkermanite	Dol, Dolomite	
Alm, Almandine		Kfs, K-feldspar
Aln, Allanite	Eck, Eckermannite	Kls, Kalsilite
Amp, Amphibole supergroup	En, Enstatite	Krs, Kaersutite
An, Anorthite	Ep, Epidote	Ktp, Katophorite
And, Andalusite		Ky, Kyanite
Anl, Analcime	Fa, Fayalite	
Ano, Anorthoclase	Fac, Ferro-actinolite	Lct, Leucite
Ap, Apatite	Fi, Fibrolite (fibrous sillimanite)	Lmt, Laumontite
Arf, Arfvedsonite	Fkrs, Ferro-kaersutite	Lpd, Lepidolite
Arg, Aragonite	Fl, Fluorite	Lws, Lawsonite
Atg, Antigorite	Fo, Forsterite	Lz, Lizardite
Ath, Anthophyllite	Fprg, Ferro-pargasite	
Aug, Augite	Frct, Ferro-richterite	Mag, Magnetite
	Fs, Ferrosilite	Marf, Magnesio-arfvedsonite
Bdy, Baddeleyite	Fsp, Feldspar	Mc, Microcline
Brc, Brucite		Mgh, Maghemite
Brl, Beryl	Ged, Gedrite	Mgs, Magnesite
Brt, Barite	Gh, Gehlenite	Mll, Melilite series
Bt, Biotite	Gln, Glaucophane	Mnz, Monazite
	Glt, Glauconite	Mor, Mordenite
Cal, Calcite	Gn, Galena	Mrbk, Magnesio-riebeckite
Cbz, Chabazite	Grs, Grossular	Ms, Muscovite
Ccp, Chalcopyrite	Grt, Garnet supergroup	Mtc, Monticellite
Chl, Chlorite	Gru, Grunerite	Mw, Merwinite
Chr, Chromite	Gth, Goethite	
Cld, Chloritoid		Nph, Nepheline
Coe, Coesite	Hbl, Hornblende series	Nsn, Nosean
Cpx, Clinopyroxene	Hd, Hedenbergite	Ntr, Natrolite
Crd, Cordierite	Hem, Hematite	
Crn, Corundum	Hgr, Hydrogrossular	Ol, Olivine group
Crs, Cristobalite	Hst, Hastingsite	Omp, Omphacite

Table 1.2 *(cont.)*

Opl, Opal	Qz, Quartz	Thr, Thorite
Opq, Opaque mineral		Tlc, Talc
Opx, Orthopyroxene	Rbk, Riebeckite	Tpz, Topaz
Or, Orthoclase	Rct, Richterite	Tr, Tremolite
	Rdn, Rhodonite	Trd, Tridymite
Pcl, Pyrochlore	Rt, Rutile	Ttn, Titanite
Per, Periclase		Tur, Tourmaline supergroup
Pg, Paragonite	Sa, Sanidine	
Pgt, Pigeonite	Sch, Scheelite	Urn, Uraninite
Ph, Phengite series	Scp, Scapolite	Usp, Ulvöspinel, Ulvospinel
Phl, Phlogopite	Sd, Siderite	Uv, Uvarovite
Php, Phillipsite	Sdl, Sodalite	
Pl, Plagioclase series	Ser, Sericite	Vtr, Vaterite
Pmp, Pumpellyite	Sp, Sphalerite	
Pmt, Piemontite	Spd, Spodumene	Wo, Wollastonite
Po, Pyrrhotite	Spl, Spinel	Wrk, Wairakite
Prh, Prehnite	Spr, Sapphirine	
Prg, Pargasite	Sps, Spessartine	Xtm, Xenotime-(Y)
Prl, Pyrophyllite	Srp, Serpentine	
Prp, Pyrope	St, Staurolite	Zeo, Zeolite family
Prv, Perovskite	Stb, Stilbite series	Zo, Zoisite
Psb, Pseudobrookite	Sti, Stishovite	Zrn, Zircon
Px, Pyroxene supergroup	Stp, Stilpnomelane	
Py, Pyrite		

1.2 Framework Silicates

Plagioclase Feldspar $NaAlSi_3O_8 - CaAl_2Si_2O_8$

Optical Properties **THE FELDSPARS**

Color in thin-section: Colorless
Pleochroism: Absent
Birefringence: 0.007–0.013
Relief: Low
Cleavage: Perfect {001} and {010} meeting at right-angle, poor {110}
Crystal system/optic sign: Triclinic / Albite (+) or (–), Anorthite (–)
Other features: Diagnostic polysynthetic twins parallel to prism faces; rarely albitic plagioclase in low-grade rocks lack twinning. Untwinned plagioclase can be confused with cordierite, but the latter may have yellowish alteration (pinite); distinguished from quartz by its biaxial nature and good cleavage; sericite alteration, sieve texture, and/or glass inclusions are common.

Paragenesis

Plagioclase is an abundant constituent of most igneous rocks, from ultrabasic–basic to acid, through intermediate and alkaline. It is scarce or absent only in some rare ultrabasic and alkaline rocks. Plagioclase is common in low- to high-grade (granulite facies) metamorphic rocks but becomes unstable at high P–T conditions (e.g. eclogite facies).

Igneous

Plagioclase is typically present in both extrusive and intrusive rocks, forming up to 50% of the total rock volume. In volcanic rocks, most plagioclase has disordered, high-temperature structure and occurs as phenocrysts and/or groundmass; phenocrysts are often compositionally zoned (Ca-rich cores with Na-rich rims). In plutonic rocks, slow cooling and diffusion erase compositional zoning. In layered mafic intrusions, plagioclase forms cumulate layers due to crystal settling processes.

Metamorphic

Plagioclase becomes more Ca-rich with increasing metamorphic grade. During low-grade regional metamorphism (chlorite and biotite zone) albite is stable, whereas oligoclase is more common in the garnet zone. The transition between amphibolite to granulite facies is marked by the reaction:

$$\underset{\text{amphibole}}{Ca_2Mg_3Al_4Si_6O_{22}(OH)_2} + SiO_2 \rightarrow 2\underset{\text{anorthite}}{CaAl_2Si_2O_8} + 3\underset{\text{orthopyroxene}}{MgSiO_3} + H_2O$$

During contact metamorphism (e.g. the pyroxene hornfels facies), plagioclase can form via the breakdown of hornblende:

$$\underset{\text{amphibole}}{NaCa_2Mg_3Fe^{3+}Al_3Si_6O_{22}(OH)_2} + 4SiO_2 \rightarrow \underset{\text{plagioclase}}{CaAl_2Si_2O_8 + NaAlSi_3O_8} + \underset{\text{diopside}}{CaMgSi_2O_6} + \underset{\text{orthopyroxene}}{Mg_2Fe^{2+}Si_3O_8} + H_2O$$

At eclogite facies, plagioclase is unstable and the anorthite component is taken up in omphacitic pyroxene and/or in garnet via the reactions:

$$\underset{\text{albite}}{NaAlSi_3O_8} + \underset{\text{olivine}}{(Mg,Fe)_2SiO_4} \rightarrow \underset{\text{omphacite}}{NaAlSi_2O_6(Mg,Fe)SiO_3} \; \underset{\text{anorthite}}{CaAl_2Si_2O_8} +$$

$$\underset{\text{olivine}}{(Mg,Fe)_2SiO_4} \rightarrow \underset{\text{garnet}}{Ca(Mg,Fe)_2Al_2Si_3O_{12}}$$

In numerous retrogressed eclogites, plagioclase commonly occurs as symplectic vermicular intergrowths with hornblende in kelyphitic rims around garnet.

Figure 1.3

A Plagioclase in basalt from Etna (Italy). Large plagioclase phenocrysts (polysynthetic twinning with numerous black glass inclusions). Plagioclase is also present in the groundmass as microlites that define a magmatic foliation enveloping the phenocrysts. Olivine (high birefringence) is also present. Cross-polarized light, 2× magnification, field of view = 7 mm.

B Plagioclase in latite from Latium (Italy). Euhedral plagioclase phenocrysts in a brown glassy groundmass. The phenocrysts contain numerous pale brown glass inclusions arranged concentrically. Some colorless plagioclase crystals are anhedral, and dark brown crystals with dark opacitic rims are hornblende. Plane-polarized light, 2× magnification, field of view = 7 mm.

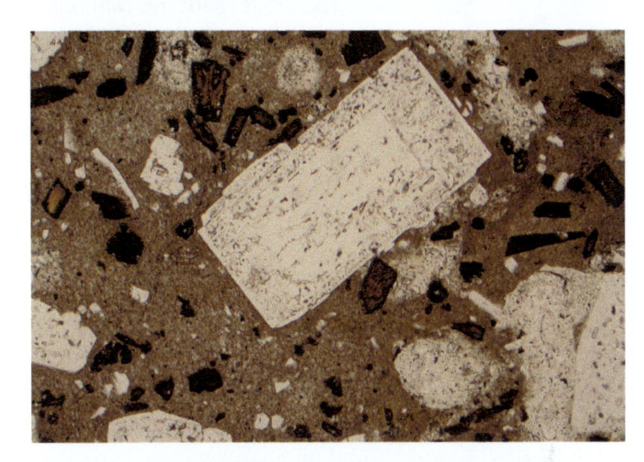

C Plagioclase in gabbro from Mutoko (Zimbabwe). Large plagioclase crystal with simple Carlsbad twinning and minor pericline twinning (right side of the crystal intersecting the albite lamellae at high angles). Pyroxene phenocrysts also present (low–medium birefringence; bottom center of image). Cross-polarized light, 2× magnification, field of view = 7 mm.

D Plagioclase in a retrogressed eclogite from Černín (Czech Republic). Symplectic intergrowth of plagioclase (gray with polysynthetic twinning) and hornblende (low birefringence) surrounds garnet (at extinction). Cross-polarized light, 10× magnification, field of view = 2 mm.

Anorthoclase (K,Na)$AlSi_3O_8$

THE FELDSPARS

ALKALI FELDSPAR

Optical Properties

Color in thin-section:	Colorless
Pleochroism:	Absent
Birefringence:	0.007–0.008
Relief:	Low
Cleavage:	Perfect on {001} and less perfect on {010}
Crystal system/optic sign:	Triclinic / (–)
Other features:	Anorthoclase is distinguished by its rhomb-shaped cross-section and cross-hatched twinning. The albite–pericline twin associated with anorthoclase may look similar to that in microcline, but anorthoclase twins have straight intersections and are parallel-sided, whereas microcline twins have curving intersections, are not straight (spindle-shaped), and microcline is only found in plutonic rocks.

Paragenesis

Anorthoclase is a common phenocryst and groundmass phase in Na-rich volcanic and shallow plutonic (sub-volcanic) rocks (e.g. alkali rhyolite [pantellerite], alkali trachyte, latite, mugearite, and phonolite). Anorthoclase represents an intermediate composition of the high-albite–sanidine series, and typically consists of 10–36% $KAlSi_3O_8$ and 64–90% $NaAlSi_3O_8$. Due to rapid cooling of volcanic rocks, high-temperature solid solution phases such as anorthoclase and sanidine are quenched in their disordered state; more ordered alkali feldspars are very rare in volcanic rocks. Rapid cooling also prevents development of exsolution/perthitic texture, although it may be present at a submicroscopic scale.

Figure 1.4

A Anorthoclase in trachyte from Colli Euganei (Italy). This anorthoclase phenocryst shows Carlsbad twinning along the central part of the crystal and cross-hatched twinning on its top half. Cross-polarized light, 2× magnification, field of view = 7 mm.

B Anorthoclase in trachyte from Colli Euganei (Italy). The higher-magnification image shows a close-up of the characteristic cross-hatched twinning in anorthoclase. It is similar to that of microcline, but less regular and more fine. Cross-polarized light, 10× magnification, field of view = 2 mm.

Microcline $KAlSi_3O_8$

Optical Properties

Color in thin-section:	Colorless; often cloudy due to alteration
Pleochroism:	Absent
Birefringence:	0.006–0.012
Relief:	Low
Cleavage:	Perfect {001} and {010} meeting at right angles
Crystal system/optic sign:	Triclinic / (–)
Other features:	Distinctive combination of cross-hatched twinning (albite and pericline twin laws), often diffuse and spindle-shaped; twinning is present at the sub-optical scale even when apparently absent, and can be distinguished from orthoclase by its extinction angle.

Paragenesis

Microcline is common in plutonic felsic rocks (e.g. granite, syenite, pegmatite) and in hydrothermal veins. It occurs in greenschist and amphibolite facies metamorphic rocks. Microcline is the triclinic low-temperature ($< 500\,°C$) K-feldspar; during slow cooling there is a gradual transition from high-temperature monoclinic sanidine, with its disordered internal structure, to the more internally ordered triclinic orthoclase or microcline. Compared to the sanidine → orthoclase transition, the orthoclase → microcline transition involves a change of symmetry from monoclinic to triclinic, in addition to the Al–Si reorganization. Such transitions are difficult to achieve, so orthoclase often persists over long timescales as a metastable phase. The mechanism by which orthoclase transforms to microcline is not well understood and has never been achieved experimentally; it is perhaps facilitated by fluid–feldspar reactions, dissolution–reprecipitation reactions, or deformation processes. It is common to find microcline subgrains in orthoclase crystals in plutonic rocks.

Figure 1.5

A Microcline in granite from Baveno (Italy). Large subhedral microcline crystal shows cross-hatched twinning and perthite unmixing (irregular, light-gray zones in the right corner of the crystal). Quartz (uniform gray, anhedral crystals) and biotite (high birefringence) are also present. Cross-polarized light, 2× magnification, field of view = 7 mm.

B Microcline in granite from Amparo (Brazil). Close-up of microcline twinning in which spindle-shaped (not parallel) twin lamellae can be seen. Cross-polarized light, 10× magnification, field of view = 2 mm.

Orthoclase $KAlSi_3O_8$

Optical Properties

Color in thin-section:	Colorless; often cloudy due to alteration
Pleochroism:	Absent
Birefringence:	0.0005–0.007
Relief:	Low
Cleavage:	Perfect {001} and {010} meeting at right angles
Crystal system/optic sign:	Monoclinic / (–)
Other features:	Orthoclase is recognized by its low birefringence, simple twinning (Carlsbad is very common), commonly cloudy appearance, its biaxial character, and the presence of perthite exsolution lamellae. These distinguish it from quartz and nepheline (both uniaxial).

Paragenesis

Orthoclase is a widespread mineral in granitic to syenitic igneous rocks and is common in the sillimanite zone of both contact and regionally metamorphosed rocks.

Igneous

Compared to its polymorphs sanidine and microcline, monoclinic orthoclase is the medium-temperature (500–900 °C) K-feldspar that forms at temperatures intermediate to the stability fields of sanidine and microcline. During the slow cooling that characterizes plutonic rocks, high-temperature monoclinic sanidine inverts to monoclinic orthoclase, then to triclinic microcline. This involves the slow reorganization of Si–Al atoms in which orthoclase assumes a "tweed" structure characterized by repetition of ordering domains on the scale of a few unit cells. Once the orthoclase "tweed" structure is formed, further internal ordering, leading to stable triclinic microcline, is blocked. This implies that the orthoclase "tweed" can persist for a long time in a metastable state such that in many plutonic rocks the alkali feldspars are mixtures of orthoclase and microcline at a sub-optical scale.

Metamorphic

In regionally metamorphosed rocks (~680 °C), orthoclase forms due to the instability of biotite and muscovite via:

$$\underset{\text{biotite}}{K(Mg,Fe)_3AlSi_3O_{10}(OH)_2} + 3SiO_2 \rightarrow \underset{\text{orthoclase}}{KAlSi_3O_8} + \underset{\text{opx}}{3(Mg,Fe)SiO_3} + H_2O$$

$$\underset{\text{muscovite}}{KAl_3Si_3O_{10}(OH)_2} + 3SiO_2 \rightarrow \underset{\text{orthoclase}}{KAlSi_3O_8} + \underset{\text{sillimanite}}{3Al_2SiO_5} + H_2O$$

During contact metamorphism, orthoclase forms in the high-grade pyroxene hornfels facies (>650–700 °C) via the breakdown of biotite:

biotite + plagioclase + quartz = K-feldspar + cordierite + orthopyroxene + melt

The important melt-producing (up to 40 vol. % melt) reaction, thought to generate significant quantities of granite in collisional orogens, also produces orthoclase via the reaction:

biotite + sillimanite = garnet + K-feldspar + melt

Figure 1.6

A Orthoclase in granite from Guaíba (Brazil). Orthoclase crystals have a cloudy appearance due to sericite alteration. Quartz is clear and free from alteration. Brown–green hornblende with an anhedral red–brown titanite crystal left of center is also present. Plane-polarized light, 2× magnification, field of view = 7 mm.

B Orthoclase in granite from Guaíba (Brazil). Same image as (A) but with crossed polarizers. Orthoclase crystals show a “zebra stripe” appearance (upper and lower-left) due to perthite unmixing. An orthoclase crystal (central left) shows a simple Carlsbad twin. Hornblende has medium interference colors and titanite is extinct. Cross-polarized light, 2× magnification, field of view = 7 mm.

C Orthoclase in quartz syenite from Biella (Italy). The central orthoclase crystal has both Carlsbad twinning and concentric growth zoning. Smaller orthoclase and hornblende crystals (brown–green interference colors) surround the large crystal. Cross-polarized light, 2× magnification, field of view = 7 mm.

Sanidine $KAlSi_3O_8$

Optical Properties

Color in thin-section:	Colorless; often "glass-clear"
Pleochroism:	Absent
Birefringence:	0.0005–0.007
Relief:	Low
Cleavage:	Perfect {001} and {010} but fresh, young crystals with poor or no cleavage
Crystal system/optic sign:	Monoclinic / (–)
Other features:	Sanidine forms clear, lath-shaped crystals with simple Carlsbad twins. If fragmented, it can be confused with quartz but the latter lacks cleavage.

Paragenesis

Sanidine is only present in extrusive volcanic rocks and is common in silica-saturated to silica-undersaturated compositions, such as rhyolite, trachyte, dacite, and phonolite. The three alkali feldspars sanidine, orthoclase, and microcline have different crystal structures which are determined by the degree of order in Al–Si within the crystal lattice; this in turn is a consequence of the rate of cooling during crystallization. Sanidine is the high-temperature, fully disordered K–Na feldspar; it is stable only above 700 °C. Sanidine commonly contains up to 30% sodium (Ab component) and at high temperatures it forms a complete solid solution with albite. Being a high-temperature phase, during slow cooling the Si–Al ordering exchange position by diffusion, gradually converting sanidine to orthoclase. Sanidine is also diagnostic of high-temperature, low-pressure contact metamorphism: the sanidinite hornfels facies. Sanidinite facies is characterized by the absence of hydrous minerals and, although relatively rare, is documented by xenoliths entrained in igneous rocks.

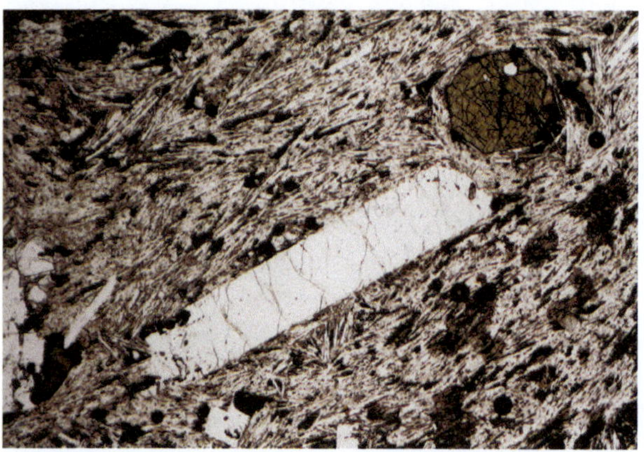

Figure 1.7

A Sanidine in trachyte from Ustica (Italy). Euhedral, glass-clear sanidine crystals set in a sanidine-rich groundmass with trachytic texture. Also present in the groundmass are pyroxene (brownish, good basal cleavage; upper-right corner) and opaques (black). Plane-polarized light, 2× magnification, field of view = 7 mm.

B Sanidine in trachyte from Ustica (Italy). Same image as (A) but with crossed polarizers. Simple Carlsbad twin is seen in the sanidine phenocryst (typical gray birefringence) and pyroxene shows typical low to medium birefringence. Cross-polarized light, 2× magnification, field of view = 7 mm.

BOX 1.1 Feldspar Solid Solution

Feldspars have both high- and low-temperature structural states that are controlled by the ordering of Al–Si atoms in the crystal lattice. Feldspar has a disordered structure at high temperature and becomes more ordered at low temperature. There is only limited solid solution between the three compositional end-members of the feldspar series:

$$NaAlSi_3O_8 - CaAl_2Si_2O_8 - KAlSi_3O_8$$

albite – anorthite – orthoclase

There is, however, complete solid solution between the Na- (albite) and Ca-rich (anorthite) plagioclase end-members, and this forms the basis for their compositional subdivision, which also varies with rock type.

Anorthite composition	Compositional name	Rock type volcanic	plutonic
An_{0-10}	Albite	Rhyolite (An_{20-40})	
An_{10-30}	Oligoclase	Dacite (An_{30-50})	Granite ($An_{<30}$)
An_{30-50}	Andesine	Andesite (An_{40-60})	Diorite ($An_{<50}$)
An_{50-70}	Labradorite	Basalt (An_{50-70})	Gabbro ($An_{>50}$)
An_{70-90}	Bytownite		
An_{90-100}	Anorthite		

The different ionic radii of Na^+ and K^+ in orthoclase impose large local strains on the Si–Al framework, resulting in perthitic (plagioclase) exsolution in alkali feldspars with disordered structure. The solubility of K in calcic plagioclase is low even at high temperature; it is only present in Ab-rich plagioclase and with slow cooling often results in "antiperthite" (Kfs) exsolution in plagioclase; this is common in granulites and charnockites, which crystallize at high temperature and have long, complex cooling histories.

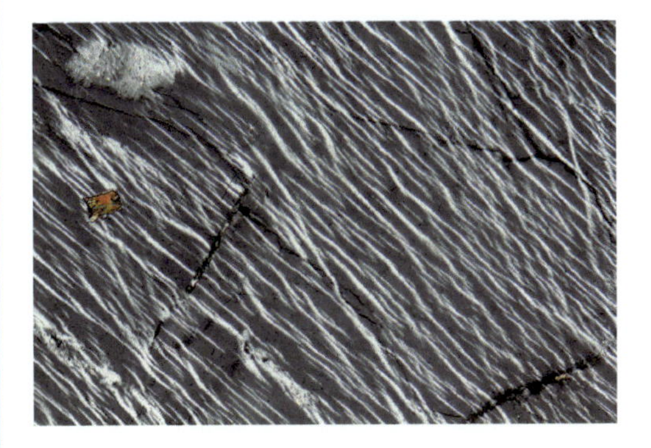

Perthite exsolution in granite from Baveno (Italy). This orthoclase shows anastomosing perthite exsolution lamellae (white birefringence). A small biotite crystal is present left of center (orange birefringence). Cross-polarized light, 10× magnification, field of view = 2 mm.

Antiperthite exsolution in charnockite from Rogaland (Sweden). Small, ovoid antiperthite exsolution in a plagioclase with polysynthetic twinning. Pyroxene (yellow–orange) is also present to the right. Cross-polarized light, 10× magnification, field of view = 2 mm.

Kalsilite $KAlSiO_4$

THE FELDSPATHOIDS

Optical Properties

Color in thin-section:	Colorless
Pleochroism:	Absent
Birefringence:	0.005–0.006
Relief:	Low
Cleavage:	Poor/indistinct on $\{10\bar{1}0\}$ and {0001}
Crystal system/optic sign:	Hexagonal / (–)
Other features:	Kalsilite is often confused with nepheline, with which it is commonly associated. While these two minerals are difficult to distinguished in thin-section, nepheline often appears cloudy due to the presence of minute inclusions, whereas kalsilite is more transparent.

Paragenesis

The presence of kalsilite is almost exclusively confined to the groundmass of K-rich extrusive rocks such as the kamafugitic series (mafurite, katungite, venanzite, coppaelite, ugandite). In rare cases kalsilite may form phenocrysts, as in the lavas of Mt. Nyiragongo (Congo); more commonly it forms intergrowths with K-feldspar due to the decomposition of leucite. Kalsilite has been found in volcanic ejecta of the Sabatini volcanic system near Rome (Italy); it is extremely rare in plutonic rocks and few cases of its occurrence are known (e.g. syenites from USSR and Alaska).

Figure 1.8

A Kalsilite in nepheline syenite from the Khibiny complex (Russia). Anhedral orthoclase crystals with rod-like exsolution of kalsilite. The orthoclase crystals are poikilitically enclosed by pyroxene (dark red–brown birefringence). Cross-polarized light, 2× magnification, field of view = 7 mm.

B Kalsilite in syenite from Synnyr (Russia). Vermicular (worm-like) intergrowths of kalsilite and orthoclase after leucite. Cross-polarized light, 10× magnification, field of view = 2 mm.

Leucite $KAlSi_2O_6$

Optical Properties

Color in thin-section: Colorless

Pleochroism: Absent

Birefringence: 0.001

Relief: Low

Cleavage: {110} very poor, not visible under the microscope

Crystal system/optic sign: Tetragonal (pseudocubic) / (+)

Other features: Leucite occurs commonly as euhedral crystals, in trapezohedron form with eight-sided sections; skeletal forms are not uncommon. Leucite can be distinguished from analcime and sodalite group minerals by its distinctive twinning, and from microcline which has higher birefringence and is often associated with quartz.

Paragenesis

Leucite is one of the most common and characteristic feldspathoid minerals. It occurs in numerous K-rich and silica-deficient volcanic rocks (e.g. basanite, tephrite, nephelinite, and phonolite). In these rocks it constitutes the main mineral, as both phenocrysts and microphenocrysts in the groundmass, and often there are two generations of leucite: large, euhedral phenocrysts and small, rounded microphenocrysts. Leucite is a common phase in some K-rich ultrabasic rocks such as ugandite and katungite (kamafugitic rocks) and in some lamproites and lamprophyres, where it occurs mainly in the matrix. If leucite occurs as phenocrysts, it may have small inclusions of pyroxene or glass. Leucite in volcanic ejecta is variously associated with kalsilite, K-feldspar, nepheline, and haüyne/nosean. Some leucites crystallized from hydrous melts display a lamellar structure which likely represents an exsolved "analcimic" or "sodic leucite" phase.

The presence of leucite in plutonic rocks is extremely rare, although it sometimes occurs in foidolite xenoliths. In this occurrence it is generally attributed to increasing P_{H_2O}, which reduces the leucite stability field until it disappears entirely. In slow-cooling plutons, the presence of water facilitates the breakdown of leucite via the reaction:

$$\text{leucite} \leftrightarrow \text{kalsilite} + \text{K-feldspar}$$

The presence of leucite is restricted to rocks of Tertiary or younger age; its absence in older rocks is attributed to the ease with which leucite alters to analcime:

$$\underset{\textit{leucite}}{KAlSi_2O_6} + H_2O \leftrightarrow \underset{\textit{analcime}}{NaAlSi_2O_6 \cdot H_2O} + K^+_{(aq)}$$

This kind of reaction occurs by ion exchange in subsolidus reactions and, as experimentally demonstrated, it is a fast exchange reaction that requires a very small activation energy.

Figure 1.9

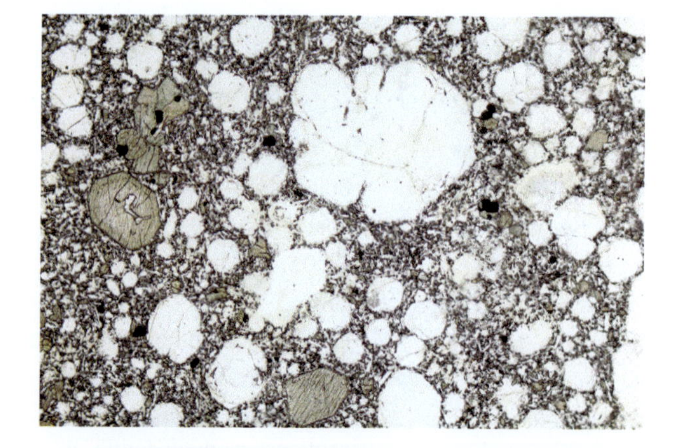

A Leucite in tephrite from Roccamonfina (Italy). Leucite phenocrysts (colorless) are seriate (of different sizes) and form rounded crystals in the groundmass. Some brown–green pyroxene phenocrysts are also present. Plane-polarized light, 2× magnification, field of view = 7 mm.

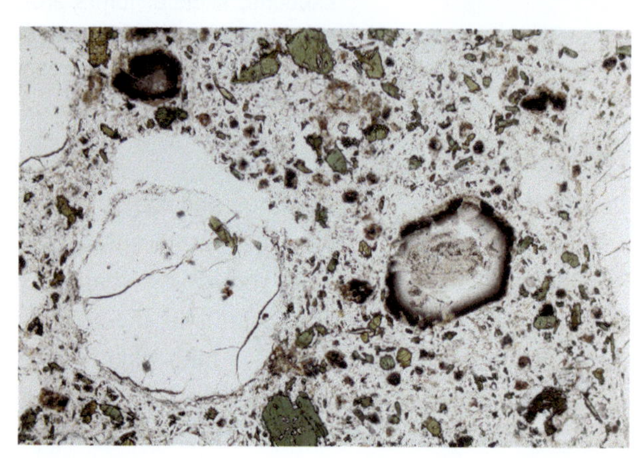

B Leucite in tephrite from Rieden (Germany). Large, euhedral but fractured leucite phenocryst, surrounded by smaller leucite crystals, green pyroxene, and some haüyne (phenocrysts with dark mantles). Plane-polarized light, 2× magnification, field of view = 7 mm.

C Leucite in tephrite from Rieden (Germany). Same image as (B) but with crossed polarizers. Leucite shows diagnostic complex twinning and low birefringence. Pyroxene has higher birefringence and haüyne is isotropic. Cross-polarized light, 2× magnification, field of view = 7 mm.

Nepheline $Na_3(Na,K)[Al_4Si_4O_{16}]$

Optical Properties

Color in thin-section:	Colorless
Pleochroism:	Absent
Birefringence:	0.003–0.006
Relief:	Low
Cleavage:	Imperfect {1010}, poor {0001}
Crystal system/optic sign:	Hexagonal / (–)
Other features:	Nepheline occurs as short columnar crystals with hexagonal and rectangular cross-sections in volcanic rocks; its optic sign distinguishes it from quartz; apatite and beryl have much higher birefringence; sanidine has oblique extinction and is twinned.

Paragenesis

Nepheline is a common mineral in silica-undersaturated igneous rocks such as nepheline syenites to basic and ultrabasic alkaline rocks. It is common in alkaline volcanic rocks such as phonolites and tephrites and in K-rich volcanic rocks, where it is subordinate to leucite. Nepheline tends to form in the last stages of crystallization (together with the alkali feldspars), hence its common hypidiomorphic to allotriomorphic or large poikilitic intercumulus form. Nepheline commonly alters to analcime, cancrinite, sodalite, muscovite, and zeolite.

Igneous

At high magmatic temperatures, the unmixing of the nepheline–kalsilite solid solution is comparable to the formation of alkali feldspar intergrowths; exsolution is rapid due to the high diffusivity of alkali ions. In volcanic rocks exsolution textures vary from cryptoperthitic to coarse-grained intergrowths, whereas in intrusive rocks exsolution tends to be complete and exsolution textures absent. Quickly cooled volcanic nepheline is often compositionally zoned, possibly related to changes in Al ↔ Fe^{3+} replacement, or Si or Ca content. Nepheline can exhibit patchy extinction (incipient unmixing of an originally homogeneous phase to a less silica-rich nepheline and alkali feldspar).

Metamorphic

Nepheline can also form by metasomatic processes, especially along contacts between country rocks and nepheline-bearing rocks. This process, commonly associated with igneous complexes, is called nephelinization. The distinction between magmatic and metasomatic nepheline-bearing rocks is mainly based on textural observations (e.g. poikilitic nepheline with rounded inclusions of feldspar, or vermicular nepheline–albite intergrowths). In late-stage magmatic hydrothermal and pneumatolytic processes, nepheline commonly alters to zeolites such as natrolite or analcime:

$$\underset{\text{nepheline}}{2NaAlSiO_4} + SiO_2 + 2H_2O \rightarrow \underset{\text{natrolite}}{Na_2Al_2Si_3O_{10}} + 2H_2O$$

$$\underset{\text{nepheline}}{NaAlSiO_4} + SiO_2 + 2H_2O \rightarrow \underset{\text{analcime}}{NaAl_2Si_2O_6} + 2H_2O$$

Figure 1.10

A Nepheline in foidite from an unknown locality (Cape Verde). Eu- to subhedral nepheline phenocryst (hexagonal outline) in a nepheline- and pyroxene-rich groundmass. The surrounding brown crystals are augite. Plane-polarized light, 10× magnification, field of view = 2 mm.

B Nepheline in foidite from an unknown locality (Cape Verde). Euhedral nepheline crystal (rectangular outline) with characteristic gray birefringence, in a groundmass rich in nepheline and pyroxene. Nepheline's imperfect prismatic cleavage is also visible. Cross-polarized light, 10× magnification, field of view = 2 mm.

C Nepheline in nephelinolite from Saxony (Germany). Nepheline is allotriomorphic and encloses augite (high birefringence). Cross-polarized light, 2× magnification, field of view = 7 mm.

Scapolite $(Na,Ca,K)_4[Al_3(Al,Si)_3Si_6O_{24}](Cl,CO_3,SO_4)$

Optical Properties

Color in thin-section:	Colorless; inclusions often cause clouding
Pleochroism:	None
Birefringence:	Marialite 0.004–0.005; meionite 0.024–0.034
Relief:	Low
Cleavage:	Good {100} and {110}
Crystal system/optic sign:	Tetragonal / (–)
Other features:	Scapolite is distinguished from quartz as the latter is uniaxial and lacks cleavage; nepheline and cancrinite occur in a different paragenesis; cordierite, andalusite, and wollastonite are optically biaxial; feldspars tend to be twinned. Scapolite commonly forms large poikiloblastic prismatic crystals, but granular or fibrous-looking aggregates are especially common in skarn.

Paragenesis

Scapolite forms a solid solution between the end-members, marialite (Na-rich) and meionite (Ca-rich), but pure end-members do not occur in nature. Scapolite may occur as a primary phenocryst phase in latites, granites, and nepheline syenites, and in associated pegmatite; however, its main occurrence is largely restricted to metamorphic and metasomatic environments. Scapolite is a common mineral in regionally metamorphosed greenschist to granulite facies calcareous rocks, in metasomatic rocks, and skarns – particularly at the contacts between impure calcareous rocks and igneous bodies. The presence of scapolite is also common in hydrothermally altered basic igneous rocks, due to the action of CO_2 and Cl-rich fluids.

Figure 1.11

A Scapolite in thermally metamorphosed limestone from Bazena (Italy). Euhedral scapolite crystals with typical interference colors set in a fine-grained recrystallized matrix of calcite. Cross-polarized light, 2× magnification, field of view = 7 mm.

B Scapolite in granitic pegmatite from Pizzo Bresciadega (Italy). A large, euhedral, and fractured scapolite crystal dominates the field of view; it is surrounded by smaller scapolite crystals and is associated with quartz (lower-left) and feldspar with sericite alteration (right of large scapolite crystal). Cross-polarized light, 2× magnification, field of view = 7 mm.

Sodalite $Na_8[Al_6Si_6O_{24}]Cl_2$
Nosean $Na_8[Al_6Si_6O_{24}]SO_4 \cdot H_2O$
Haüyne $Na_6Ca_2Al_6Si_6O_{24}(SO_4)_2$

Optical Properties

Color in thin-section:	Typically colorless to rare pale pink or blue; nosean and haüyne are colorless to pale blue
Pleochroism:	None
Birefringence:	Isotropic
Relief:	Moderate to low negative
Cleavage:	Weak parallel to {110}
Crystal system/optic sign:	Cubic
Other features:	Sodalite group minerals have 6-, 8-, or 10-sided cross-sections with corroded or embayed faces. They are isotropic or, more rarely, weakly anisotropic (nosean and haüyne may show weak birefringence). They are not easy to distinguish from each other, but nosean and haüyne may have dark rims and/or distinctive exsolution; they are distinguished from leucite by its weak birefringence and characteristic twinning.

Paragenesis

These three minerals form the sodalite group. The sodalite group minerals are characteristic of alkaline igneous rocks and are commonly associated with aegirine–augite, melanite garnet, cancrinite, nepheline, leucite, and other SiO_2-undersaturated minerals; they are never found together with quartz. Late-stage hydrothermal alteration of sodalite group minerals often results in the formation of fibrolitic zeolite (commonly natrolite).

Of the minerals in the sodalite group, sodalite is the most common. It is a common rock-forming mineral in SiO_2-undersaturated, Na- and NaCl-rich plutonic and volcanic rocks, such as syenite, monzosyenite, and phonolite. It is also found in metasomatized calcareous rocks at the intrusive contact with alkaline igneous rocks. Sodalite is typically associated with nepheline, cancrinite, melanite garnet, and fluorite.

Nosean and haüyne are restricted to phonolites and related rock types, and are common in volcanic bombs and ejected volcanic blocks. Haüyne occurs in some alnöites. Haüyne and nosean form a complete solid solution at 600 °C and 100 MPa, while the sodalite–nosean and sodalite–haüyne solid solution is more limited due to the difficulty replacing Ca^{2+} and Na^+ with Cl^- and SO_4^{2-}. Nosean and haüyne are volatile-rich minerals and accommodate S, Cl, H_2O, and CO_2; the presence of these elements is directly related to the composition of the magma from which they crystallize. Thus, these minerals are very important as their composition can be used to determine, and constrain, the composition of the magma from which they crystallized, as well as to determine the concentration of volatiles under magmatic conditions.

Figure 1.12

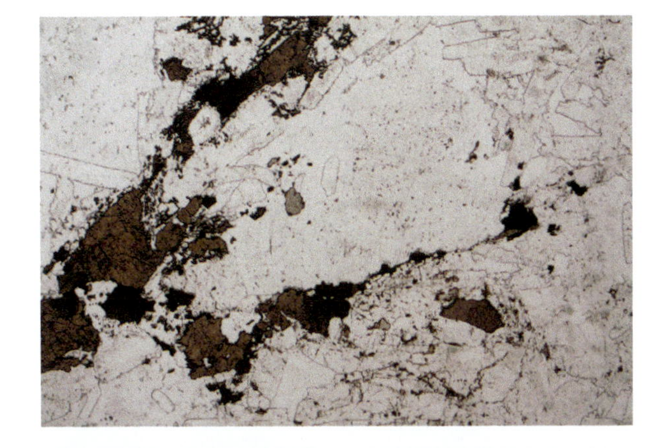

A Sodalite in syenite from Ditrău (Romania). Sodalite crystals (center and upper-left) are colorless and have low relief; they are associated with feldspar (also colorless) and pyroxene (brown). Plane-polarized light, 2× magnification, field of view = 7 mm.

B Sodalite in syenite from Ditrău (Romania). Same image as (A) but with crossed polarizers. Sodalite is isotropic and is often rich in inclusions (as seen here). Feldspar shows typical interference colors and twinning, while pyroxene has high-interference colors. Cross-polarized light, 2× magnification, field of view = 7 mm.

C Haüyne in phonolite (haüynophyre) from Melfi (Italy). Subeuhedral haüyne phenocrysts are corroded, embayed, and have dark, alteration rims. The central phenocryst shows characteristic exsolution of inclusions in linear and intersecting patterns. The green mineral is pyroxene. Plane-polarized light, 2× magnification, field of view = 7 mm.

D Nosean in phonolite from Schellkopf (Germany). Subhedral nosean phenocryst sits in a brown glassy groundmass. It has a corroded crystal face showing a typical fluid inclusion-rich core and alteration along fractures. Plane-polarized light, 10× magnification, field of view = 2 mm.

Coesite SiO_2

Optical Properties

Color in thin-section:	Colorless
Pleochroism:	None
Birefringence:	0.0050–0.0060
Relief:	Low
Cleavage:	None
Crystal system/optic sign:	Monoclinic / (+)
Other features:	High relief, low birefringence, and radial fractures in the host are diagnostic; it has slightly higher relief than quartz and radial fractures are distinctive.

Paragenesis

Coesite is a high-pressure polymorph of SiO_2. Coesite forms under the ultra-high pressure associated with meteorite impact events and the burial of crustal rocks to great depths (>90 km). It is not stable at surface conditions, requires at least 450–800 °C and ≥2.8 GPa to form, and is an indicator of ultra-high-pressure metamorphism.

At impact sites (meteor craters) it is associated with quartz and silica glass from impact-related metamorphism (e.g. at Meteor Crater in Arizona, or Ries crater in Bavaria), where it forms as a direct result of high-pressure impact events. In convergent orogens such as the Alpine Belt, the Central Orogenic Belt of China, and the Norwegian Caledonides, coesite forms during the subduction of continental crust to great depths. Coesite is preserved as inclusions in other minerals such as garnet, or reverts to quartz during exhumation and is preserved as non-annealed polycrystalline quartz pseudomorphs after coesite; the latter occur with diagnostic radial cracks resulting from the volume expansion associated with the coesite–quartz transition. Coesite is typically associated with other ultra-high-pressure indicators, such as microdiamonds, xenoliths from diamond-bearing kimberlites, and majoritic garnet.

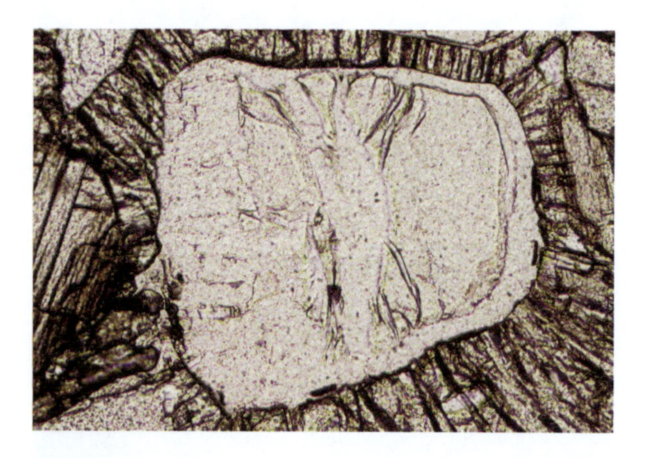

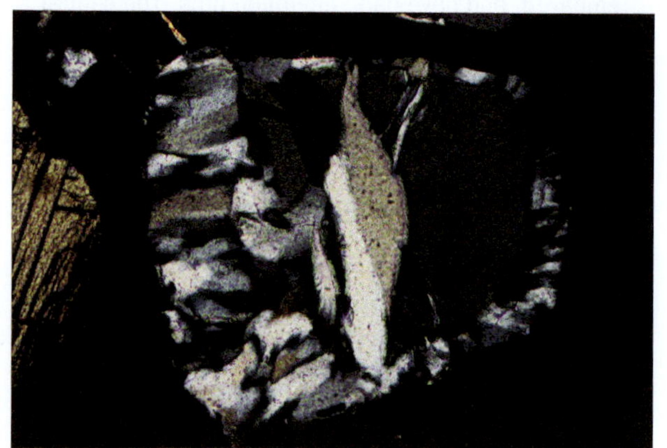

Figure 1.13

A Coesite-bearing quartzite from Dora Maira (Italy). Coesite and quartz enclosed in large pyrope crystal with numerous radial fractures. Coesite has higher relief than the quartz. Kyanite with good cleavage is also present (left of center). Plane-polarized light, 10× magnification, field of view = 2 mm.

B Coesite-bearing quartzite from Dora Maira (Italy). Same thin-section as (A) but with crossed polarizers. Coesite crystal is surrounded by a "palisade" quartz rim (thin radially, elongate grains). Cross-polarized light, 10× magnification, field of view = 2 mm.

Cristobalite SiO_2

Optical Properties

Color in thin-section:	Colorless
Pleochroism:	None
Birefringence:	0.003
Relief:	Absent
Cleavage:	None
Crystal system/optic sign:	Tetragonal / (–)
Other features:	Minute, platy crystals or spherulitic fibrous needles together with alkali feldspar grown in cavities of volcanic rock; chalcedony has higher birefringence, quartz is uniaxial (+), and chabazite has better cleavage.

Paragenesis

Cristobalite is usually found in the cavities of silica-oversaturated volcanic rocks or as a devitrification product of highly glassy rocks. It occurs in sanidine facies rocks as a contact metamorphic mineral (e.g. associated with thermally metamorphosed sedimentary xenoliths in basaltic rocks). Cristobalite has low- and high-temperature polymorphs (α- and β-cristobalite), but α-cristobalite is never stable and is metastable up to 200–275 °C. β-cristobalite is metastable above 200–275 °C and stable from 1,470–1,713 °C.

Igneous

Cristobalite is usually found in acid volcanic rocks (e.g. trachyte, dacite, rhyolite, pitchstone) and may be associated with tridymite. Both cristobalite and tridymite fill cavities of effusive rocks.

Metamorphic

Cristobalite is generated during impact events and also occurs during the thermal metamorphism of sandstone in contact with basaltic melt and in sandstone xenoliths within basic rocks.

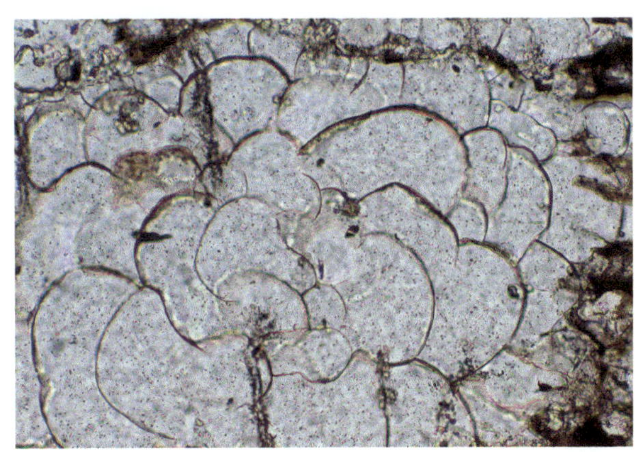

Figure 1.14

A Cristobalite in an ignimbrite from Ponza island (Italy). Fibrous cristobalite (gray interference colors, center) mantled by fibrous chalcedony (anomalous beige birefringence due to impurities). Cross-polarized light, 10× magnification, field of view = 2 mm.

B α-cristobalite in an impactite from the Bosumtwi crater (Ghana). α-cristobalite, a high-temperature polymorph of SiO_2, formed from a meteorite impact. This sample (suevite BH1-0790) shows distinctive "ballen" structure. Plane-polarized light, 100× magnification, field of view = 1.2 mm.

Image courtesy of Ludovic Ferrière (Natural History Museum Vienna).

Quartz SiO_2

Optical Properties

Color in thin-section:	Colorless
Pleochroism:	None
Birefringence:	0.0091
Relief:	Low
Cleavage:	None, only irregular, curved fractures (conchoidal fracture)
Crystal system/optic sign:	Trigonal / (+)
Other features:	The absence of cleavage, lack of alteration, low relief, and conchoidal fracture are diagnostic. May be mistaken with nepheline and cordierite, but nepheline never forms such clear crystals and is optically (–), while cordierite has higher relief and is often altered. Undulatory extinction and subgrains are common.

Paragenesis

Together with feldspar, quartz is the most abundant mineral in the Earth's crust. Quartz is a common mineral in many igneous, metamorphic, and sedimentary rocks and is also the most common polymorph of SiO_2. Quartz has low- and high-temperature polymorphs, respectively known as α-quartz and β-quartz. At ambient pressure, α-quartz is stable up to 573°C, whereas β-quartz is stable up to 870 °C and metastable above 870 °C. All quartz observed under the microscope is α-quartz.

Igneous

Quartz is ubiquitous in all acid, silica-saturated, magmatic rocks such as granites and granodiorites. Quartz is absent in ultramafic rocks and can be present up to 5 vol. % in mafic rocks. In plutonic rocks, quartz commonly forms anhedral crystals, but in rapidly cooled extrusive and hypabyssal rocks like rhyolites, dacites, and porphyries, quartz may occur as euhedral phenocrysts. Quartz phenocrysts may show signs of magmatic resorption, the so-called "embayed" texture. Since feldspathoid minerals and forsteritic olivine are incompatible in a high-silica environment, they are never found in equilibrium with quartz; nevertheless, it is very common to find quartz xenocrysts in silica-poor rocks. The presence of quartz in these rocks gives rise to various chemical reactions such as:

$$Mg_2SiO_4 + SiO_2 = Mg_2Si_2O_6$$

forsterite quartz enstatite

In this scenario, when all the orthopyroxene is crystallized, if the melt is still silica-undersaturated, the silica must be reassigned to create feldspar and feldspathoids. Similarly, the presence of quartz in feldspathoid-rich rocks is incompatible and quartz will react with feldspathoids to form feldspar:

$$(Na,K)AlSiO_4 + SiO_2 = NaAlSi_3O_8$$

nepheline quartz albite

$$KAlSi_2O_6 + SiO_2 = KAlSi_3O_8$$

leucite quartz alkali feldspar

Metamorphic

Quartz is also a common mineral in many metamorphic rocks derived from metamorphism of quartz-bearing sedimentary and igneous rocks. Quartz crystals from magmatic rocks and quartz grains from sedimentary rocks are deformed or recrystallized under most metamorphic conditions, and with increasing metamorphic grade reactions involving the release of SiO_2 can crystallize quartz.

Figure 1.15

A Quartz in granitic aplite from Elba island (Italy). Quartz (gray interference colors, center) shows undulatory extinction and subgrains. It is surrounded by plagioclase with polysynthetic twinning and biotite (high birefringence). Cross-polarized light, 2× magnification, field of view = 7 mm.

B Quartz in a rhyolitic dike from Nuoro (Sardinia). Euhedral β-quartz (now converted in α-quartz) phenocryst (hexagonal outline) set in a sericite-rich groundmass in an altered rhyolite dike. In the upper-left corner, biotite is present. Cross-polarized light, 10× magnification, field of view = 2 mm.

C Quartz in a basalt from Volsini volcanic complex (Italy). Quartz xenocryst (rounded and fractured) with a well-developed pyroxene corona, set in a pyroxene- and plagioclase-rich groundmass. Cross-polarized light, 10× magnification, field of view = 2 mm.

D Granitic gneiss from Siniscola (Sardinia). Large quartz crystal dominates the image and shows pronounced undulatory extinction; it is surrounded by plagioclase with polysynthetic twinning. Cross-polarized light, 10× magnification, field of view = 2 mm.

BOX 1.2 Polymorphism

Polymorphism – literally "many shapes" – is the ability of a mineral to crystallize in more than one form in response to changes in temperature and/or pressure. Common natural polymorphs include SiO_2, kyanite–andalusite–sillimanite, and calcite–aragonite. A mineral with a constant chemical composition can have more than one form due to the energy state within the crystal lattice; some configurations have higher (or lower) energy than others. The internal energy of a specific polymorph is a function of temperature and/or pressure. As temperature increases, the atomic vibrational energy within the lattice increases and results in a more loosely bound crystal structure. Increasing pressure, on the other hand, favors a more compact ("tighter") atomic structure and results in increased density.

A given polymorph structure is stable over a particular range of P–T conditions; this implies that a particular polymorph can transform into another when its P–T conditions change. Since polymorphs share the same chemical formula and polymorph transformation depends on temperature and pressure only, theoretically polymorph stability is unaffected by other chemical components (i.e. polymorphism is only affected by the chemical components that can be incorporated into the crystal lattice). Therefore, the presence of a mineral polymorph should reflect the pressure and temperature conditions of its host rock. It is unusual for multiple polymorphs to coexist in the same rock; in such cases, one polymorph is typically a metastable relict. This occurs because most polymorph transformations are characterized by small changes in entropy and volume.

This image illustrates the metastability of aluminosilicates in a metapelite from Mull (England). Pink andalusite pseudomorphs blue kyanite and represents a different P–T environment. In this case, kyanite is metastable, existing under andalusite P–T conditions.

Image courtesy of J. Wheeler

Tridymite SiO_2

Optical Properties

Color in thin-section:	Colorless
Pleochroism:	None
Birefringence:	0.004
Relief:	Low
Cleavage:	Rarely visible
Crystal system/optic sign:	Orthorhombic / (+)
Other features:	Forms thin, platy to flaky aggregates of six-sided tabular crystals in cavities or veins; twinning and wedge-shaped crystals are characteristic. Compared to tridymite, sodalite is isotropic, zeolites have higher refractive indices and higher birefringence, and nepheline and cristobalite are optically negative.

Paragenesis

Tridymite typically occurs in the cavities and fractures of acidic to intermediate volcanic rocks such as rhyolite, andesite, trachyte, latite, and obsidian, and is found in lunar basalts and meteorites. The presence of tridymite in many igneous rocks is due to pneumatolytic metamorphism, rather than magmatic crystallization. Tridymite together with feldspar is common in spherulites of devitrified glassy rocks. Tridymite has been recorded from high-temperature (sanidinite facies), thermally metamorphosed impure limestones.

Figure 1.16

A Tridymite in andesite from Capraia island (Italy). Platy tridymite crystals fill a large vesicle. Smaller tridymite-filled vesicles are seen throughout the section. The higher birefringence crystal (upper-right) is pyroxene. Cross-polarized light, 10× magnification, field of view = 2 mm.

B Tridymite in tuff from Eifel (Germany). Platy tridymite crystals fill two vesicles set in a fine-grained, dark, cineritic groundmass. Cross-polarized light, 10× magnification, field of view = 2 mm.

APPLICATION 1.1

Common Polymorphs

SiO_2 has six (three low-temperature and three higher-temperature) principal polymorphs. Three structural categories define the low-temperature polymorphs:

α-quartz (lowest symmetry, densest structure)

α-tridymite (higher symmetry, less dense structure)

α-cristobalite (highest symmetry, lowest density structure)

Each polymorph defines a specific P–T regime; they are linked through reconstructive transformations which require large amounts of energy and time. The slow speed of such transformations allows the polymorphs in these structural categories to persist in a metastable state for long periods of time. Each polymorph has a high-temperature phase (β-quartz, β-tridymite, and β-cristobalite). The transition from the high-temperature (β) to the low-temperature (α) phase is linked to faster and totally reversible displacive transformations.

Among the silica polymorphs, quartz is the most abundant because thermodynamically it is stable under most geological conditions. It occurs as one of two polymorphs: α-quartz and β-quartz. β-quartz is stable at temperatures above 573 °C and readily converts to α-quartz upon cooling. Tridymite and cristobalite are high-temperature/low-pressure polymorphs of silica. Tridymite is stable above 870 °C, while cristobalite is stable above 1,470 °C. Tridymite and cristobalite are common products of the devitrification of siliceous volcanic glass. Coesite and stishovite are the high-pressure polymorphs, and though neither is stable at lower P–T conditions, they remain as metastable phases for long periods of time. The transformation to a low-pressure silica polymorph is reconstructive, involving the slow and complete rearrangement of atoms in the crystal lattice.

Amorphous varieties of silica are common in nature but are not considered to be polymorphs. For example, chalcedony includes several varieties of

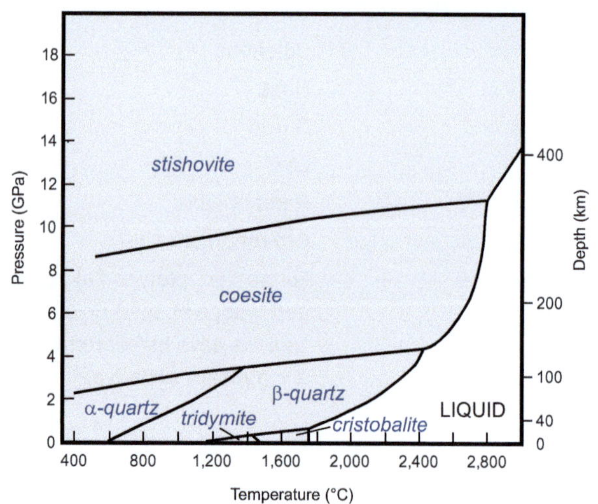

From Best (2013), with permission.

compact, microcrystalline silica. *Agate* is a variety of chalcedony that consists of alternating or concentric bands of various color. Massive or stratified opaque chalcedony is called *chert*, while nodular forms common in limestone and chalk are called *flint*. The red opaque massive form of chalcedony is called *jasper*. Amorphous silica is known as *opal*.

Al_2SiO_5 has three polymorphs: andalusite, kyanite, and sillimanite. All three occur in Al-rich metapelitic rocks and they are widely used as metamorphic *index minerals* to define metamorphic facies and P–T conditions. The Al_2SiO_5 phase diagram defines the stability fields for these polymorphs, within which a single mineral phase is stable. Within each field, pressure and temperature may vary (*divariant*), but each field is separated from the others by a *univariant* boundary along which the stable coexistence of two phases may occur (e.g. andalusite + sillimanite, andalusite + kyanite, and kyanite + sillimanite). In this phase diagram there is an *invariant* point where all three Al_2SiO_5 polymorphs may exist in equilibrium. This is one of the most important invariant points in metamorphic petrology and we define it at approximately 500 °C at 3.8 kbar and 550 °C at 4.5 kbar. Its exact location in P–T space is not precisely known, because of (1) the sluggish kinetics associated with polymorph transitions; and (2) the presence of chemical impurities such as Fe^{3+} and Mn^{3+}. The Al_2SiO_5 polymorph transformations are extremely slow and crystals of one polymorph may remain as a metastable phase in the stability field of another. The presence of more than one polymorph in a rock usually reflects metastability, and particular attention should be paid to distinguishing secondary textures (partial replacement, overgrowth, etc.) from true stable equilibrium grain boundaries.

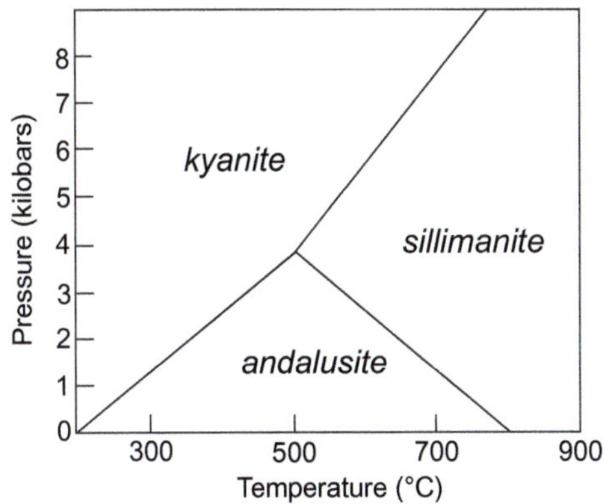

From Winter (2014), with permission.

CO_3 is found in three structural forms: rhombohedral calcite, orthorhombic aragonite, and hexagonal vaterite. Calcite is stable at atmospheric pressure, whereas aragonite is stable under high-pressure, low-temperature conditions, such as blueschist facies. The stability field of vaterite is not well established. Despite aragonite being a stable high-pressure phase, it is commonly a metastable phase within the stability field of calcite, coexisting with calcite. Aragonite is also a common phase in marine environments, where it is made from biological activity. Biological aragonite is also metastable and can convert to calcite. At lower pressure, fluids can precipitate aragonite instead of calcite. The structural parameters of biogenic aragonite are slightly different from those of metamorphic aragonite due to organic macromolecules. The calcite–aragonite phase diagram shows a univariant boundary whose slope varies with increasing pressure and temperature. The curvature of the calcite–aragonite boundary reflects the different states of internal ordering of the two phases.

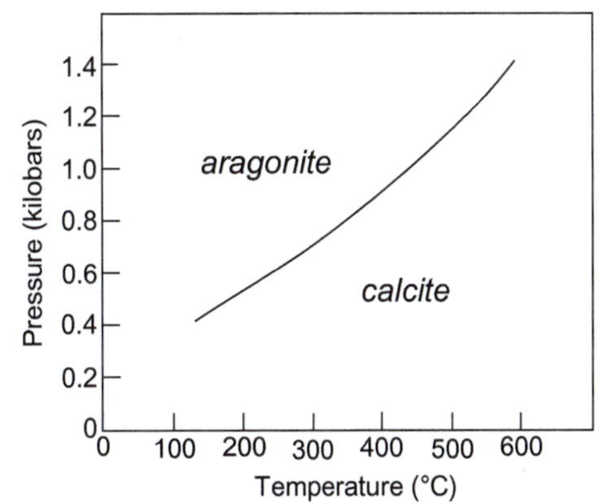

From Johannes and Puhan (1971), with permission.

BOX 1.3 The Zeolites

Zeolites are microporous, aluminosilicate minerals that form a linked framework tetrahedra with an unusually open structure and the formula $(Na_2,K_2Ca,Ba)[(Al,Si)O_2]_x\ nH_2O$. Zeolites are known as "molecular sieves" because of their ability to sort molecules based on size. Their molecular structures and dimensions determine the maximum size of the molecular or ionic species that can enter zeolite pores (as shown by the blue area below).

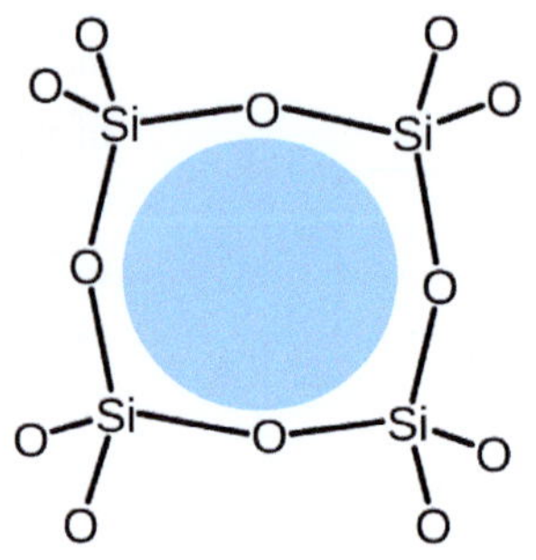

This "open" structure can consist of four-, six-, and even eight-membered pores. These "channels" and "cavities" accept diverse ions and water molecules. Consequently, the zeolite group is large and diverse; in 2019 there were >250 known species. Today, zeolites are classified within the following subgroups according to their dominant cation, e.g. chabazite (Ca).

Zeolite subgroups
Analcime (Na)
Natrolite (Na)
Harmotome (Ba)/phillipsite (K, Ca, Na)
Laumontite (Ca) and others
Heulandite (Ca, Na, K), includes stilbite and brewsterite
Chabazite (Ca)
Mordenite (Na, K, Ca)

Zeolite commonly occurs as euhedral crystals in amygdales and fissures of basic volcanic rocks, where it forms from late-stage fluids permeating basalt after extrusion. These amygdales often contain more than one species of zeolite. Zeolite is associated with tuffs/tuffaceous sediments in marine and lake environments. Zeolite occurs in low-grade metamorphic rocks as the result of hydrothermal and/or burial metamorphism. Zeolites, as hydrous aluminosilicates, are relatively sensitive to temperature and pressure, and the sequence chabazite → analcime → heulandite documents increasing temperature. Analcime is a zeolite, but has close affinities with the feldspathoids and can be associated with a higher-temperature paragenesis.

Pure zeolite is colorless or white, but many specimens are colored due to the presence of impurities. The radiating habit of natrolite is fairly distinctive, while analcime, chabazite, heulandite, and stilbite typically form somewhat trapezohedral or rhombohedral tabular crystals.

BOX 1.3 The Zeolites (*cont.*)

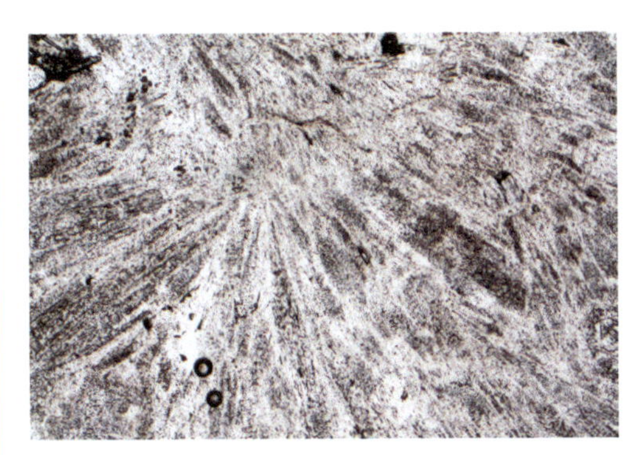

Natrolite in alkaline pegmatite from Tvedalen (Norway). The entire view is dominated by natrolite crystals of radiating habit. Natrolite has a dusty appearance due to numerous inclusions in plane-polarized light (*left*) and shows typical low first-order birefringence in cross-polarized light (*right*). Both images: magnification = 2×, field of view = 7 mm.

Analcime in analcime-gabbro (teschenite) from Queensferry (Scotland). In plane-polarized light (*left*) analcime is white with low relief and surrounded by higher-relief titanium augite (brownish) and dusty plagioclase. In cross-polarized light (*right*), analcime is isotropic, while augite shows high birefringence and plagioclase shows first-order birefringence. Both images: 10× magnification, field of view = 2 mm.

1.3 Sheet Silicates

Biotite $K(Mg,Fe^{2+})_3(Al,Fe^{3+})Si_3O_{10}(OH,F)_2$

THE MICAS

Optical Properties

Color in thin-section:	Pale yellows, greens, browns
Pleochroism:	Strong from pale to dark yellow, green, or brown
Birefringence:	0.008–0.028
Relief:	High
Cleavage:	Perfect basal cleavage on {001}
Crystal system/optic sign:	Monoclinic/ (–)
Other features:	Biotite is distinguished by its strong pleochroism and its so-called bird's eye extinction; phlogopite has a weaker pleochroism in yellowish colors. The relief is commonly higher than white mica. Pleochroic halos around inclusions such as zircon are common in biotite. Biotite commonly alters to chlorite.

Paragenesis

Biotite includes the trioctahedral micas and has a wide compositional range between phlogopite (Mg-rich end-member), annite (Fe-rich end-member), and siderophyllite (Al-rich end-member). It is a common mineral in many igneous and metamorphic rocks. In igneous rocks, biotite occurs predominantly in granites and peralkaline rocks, as well as some lamprophyres. Phlogopite occurs in ultrabasic rocks, such as kimberlites. In metamorphic rocks, biotite is a characteristic mineral of the greenschist to amphibolite facies, occurring in metapelite and metabasite; phlogopite is associated with the metamorphism of impure limestone and dolomite.

Igneous

Biotite is common in many acid to intermediate igneous rocks. In general, biotite tends to be more Fe-rich in acid igneous rocks and more Mg-rich in intermediate or mafic rock types. Biotite is rare in mafic rocks, but some biotite-bearing norite and gabbro exist in nature. Phlogopite has been reported from kimberlites and some leucite-bearing rocks. As a hydrous mineral, biotite is stable at depth and in volcanic rocks commonly occurs as partially or completely resorbed phenocrystals with a dark "opacitic" rim.

Metamorphic

In metapelites, metabasites, and metagranitoids biotite appears at upper greenschist facies from continuous reactions between white mica and/or K-feldspar with chlorite. The metamorphic reactions that lead to the formation of biotite are numerous; for example:

$$\text{muscovite} \longrightarrow \text{biotite} + \text{K-feldspar} + \text{quartz} + H_2O$$
$$\text{muscovite} + \text{chlorite} \longrightarrow \text{biotite}$$

At medium grade, biotite may coexist with different ferromagnesian minerals (garnet, staurolite, cordierite, amphibole, etc.). At high grade (e.g. granulite facies), biotite is consumed by continuous and discontinuous reactions that may produce melt (i.e. biotite dehydration melting) or cause the breakdown of biotite to K-feldspar and orthopyroxene. Biotite is not stable at blueschist to eclogite facies, and at these conditions it is replaced by white mica, amphibole, garnet, and/or clinopyroxene.

Figure 1.17

A Biotite in ignimbrite from Viterbo (Italy). Brown, prismatic biotite phenocrysts showing {001} cleavage (left of center) with apatite inclusions (colorless, prismatic crystals), set in a glassy groundmass. Plane-polarized light, 2× magnification, field of view = 7 mm.

B Phlogopite in lamproite from Corsica (France). Phlogopite phenocrysts with characteristic yellowish to pale brown colors, set in a fine-grained sanidine- and pyroxene-rich groundmass. Plane-polarized light, 2× magnification, field of view = 7 mm.

C Biotite in granite from Amparo (Brazil). Large biotite crystal with bird's eye extinction: The surface of the crystal has a "pebbly" appearance. The dark area surrounding biotite is quartz at extinction. Cross-polarized light, 10× magnification, field of view = 2 mm.

D Biotite in sillimanite–gneiss from Lombardy (Italy). Biotite crystals (orange–brown) occur with hypidiomorphic garnet crystals (high relief, brown), plagioclase (colorless), and small, elongate sillimanite crystals (mixed with biotite crystals). Plane-polarized light, 10× magnification, field of view = 2 mm.

Muscovite $K_2Al_4[Si_6Al_2O_{20}](OH,F)_4$

Optical Properties

Color in thin-section:	Colorless
Pleochroism:	Weak (higher absorption parallel to cleavage planes)
Birefringence:	0.035–0.042; sections parallel to cleavage have low-interference colors of the first order
Relief:	Moderate (changes with rotation)
Cleavage:	Perfect {001}
Crystal system/optic sign:	Monoclinic / (–)
Other features:	Muscovite can be distinguished from other phyllosilicate minerals by its high birefringence. Parallel intergrowth between biotite and muscovite is common. Muscovite has straight or nearly straight extinction in sections perpendicular to cleavage.

Paragenesis

Muscovite is common and occurs in both igneous and metamorphic rocks, in a wide variety of geological environments, from lower greenschist to upper amphibolite facies. Muscovite is not readily altered, but can be converted to clay minerals with weathering.

Igneous

Muscovite is less common than biotite in plutonic rocks, but common in Al-rich granitoids where it may occur together with cordierite. The stability curve of muscovite intersects the minimum melting curve of granite (P = 0.35 GPa; T = 700 °C); therefore it can crystallize from liquid of granitic composition. Below 0.35 GPa muscovite forms mainly as the result of hydrothermal alteration of feldspar or through reactions such as:

andalusite (any aluminosilicates) + K-feldspar + vapor ↔ muscovite + quartz

Muscovite is one of the most problematic minerals of the peraluminous granitoids because in many rocks muscovite may have been derived in part or entirely from the alteration of peraluminous minerals (cordierite, aluminosilicates, corundum, etc.). Muscovite is also a common constituent of pegmatites, where it occurs together with biotite, feldspar, quartz, and many other pegmatitic phases.

Metamorphic

Muscovite is a common K-bearing phase in several metamorphic rocks. At increasing temperature, muscovite grows from clay minerals or due to the destabilization of K-feldspar in metapelites, metasandstones, metamarls, and metagranitoids. At low metamorphic grade, white mica is phengitic and occurs commonly associated with chlorite, albite, K-feldspar, biotite, epidote, quartz, and/or calcite. At amphibolite facies, muscovite destabilizes, producing K-feldspar and Al-silicates:

muscovite + quartz ⟶ K-feldspar + sillimanite

At blueschist facies conditions, muscovite tends to become increasingly phengitic, substituting Al for (Fe, Mg); at these conditions it commonly coexists with glaucophane, lawsonite, and chloritoid. Paragonite occurs in Na- and Al-rich metamorphic rocks and may be substituted by albite or plagioclase during regional metamorphism or by glaucophane and omphacite at high-pressure conditions.

Figure 1.18

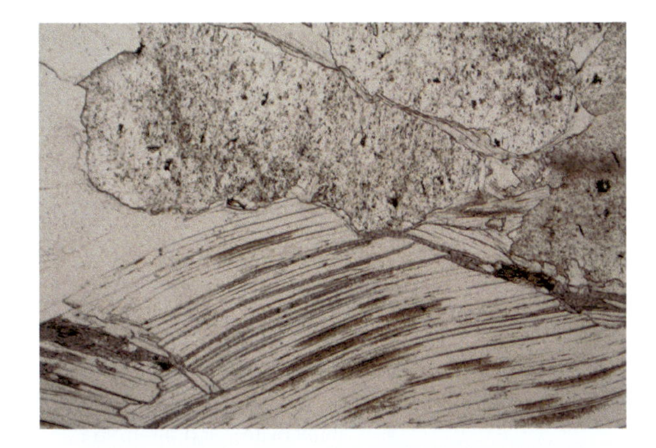

A Muscovite in granite pegmatite from Adamello (Italy). Large muscovite crystal (colorless, bottom) with good cleavage, in association with orthoclase (top-center) cloudy due to sericite alteration. Muscovite parallel to the basal section {110} is also present (upper-left) and therefore no cleavage is visible. Plane-polarized light, 2× magnification, field of view = 7 mm.

B Muscovite in granite pegmatite from Adamello (Italy). Same image as (A) but with crossed polarizers. Muscovite shows its characteristic birefringence, and the muscovite crystal cut parallel to {110} (upper-left) has low-interference colors (yellow to white and dark gray). Orthoclase at extinction. Cross-polarized light, 2× magnification, field of view = 7 mm.

C Muscovite in gneiss from Alpe Arami (Switzerland). Colorless muscovite alternates with brown biotite. In both cases the {110} cleavage is clearly visible. Muscovite and biotite together define the coarse-grained foliation. Plane-polarized light, 2× magnification, field of view = 7 mm.

D Muscovite in gneiss from Alpe Arami (Switzerland). Same image as (C) but with crossed polarizers. Note the variation in birefringence between muscovite and biotite. Cross-polarized light, 2× magnification, field of view = 7 mm.

Sericite $K_2Al_4[Si_6Al_2O_{20}](OH,F)_4$

Optical Properties

Color in thin-section:	Colorless
Pleochroism:	Absent
Birefringence:	0.036–0.054
Relief:	Low to moderate
Cleavage:	Not visible (due to fine-grained character)
Crystal system/optic sign:	Monoclinic / (–)
Other features:	Sericite is identified by its fine grain size (typically <50 μm), birefringence (identical to muscovite), and alteration association with feldspar.

Paragenesis

Sericite is a general term for fine-grained mica (muscovite or paragonite), commonly found as the alteration product of feldspar. Sericite is chemically indistinguishable from muscovite, although it can show high SiO_2, MgO, and H_2O, and low K_2O. Sericite has a wide paragenesis, but is always linked to fluid circulation and hydrothermal processes. Sericite is a common product associated with contact-metamorphosed rocks and abundant in the outer zones of contact aureoles; at the intrusion contact, the crystallinity of sericite increases with increasing temperature and muscovite forms. In some low-grade metamorphic rocks, sericite can be a rock-forming mineral giving rise to sericite–phyllites, sericite–greenschist, and sericite–quartzite. Sericite is a common product in porphyry ore deposits, where it is a dominant mineral in sericitic and phyllitic alteration zones.

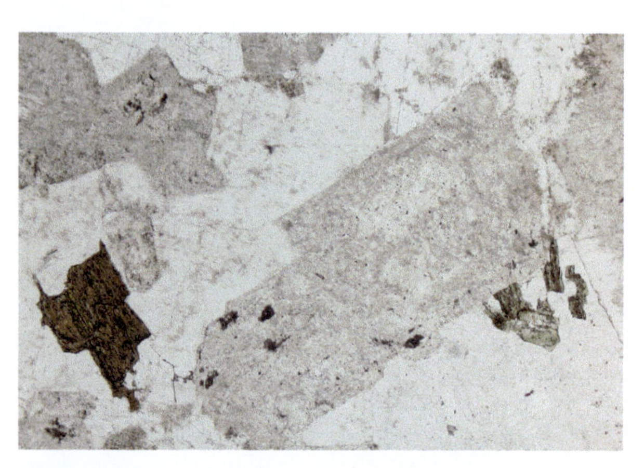

Figure 1.19

A Sericitized plagioclase in granite from Gavorrano (Italy). Strong sericite alteration makes plagioclase crystals clearly visible; they are surrounded by quartz (clear, colorless) and highly altered biotite crystals (green–brown). Plane-polarized light, 2× magnification, field of view = 7 mm.

B Sericitized plagioclase in granite from Gavorrano (Italy). Same image as (A) but with crossed polarizers. Sericitization of plagioclase (center) has the high birefringence of white mica; it is surrounded by quartz, mostly at extinction. Cross-polarized light, 2× magnification, field of view = 7 mm.

Chlorite Group $(Mg, Fe^{2+})_{10}Al_2[Al_2Si_6O_{20}](OH)_{16}$

Optical Properties

Color in thin-section:	Colorless or green
Pleochroism:	Green varieties are pale green to colorless
Birefringence:	0.00–0.01
Relief:	Moderate
Cleavage:	Perfect {001} basal cleavage
Crystal system/optic sign:	Monoclinic / Mg-rich (+), Fe-rich (–)
Other features:	Chlorite is distinguished by its lower birefringence and anomalous interference colors. Although some serpentine minerals can have a similar platy morphology, they have a lower birefringence. Green biotite, glauconite, and celadonite can have a similar color and pleochroism as chlorite, but all have higher birefringence. Mg-rich chlorite has brown anomalous interference colors, while Fe-rich chlorite is violet to blue.

Paragenesis

The chlorite group represents a solid solution between the Mg- and Fe-rich end-members clinochlore and chamosite, respectively. "Chlorite" is often used to describe any member in this series. Its optical properties are controlled by the Fe/Mg ratio and the replacement of Si by Al. There is widespread solid-solution between Mg varieties (clinochlore) and Fe varieties (chamosite). Chlorite is a common product of weathering and occurs in many argillaceous rocks and in some iron-rich sediments.

Igneous

In igneous rocks the presence of chlorite is linked to the hydrothermal alteration of ferromagnesic minerals such as pyroxenes, amphiboles, and biotite. It is also a relatively common mineral filling vesicles, forming amygdales in many volcanic rocks.

Metamorphic

Chlorite is a widespread and very common mineral, from the lower part of zeolite facies up to medium-grade regional metamorphic rocks (~400 °C and ~0.3 GPa). As metamorphism increases, chlorite can be involved in numerous reactions, such as the transition between the zeolite to prehnite–pumpellyite facies:

$$\text{chlorite} + \text{laumontite} + \text{prehnite} \longrightarrow \text{pumpellyite} + H_2O$$
$$\text{laumontite} + \text{chlorite} + \text{calcite} \longrightarrow \text{pumpellyite} + H_2O + CO_2$$

As pressure and temperature increase, chlorite gives way to biotite via the reaction:

$$\text{chlorite} + \text{muscovite} + \text{quartz} \longrightarrow \text{andalusite} + \text{biotite} + \text{cordierite} + H_2O$$

and at pressures in excess of 0.3 GPa, its upper thermal stability is controlled by the reaction:

$$\text{chlorite} \longrightarrow \text{enstatite} + \text{spinel} + \text{forsterite} + H_2O$$

In high-grade metamorphic rocks, chlorite is absent or present only in minor amounts; if present, it is linked to the alteration of ferromagnesian phases such as garnet, staurolite, and biotite.

Figure 1.20

A Chlorite in granite from Elba island (Italy). Chlorite, with typical green pleochroism, partially substitutes for brown biotite. The elongate, high-relief mineral enclosed in biotite (lower-center) is apatite. Other phases include quartz (clear) and plagioclase (partially altered by "dusty" sericite). Plane-polarized light, 2× magnification, field of view = 7 mm.

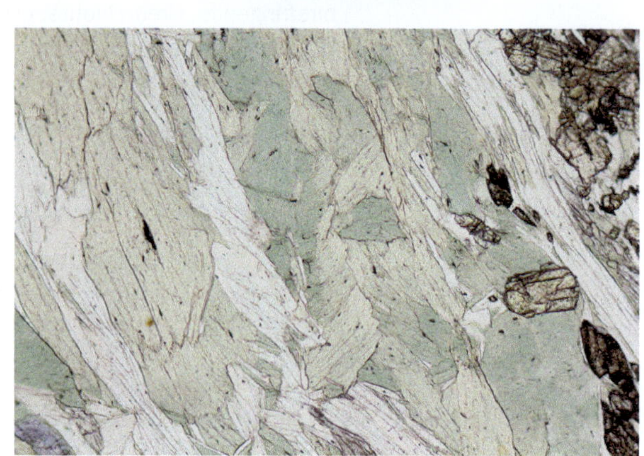

B Chlorite schist from Val Vigezzo (Italy). Chlorite aggregate with typical green pleochroism associated with colorless muscovite. The high-relief, pale yellowish minerals are epidote. Plane-polarized light, 2× magnification, field of view = 7 mm.

C Chlorite schist from Val Vigezzo (Italy). Same image as (B) but with crossed polarizers. Chlorite shows anomalous low-interference colors while muscovite has higher birefringence. Cross-polarized light, 2× magnification, field of view = 7 mm.

Prehnite $Ca_2(Al,Fe^{3+})[AlSi_3O_{10}](OH)_2$

Optical Properties

Color in thin-section:	Colorless
Pleochroism:	None
Birefringence:	0.022–0.051 (increases with increasing Fe)
Relief:	Medium to high
Cleavage:	Good {001}, weak {010}
Crystal system/optic sign:	Orthorhombic / (+)
Other features:	Colorless with bright, pure second- and third-order interference colors is diagnostic. Fine lamellar twinning may occur. Radial aggregates and barrel-shaped clusters often give the characteristic imperfect "bow tie" or "hour glass" appearance. Lawsonite has lower birefringence.

Paragenesis

Prehnite is a secondary or hydrothermal mineral in veins and amygdales in basic volcanic rocks and in contact metamorphism of calcareous rocks. It indicates low-grade regional and burial metamorphism, and defines the prehnite–pumpellyite facies. At prehnite–pumpellyite facies, the transition from the prehnite to the pumpellyite zone occurs at 260 °C at 0.7 GPa via the reaction:

$$\text{chlorite} + \text{prehnite} \longrightarrow \text{pumpellyite} + \text{quartz}$$

At higher temperatures prehnite breaks down to anorthite + wollastonite + H_2O. The low-temperature alteration of prehnite may produce a zeolite or a chloritic material.

Igneous

Prehnite in basic volcanic rocks is often associated with calcite and zeolites. In plutonic rocks it occurs in veins and as pseudomorphs (e.g. after laumontite and clinozoisite).

Metamorphic

Prehnite occurs in contact-altered impure limestones and marls, and in rocks which have undergone calcium metasomatism, such as rodingites or garnet-bearing gabbros; in the latter it is often associated with diopside and hydrogrossular garnet. The definitive reaction separating the prehnite–pumpellyite facies from the prehnite–amphibolite facies is:

$$\text{prehnite} + \text{chlorite} + \text{quartz} \longrightarrow \text{pumpellyite} + H_2O$$

With increasing temperatures, prehnite–actinolite assemblages give way directly to the greenschist facies by the reaction:

$$\text{prehnite} + \text{chlorite} + \text{quartz} \longrightarrow \text{clinozoisite} + \text{tremolite} + H_2O$$

The upper limit of prehnite stability occurs at about 400 °C at 0.2–0.4 GPa ($P_{H_2O} = P_{Total}$) and is defined by the reaction:

$$5\ \text{prehnite} \longleftrightarrow 2\ \text{grossular} + 2\ \text{zoisite} + 3\ \text{quartz} + 4\ H_2O$$

The temperature range of the prehnite–pumpellyite facies in natural environments at 0.3 GPa is about 250–380 °C.

Figure 1.21

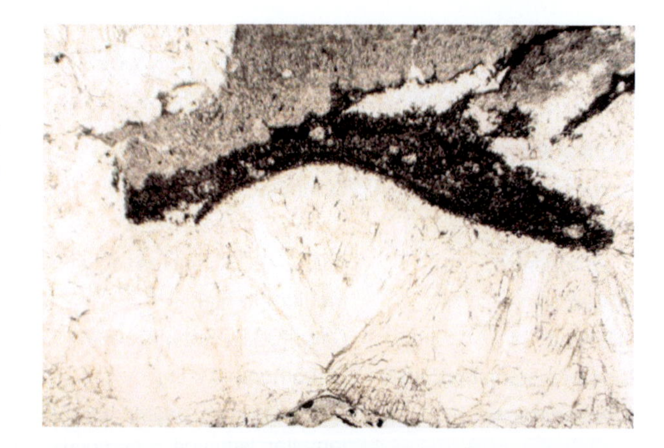

A Prehnite in altered basalt, Loanhead Quarry (Scotland). Colorless prehnite crystals occupy most of the image; original basaltic material (black) occurs in the center of the image. Plane-polarized light, 2× magnification, field of view = 7 mm.

B Prehnite in altered basalt, Loanhead Quarry (Scotland). Same image as in (A) but with crossed polarizers. Note the radial habit and second-order interference colors. Cross-polarized light, 2× magnification, field of view = 7 mm.

C Prehnite in metalimestone from Orciatico (Italy). Prehnite occurs between calcite crystals (it has lower relief than calcite). Along the left edge of the image its tabular form can be seen. Cross-polarized light, 2× magnification, field of view = 7 mm.

D Prehnite in metalimestone from Orciatico (Italy). Same image as in (C) but with crossed polarizers. Note the clarity and brightness of the interference colors, which is typical. Cross-polarized light, 2× magnification, field of view = 7 mm.

Pyrophyllite $Al_4(Si_8O_{20})(OH)_4$

Optical Properties

Color in thin-section:	Colorless
Pleochroism:	None
Birefringence:	0.043–0.05
Relief:	Low to moderate
Cleavage:	Perfect {001}, {100}, but moderate {010} and poor {100} in Fe-rich varieties
Crystal system/optic sign:	Monoclinic or triclinic / (–)
Other features:	Occurs as fine-grained foliated lamellae with platy cleavage, radiating needles/granules, or massive spherulitic aggregates of smaller crystals; fine grain size and high interference colors make it difficult to distinguish from muscovite and talc.

Paragenesis

Pyrophyllite is a relatively uncommon mineral that occurs in relatively low-grade Al-rich metamorphic rocks (e.g. metapelite, metabauxite, metaquartzite up to greenschist facies), and also in blueschist high-pressure/low-temperature rocks. It is also formed by hydrothermal alteration of aluminous minerals such as feldspar and muscovite in moderately to highly siliceous rocks. Pyrophyllite has the dioctahedral layered structure of muscovite and in natural systems forms via various reactions:

$$\text{pyrophyllite} + \text{chlorite} \longrightarrow \text{chloritoid} + \text{quartz} + H_2O$$

$$\text{pyrophyllite} + \text{calcite} \longrightarrow \text{margarite} + \text{quartz} + H_2O + CO_2$$

Experimental studies in the Al_2O_3–SiO_2–H_2O system indicate pyrophyllite may form as follows:

at c. 300 °C (0.1-1 GPa): kaolinite + quartz $\longrightarrow$ pyrophyllite + H_2O

at c. 350-420 °C (and increasing P):
pyrophyllite $\longrightarrow$ quartz + andalusite/kyanite + H_2O

Figure 1.22

A Pyrophyllite in metapelite from California (USA) in cross-polarized light. Radiating fan-shaped crystals with high birefringence are diagnostic of pyrophyllite and occur with altered portions of the original rock (brownish material in top image). (*Top*) 2× magnification, field of view = 7 mm. (*Bottom*) Close-up, 10× magnification, field of view = 2 mm.

Antigorite $Mg_3[Si_2O_5](OH)_4$

Optical Properties **THE SERPENTINE MINERALS**

Color in thin-section:	Colorless to pale green
Pleochroism:	Fe-bearing varieties have a weak pleochroism from pale greenish-yellow to pale green
Birefringence:	0.004–0.007
Relief:	Low
Cleavage:	Perfect on {001}, partings {100} and {010}
Crystal system/optic sign:	Monoclinic / (–)
Other features:	Antigorite forms small platy crystals and is distinguished from Mg-chlorite by the anomalous colors and positive optic sign of the latter, and from talc which has higher birefringence. Antigorite can only be differentiated from fine-grained lizardite by X-ray analysis.

Paragenesis

Antigorite is the high-pressure variety of serpentine, and is commonly found in low-temperature, high-pressure environments; it is a common mineral in paleo-subduction zones. Compared to the other two main varieties of the serpentine group, lizardite and chrysotile, antigorite is the less common and its presence indicates prograde metamorphism or that the host was serpentinized in a higher P–T regime than lizardite and chrysotile. Antigorite is very rare in oceanic serpentinites, and in these rocks is generally restricted to veins or shear zones. Serpentine minerals, like antigorite, contains 13 wt% H_2O and they play a key role in water transport into the deep mantle. The dehydration process of serpentine minerals during slab subduction is believed to be one of the major causes of mantle wedge hydration. Antigorite is stable up to 600–700 °C; above this temperature, antigorite → forsterite + enstatite + H_2O. It may be accompanied by lizardite and can form the major component in pseudomorphs after orthopyroxene (bastite).

Figure 1.23

A Antigorite in serpentinite from Bergamo (Italy). Antigorite crystals (colorless, platy habit) in association with magnetite (black), talc (see XPL) and calcite (high-relief crystal, right). Plane-polarized light, 2× magnification, field of view = 7 mm.

B Antigorite in serpentinite from Bergamo (Italy). Same image as (A) but with crossed polarizers. Antigorite crystals show platy habit and low-interference colors. Talc shows second-order interference colors. Cross-polarized light, 2× magnification, field of view = 7 mm.

Chrysotile $Mg_3[Si_2O_5](OH)_4$

THE SERPENTINE MINERALS

Optical Properties

Color in thin-section:	Colorless (rare Fe-bearing varieties are greenish)
Pleochroism:	Absent to weak, from greenish-yellow to pale green
Birefringence:	0.013–0.017
Relief:	Low
Cleavage:	Fibrous
Crystal system/optic sign:	Monoclinic/ (–)
Other features:	The fine grain size of serpentine minerals makes them difficult to distinguish from one another; they have lower birefringence than chlorite and fibrous amphiboles. X-ray analysis is needed for a correct and precise characterization.

Paragenesis

Lizardite, antigorite, and chrysotile are the main serpentine minerals. At temperatures below 400 °C they form from the hydrothermal alteration of ultrabasic rocks (e.g. peridotites, pyroxenites). Serpentine may also form in thermally metamorphosed siliceous dolomitic limestones. Chrysotile, together with lizardite, are the main serpentine minerals in low-grade serpentinites from the oceanic lithosphere and from low-grade metamorphic ophiolites. Chrysotile, also known as white asbestos, is commonly found as veins. Given its mechanical strength, thermal stability, and low thermal conductivity, it has wide industrial uses.

The serpentinization process generally occurs at temperatures <400 °C and leads to the progressive hydration of minerals such as olivine and pyroxene during interaction with seawater at decreasing temperature (retrograde alteration). The textural and mineralogical assemblage can be very complex and reflects multiple phases of alteration, deformation during exhumation processes, and late-stage, low-temperature weathering. Although serpentinization involves numerous reactions, it is dominated by the progressive alteration of olivine to lizardite and chrysotile:

$$\underset{\text{fosterite}}{2Mg_2SiO_4} + 3H_2O \rightarrow \underset{\text{serpentine}}{Mg_3SiO_5(OH)_4}$$

If olivine contains too much Fe to be accommodated in the serpentine structure, it is exsolved as iron oxide (magnetite). Serpentinization processes are commonly accompanied by volume expansion, which is clearly visible at the outcrop scale and in thin-section. Chrysotile is stable to 500–750 °C, but above this begins to alter to its anhydrous phase (metachrysotile) via water loss. If the temperature rises further, first forsterite is formed and then tremolite:

$$\underset{\text{chrysotile}}{Mg_3Si_2O_5(OH)_4} \rightarrow \underset{\text{metachrysotile}}{Mg_2Si_2O_7} + H_2O + \underset{\text{fosterite}}{Mg_2SiO_4} + \underset{\text{enstatite}}{MgSiO_3}$$

Figure 1.24

A Chrysotile in serpentinite from Tuscany (Italy). Massive serpentine characterized by fibrous layers. Chrysotile has low-interference colors. Cross-polarized light, 2× magnification, field of view = 7 mm.

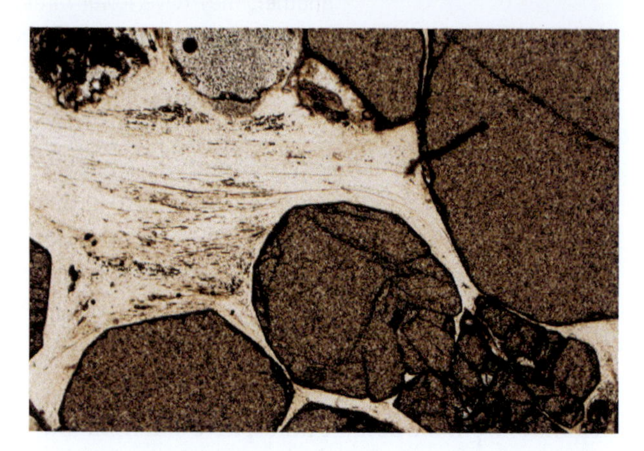

B Chrysotile (white asbestos) in a garnet-rich serpentinite from Valmalenco (Italy). Colorless, fibrous chrysotile crystals between rounded andradite garnets. Plane-polarized light, 10× magnification, field of view = 2 mm.

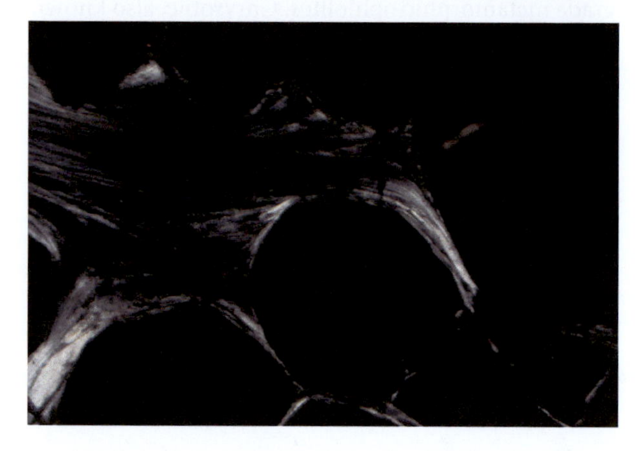

C Chrysotile (white asbestos) in a garnet-rich serpentinite from Valmalenco (Italy). Same image as (B) but with crossed polarizers. Fibrous chrysotile crystals (low birefringence) between isotropic andradite garnets (var. demantoid). Cross-polarized light, 10× magnification, field of view = 2 mm.

Lizardite $Mg_3[Si_2O_5](OH)_4$

Optical Properties

Color in thin-section:	Colorless to pale green
Pleochroism:	Absent
Birefringence:	0.006–0.008
Relief:	Low
Cleavage:	Perfect on {001}
Crystal system/optic sign:	Orthorhombic / (–)
Other features:	Lizardite forms fine-grained platy or lath-like crystals. The distinction between lizardite and antigorite usually requires X-ray analysis, but fibrous lizardite is "length-fast," compared to "length-slow" chrysotile.

Paragenesis

Lizardite together with antigorite are the most common serpentine minerals, and are formed principally by retrograde hydrothermal alteration of ultrabasic rocks such as dunite, peridotite, or pyroxenite. Lizardite dominates low-temperature serpentinites from blueschist to greenschist facies (~300–360 °C), whereas antigorite appears as temperature progressively increases. Lizardite often forms the matrix material of chrysotile veins and is commonly a retrograde product: a pseudomorph of olivine, forming the "hour glass" texture, or a pseudomorph of pyroxene, generating the "bastite" texture.

Figure 1.25

A Lizardite in serpentinite from Baden (Germany). The rock is completely composed of lizardite pseudomorphing olivine and generating the classic "hour glass" or "mesh" texture. Cross-polarized light, 10× magnification, field of view = 2 mm.

B Lizardite in serpentinite from Baden (Germany). A large pyroxene crystal (center) is completely replaced by lizardite, generating the "bastite" texture. Cross-polarized light, 10× magnification, field of view = 2 mm.

Stilpnomelane $(K,Na,Ca)_{0.6}(Mg,Fe^{2+},Fe^{3+})_6Si_8Al(O,OH)_{27}2–4H_2O$

Optical Properties

Color in thin-section:	Pale yellow, brown, or green
Pleochroism:	Yellow to golden, greenish- and reddish-brown, black
Birefringence:	0.03–0.11
Relief:	High but varies with rotation
Cleavage:	Perfect {001}, imperfect {110}
Crystal system/optic sign:	Triclinic / (–)
Other features:	Micaceous, fan-like, or radiating aggregates. $Fe^{2+,3+}$ substitution influences pleochroism. May resemble biotite but has an additional cleavage, higher birefringence, and lacks "bird's eye" extinction.

Paragenesis

Stilpnomelane is a sheet silicate that is common in low-grade regionally metamorphosed rocks (e.g. greenschists) and also occurs in blueschists and silicate–iron formations. In greenschist facies rocks it is associated with chlorite, albite, and muscovite; in blueschists with glaucophane, garnet, calcite, epidote group minerals, and pumpellyite; in the iron formations it is accompanied by minerals such as minnesotaite, greenalite, chamosite, and grunerite. It is commonly zoned from core (Fe^{2+}) to rim (Fe^{3+}), reflecting the coupled substitution of Fe, Mg, and OH, and its color and pleochroism vary with Fe content. Stilpnomelane decomposes to biotite and almandine garnet at higher temperatures, to zussmanite and chlorite at higher pressures, and alters to chlorite, iron oxides, and clay.

Figure 1.26 Stilpnomelane in meta-ironstone from Laytonville quarry (California). In plane-polarized light (*left*) stilpnomelane crystals form golden-brown, radiating aggregates in a recrystallized quartz-rich matrix. Small, fibrous dark blue–green crystals (top, left of center) are howieite. In crossed-polarized light (*right*), the birefringence of stilpnomelane has high first-order colors. 2× magnification, field of view = 7 mm.

Talc $Mg_6[Si_8O_{20}](OH)_4$

Optical Properties

Color in thin-section:	Colorless
Pleochroism:	None
Birefringence:	c. 0.050; basal sections give very low first-order colors
Relief:	Low
Cleavage:	Perfect {001}
Twinning:	None
Crystal system/optic sign:	Triclinic or monoclinic / (–)
Other features:	Tabular crystals exhibit perfect {001} cleavage but more often talc occurs as fibrous or radiating aggregates. Talc is best distinguished from pyrophyllite and sericite by chemical or X-ray analyses. Pyrophyllite and muscovite have a larger 2V; brucite is optically (+) with anomalous interference colors; gibbsite is optically biaxial (+) and has oblique extinction.

Paragenesis

Talc occurs mainly in low-grade metamorphosed ultrabasic rocks and in hydrothermally altered basic to ultrabasic rocks, in contact and regionally metamorphosed siliceous dolomites, and in some rare high-pressure metapelites called "whiteschists." Talc can form important economic deposits and monomineralic rocks formed of talc are called soapstone or steatites.

Igneous

No igneous occurrence is known.

Metamorphic

Talc is a relatively common mineral in ultrabasic Mg-rich rocks. In the lower metamorphic facies, talc is often associated with serpentine, tremolite, forsterite, and carbonates (calcite, dolomite, or magnesite). It is also an alteration product of tremolite and/or forsterite.

The formation of talc in serpentine rocks is due to numerous factors (e.g. addition of SiO_2, addition of CO_2, or removal of Mg):

$$\underset{\text{serpentine}}{2Mg_3Si_2O_5(OH)_4} + 3CO_2 \rightarrow \underset{\text{talc}}{Mg_3Si_4O_{10}(OH)_2} + \underset{\text{magnesite}}{3MgCO_3} + 3H_2O$$

In serpentinites subjected to thermal metamorphism, talc can be produced by the decomposition of antigorite: antigorite → forsterite + talc + H_2O, while in siliceous dolomites talc can be produced by the reaction between dolomite and silica:

$$\underset{\text{dolomite}}{3CaMg(CO_3)_2} + \underset{\text{quartz}}{4SiO_2} + H_2O \rightarrow \underset{\text{talc}}{Mg_3Si_4O_{10}(OH)_2} + \underset{\text{calcite}}{CaCO_3} + 3CO_2$$

Talc also forms from the high-pressure (eclogite facies) metamorphism of Mg-rich pelites, known as "whiteschists." In these rocks, talc is associated with phengite, kyanite, and garnet, and talc formation is associated with the progressive destabilization of chlorite and quartz:

$$\text{chlorite} + \text{quartz} \rightarrow \text{kyanite} + \text{talc} + H_2O$$

Figure 1.27

A Talc in serpentinite from Tyrol (Italy). Fibrous, hair-like talc crystals (very high-interference colors) coexist with antigorite crystals (low, gray interference colors). Cross-polarized light, 2× magnification, field of view = 7 mm.

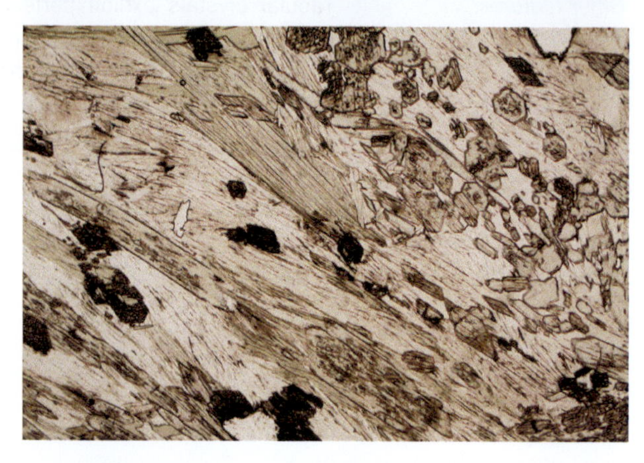

B Talc in schist from Prata Camportaccio (Italy). In this tremolite–talc schist, tabular talc crystals (colorless, with good cleavage) look similar to muscovite. Talc surrounds tremolite crystals (inclusion-rich, prismatic or basal sections, high relief) notable in the upper-right quadrant of the image. Plane-polarized light, 2× magnification, field of view = 7 mm.

C Talc in schist from Prata Camportaccio (Italy). Same image as (B) but with crossed polarizers. Talc interference colors are very high (higher than muscovite), while tremolite shows moderate to high-interference colors. Cross-polarized light, 2× magnification, field of view = 7 mm.

D Talc in whiteschist from Mautia Hills (Tanzania). Tabular talc crystals (greens) are associated with quartz (first-order colors). The talc crystal in the upper-left shows some internal deformation (small kink). Cross-polarized light, 2× magnification, field of view = 7 mm.

1.4 Chain Silicates

Anthophyllite $(Mg,Fe^{2+})_7[Si_8O_{22}](OH)_2$
Gedrite $(Mg,Fe^{2+})_5Al_2[Si_6 Al_2O_{22}](OH)_2$

Optical Properties **THE AMPHIBOLES**

Color in thin-section:	Colorless to pale brown, gray–brown, green–brown
Pleochroism:	None to pale yellow/gray–brown to brown
Birefringence:	0.013–0.021 (anthophyllite); 0.021–0.029 (gedrite)
Relief:	Moderate to high
Cleavage:	Perfect {210}, distinct {100}, {010}
Crystal system/optic sign:	Orthorhombic / anthophyllite (–) and gedrite (+)
Other features:	Anthophyllite can be columnar, bladed, acicular, or fibrous; orthorhombic amphiboles are untwinned and have parallel extinction along [001] sections, whereas monoclinic amphiboles can show twinning and have inclined extinction.

Paragenesis

These orthorhombic amphiboles belong to the Fe–Mg amphibole subgroup and are common in metamorphic or metasomatic rocks. They are typical of greenschist and lower amphibolite facies, where they are often associated with staurolite and cordierite. Anthophyllite and gedrite are common products in the reaction zone between ultramafic bodies (e.g. serpentinized peridotites) and country rock, and may occur as a hornfels constituent within the metamorphic aureoles of intrusions. Higher grades of metamorphism may result in the dehydration and breakdown of orthoamphibole to orthopyroxene, while during retrograde metamorphism of previously thermally metamorphosed rocks orthoamphibole may form rims around orthopyroxene. Orthoamphiboles alter to fine-grained serpentine, talc, and chlorite.

Igneous

Orthoamphiboles are unknown in igneous rocks.

Metamorphic

Regional metamorphism may produce orthorhombic amphiboles at lower amphibolite facies via reactions involving chlorite and quartz, with or without plagioclase. Anthophyllite is more Mg-rich, whereas gedrite has more Fe and Al. Minerals commonly associated with orthoamphiboles include staurolite + cordierite (relatively low pressure and temperature environments), garnet + cordierite (low pressure, high temperature), and aluminosilicate (high pressure). Other coexisting minerals may include talc, chlorite, spinel, olivine, orthopyroxene, hornblende, plagioclase, and quartz.

Cordierite–orthoamphibole gneisses are thought to originate via metasomatic alteration of a variety of common rock types, including granitic rocks and felsic and mafic volcanic rocks. This includes metasomatism of mafic rocks associated with seafloor hydrothermal systems.

Figure 1.28

A Anthophyllite from Südtirol (Italy). Note the typical color and pseudo-rhombohedral crystal outline. Plane-polarized light, 10× magnification, field of view = 2 mm.

B Anthophyllite from Südtirol (Italy). Same image as (A) but with crossed polarizers. Typical second-order interference colors nicely displayed. Cross-polarized light, 10× magnification, field of view = 2 mm.

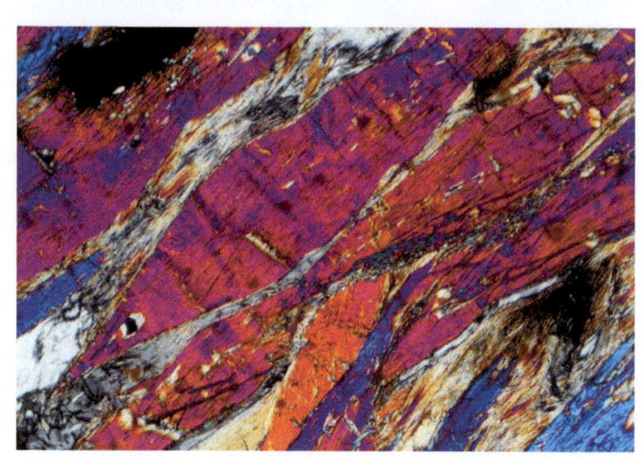

C Gedrite in a metamorphosed aplite from West Silvberg (Sweden). Fibrous, acicular gedrite crystals (gray–green, moderate relief) with cleavage roughly perpendicular to the long axis. Plane-polarized light, 10× magnification, field of view = 2 mm.

D Gedrite in a metamorphosed aplite from West Silvberg (Sweden). Same image as (C) but with crossed polarizers. Gedrite shows typical birefringence in this image. Cross-polarized light, 10× magnification, field of view = 2 mm.

Cummingtonite–Grunerite $(Mg,Fe^{2+},Mn)_7[Si_8O_{22}](OH)_2$

Optical Properties

Color in thin-section:	Colorless to pale green or brown (darker with increasing Fe)
Pleochroism:	Mg-rich cummingtonite, none; Fe-rich cummingtonite, pale green; grunerite, pale yellow or pale brown to brown
Birefringence:	0.020–0.032 (cummingtonite); 0.032–0.045 (grunerite)
Relief:	Moderate to high
Cleavage:	Good on {110}
Crystal system/optic sign:	Monoclinic / Cummingtonite (+), Grunerite (–)
Other features:	Narrow lamellar twins and an acicular or fibrous form are diagnostic; inclined extinction distinguishes it from anthophyllite–gedrite.

Paragenesis

The cummingtonite–grunerite amphibole series includes intermediate to Fe-rich amphiboles that are poor in calcium. The name cummingtonite is used for minerals containing $Mg > Fe^{2+}$ and grunerite is used for those with $Fe^{2+} > Mg$. Cummingtonite is most often found in Ca-poor amphibolites produced by regional metamorphism of basic igneous rocks, whereas grunerite is more commonly associated with metamorphosed banded ironstones.

Igneous

Minerals of this series are rare in igneous rocks. Cummingtonite is occasionally found in intermediate volcanic rocks, or in association with hornblende in diorite, gabbro, norite, and skarn.

Metamorphic

Cummingtonite–grunerite occurs in metamorphosed mafic and ultramafic rocks as either individual crystals or in crystals sharply zoned by calcic amphibole, in association with hornblende. Mg-rich cummingtonite occurs with anthophyllite in isochemically metamorphosed ultrabasic rocks, as well as in hybrid rocks of intermediate composition via the reaction:

$$\text{orthopyroxene} \longrightarrow \text{cummingtonite} \longrightarrow \text{hornblende}$$

It can occur with, and be hard to distinguish from, orthoamphibole (anthophyllite–gedrite). Other associated minerals include cordierite, garnet, plagioclase, hornblende, and biotite.

Grunerite is a common mineral in moderately metamorphosed iron formations, being stable from the biotite isograd through the garnet and staurolite isograds. It may be derived via the reaction:

$$\text{ferrodolomite} + \text{quartz} + H_2O \longrightarrow \text{grunerite} + \text{calcite} + CO_2$$

During regional metamorphism a characteristic assemblage is magnetite + grunerite + quartz. In association with contact and regional metamorphism, grunerite is commonly associated with fayalite, hedenbergite, and almandine garnet. Alteration products include chlorite, talc, or serpentine.

Figure 1.29

A Cummingtonite in amphibolite from Quabbin Reservoir, Massachusetts (United States). It occurs with hornblende (light brown in central, upper-right) and biotite (green–brown in lower-left), and is almost colorless with one good cleavage (center to left of the image). Fine exsolution lamellae of hornblende are semi-perpendicular to the cleavage. Plane-polarized light, 10× magnification, field of view = 3 mm.

Image courtesy of K. Hollocher

B Cummingtonite in amphibolite from Quabbin Reservoir, Massachusetts (United States). Image same as (A) but with crossed polarizers. Like most monoclinic amphiboles, birefringence is in the lower second order. Cross-polarized light, 10× magnification, field of view = 3 mm.

Image courtesy of K. Hollocher

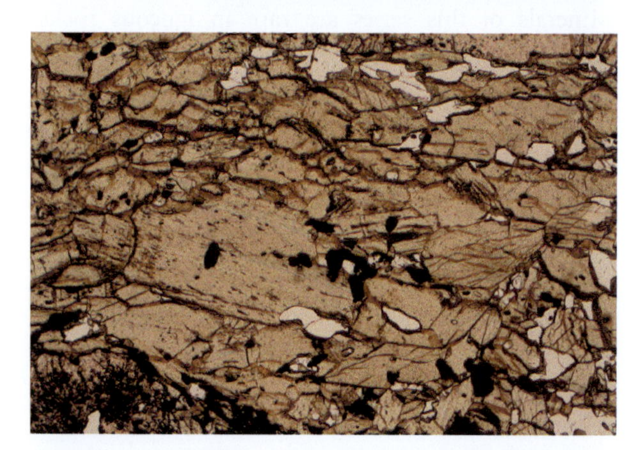

C Grunerite gneiss from Maine (United States). This thin-section is mostly grunerite (brownish) with minor plagioclase (clear), oxides (black), and garnet (cloudy, lower-left). Plane-polarized light, 4× magnification, field of view = 2.5 mm.

D Grunerite gneiss from Maine (United States). Same image as (C) but with crossed polarizers. Grunerite is highly twinned and shows typical interference colors. Garnet is at extinction, plagioclase is gray and shows twinning. Cross-polarized light, 4× magnification, field of view = 2.5 mm.

Eckermannite–Arfvedsonite $(K,Na)Na_2(Fe^{2+},Mg)_4(Fe^{3+},Al)[Si_8O_{22}](OH,F)_2$

Optical Properties

Color in thin-section:	Pale blue–green to yellow–, brown–, and gray–green, to violet
Pleochroism:	Eckermannite, blue–green to yellow–green; arfvedsonite, olive green, blue–green to dark blue
Birefringence:	0.013–0.020 (eckermannite); 0.010–0.012 (arfvedsonite)
Relief:	Moderate
Cleavage:	Perfect {110}, parting {010}, typical amphibole cleavage intersection 56°
Crystal system/optic sign:	Monoclinic / (–)
Other features:	Simple and lamellar {100} twins; birefringence is greatest in Mg-rich varieties; eckermannite has a large extinction angle, whereas arfvedsonite has a small extinction angle, has strong absorption, strong dispersion, and anomalous extinction. Eckermannite has lower refractive indices, birefringence, and extinction angle than cummingtonite and lower birefringence, larger extinction angle, and pleochroism than tremolite. Arfvedsonite is recognized by its characteristic pleochroism.

Paragenesis

The sodic amphiboles in this series are classified by four compositional end-members, with the Fe-rich members being the most common:

Eckermannite	$Na(Na_2)(Mg_4Al)[Si_8O_{22}](OH)_2$
Ferro-eckermannite	$Na(Na_2)(Fe^{2+}{}_4Al)[Si_8O_{22}](OH)_2$
Arfvedsonite	$Na(Na_2)(Fe^{2+}{}_4Fe^{3+})[Si_8O_{22}](OH)_2$
Magnesio-arfvedsonite	$Na(Na_2)(Mg_4Fe^{3+})[Si_8O_{22}](OH)_2$

Eckermannite occurs in alkaline rocks such as nepheline syenites and is often accompanied by aegirine, while arfvedsonite occurs in peralkaline granites and syenites. Conditions of formation for eckermannite are $300 < T < 400\,°C$ and $5 < P < 15$ kbar (high pressure/low temperature) and ascribed to hydrous fluids.

Igneous

Members of this group occur in silica-saturated peralkaline rocks such as lamprophyres, lamproites, syenites, alkali granites and their pegmatites, as well as in carbonatites. Aenigmatite is commonly reported as an associated mineral. Arfvedsonites have been reported to form by magmatic/subsolidus processes in reducing environments, whereas hydrothermal and oxidizing conditions favor riebeckite.

Metamorphic

True eckermannite seems to be rare, with most samples being ferro- to fluoro-eckermannite. The coupled OH–F substitution may interact with the octahedral strip containing Mg_4Al in eckermannite to repel Na, whereas the $Fe_4^{2+}Fe^{3+}$ arrangement of arfvedsonite allows Na to be accommodated. These arguments suggest that arfvedsonite should be more common than eckermannite.

Figure 1.30

A Eckermannite from Tawmaw (Myanmar). The image is dominated by eckermannite. The large crystal (lower-left) shows typical amphibole cleavage. The color variation in the image from bottom to top (colorless to pale yellow–green) relates to Cr content. Emerald green stringers are kosmochlor. Plane-polarized light, 50× magnification, field of view = 1.8 mm.
Image courtesy of F. Mazdab www.rockptx.com

B Eckermannite from Tawmaw (Myanmar). Same image as (A) but with crossed polarizers. Note the typical first-order birefringence of eckermannite. Crossed-polarized light, 50× magnification, field of view = 1.8 mm.
Image courtesy of F. Mazdab www.rockptx.com

C Arfvedsonite from Norr Kärr (Sweden). Recognizable by its distinctive blue–green color. Plane-polarized light, 2× magnification, field of view = 7 mm.

D Arfvedsonite from Norr Kärr (Sweden). Same image as (C) but with crossed polarizers showing arfvedsonite's distinctive pleochroism. Cross-polarized light, 2× magnification, field of view = 7 mm.

Glaucophane–Riebeckite $Na_2(Mg,Fe^{2+})_3[Al,Fe^{3+}]_2[Si_8O_{22}](OH)_2$

Optical Properties

Color in thin-section:	Colorless to medium blue (glaucophane); lavender blue (riebeckite)
Pleochroism:	Colorless, pale blue, yellow–blue–green, lavender, to violet (glaucophane); gray–blue, indigo, yellow–green, to yellow–brown (riebeckite)
Birefringence:	0.018–0.020 (glaucophane); 0.006–0.016 (riebeckite)
Relief:	Moderate to high
Cleavage:	Perfect {110}, parting {100}, typical amphibole cleavage intersection 56°
Crystal system/optic sign:	Monoclinic/ glaucophane (–) and riebeckite (–) or (+)
Other features:	Glaucophane lacks twins, has parallel extinction, is optically (–), and length-slow, with distinctive pale lavender pleochroism. Glaucophane may be confused with tourmaline but the latter is uniaxial. Magnesioriebeckite and riebeckite can have simple and lamellar twinning on {100}, both are length-fast and typically show intense dark blue pleochroism. Needle-like riebeckite crystals are common. Relative to arfvedsonite, riebeckite has smaller extinction angles and distinctive pleochroism.

Paragenesis

In the sodic amphiboles, Fe^{2+} substitutes for Mg and Fe^{3+} substitutes for Al, generating the series:

Glaucophane	$Na_2Mg_3Al_2Si_8O_{22}(OH)_2$
Magnesioriebeckite	$Na_2(Mg,Fe^{2+})_3(Al,Fe^{3+})_2Si_8O_{22}(OH)_2$
Riebeckite	$Na_2Fe^{2+}{}_3Fe^{3+}{}_2Si_8O_{22}(OH)_2$

Glaucophane is an indicator mineral for high-pressure (0.4–0.9 GPa), low-temperature (100–250 °C) regional metamorphism. It is derived from metamorphosed basaltic rocks (e.g. glaucophane schists), which are typically associated with orogenic belts. Glaucophane often occurs with lawsonite, pumpellyite, chlorite, albite, quartz, jadeite, and members of the epidote group. Riebeckite occurs in alkaline plutonic rocks, some schists, meta-iron formations, or can be authigenic in sediment.

Igneous

Glaucophane does not form in igneous rocks. Riebeckite is found in Na-rich rocks such as nepheline syenite. Late-oxidizing fluids can also result in the subsolidus formation of riebeckite.

Metamorphic

During regional metamorphism, glaucophane may form via the reactions:

plagioclase + chlorite + tremolite ⟶ glaucophane + lawsonite

plagioclase + chlorite + epidote + quartz + H_2O ⟶ glaucophane + lawsonite

Glaucophane may also form during retrograde metamorphism of eclogite.

Under lower-pressure conditions it may alter:

glaucophane ⟶ albite + talc + mica

Under higher-pressure conditions, the reaction is:

glaucophane ⟶ jadeite + talc

Riebeckite is found in some metamorphosed iron formations in its fibrous form "crocidolite."

Figure 1.31

A Riebeckite in quartz syenite from Mt. Cabot, New Hampshire (United States). Fresh poikilitic riebeckite (dark green–brown) encloses perthite crystals. Strong color is associated with high Fe content. Plane-polarized light, 2× magnification, field of view = 3.5 mm.

Attributed to C. Kraft, courtesy of G. Robert.

B Riebeckite in quartz syenite from Mt. Cabot, New Hampshire (United States). Same thin-section as (A) but with crossed polarizers. Although maximum interference colors range to upper first order, they are often masked by strong absorption as in this image. Cross-polarized light, 2× magnification, field of view = 3.5 mm.

Attributed to C. Kraft, courtesy of G. Robert.

C Glaucophane schist from Milos (Greece). Glaucophane has typical blue to lavender color and crystal habit. Plane-polarized light, 2× magnification, field of view = 7 mm.

D Glaucophane schist from Milos (Greece). Image same as (C) but with crossed polarizers. The large crystal (center-bottom) shows anomalous birefringence due to its natural blue color. Cross-polarized light, 2× magnification, field of view = 7 mm.

Hornblende $(K,Na)_{0-1}(Ca,Na,Fe,Mg)_2(Mg,Fe,Al)_5(Si,Al)_8O_{22}(OH)_2$

Optical Properties

THE AMPHIBOLES

Color in thin-section:	Colorless or pale green, yellow with increasing Fe
Pleochroism:	Distinctive shades of yellow–green, green, blue–green, yellow–brown, brown, red–brown (compositionally dependent)
Birefringence:	0.018–0.025 (may be distorted by mineral color)
Relief:	Moderate to high
Cleavage:	Good {110}, parting on {100}, typical amphibole cleavage intersection 56°
Crystal system/optic sign:	Monoclinic / generally (–) but Mg-rich paragasite (+)
Other features:	Simple and lamellar {100} twins are common. Hornblendes are distinguished from most other minerals by their elongate, prismatic shape, distinctive green to brown colors, distinctive cleavage, and inclined extinction. The properties of actinolite and hornblende overlap. Dark oxyhornblende may resemble biotite, but has inclined extinction.

Paragenesis

Hornblende has a complex and diverse compositional variation due to substitutions of Na, K, Fe, and Al for Si, and Ti, Mn, and Cr for (OH). Hornblende stability is affected by fO_2 and fluid composition. Oxyhornblende contains Fe^{3+}, but the principal varieties include:

Magnesiohornblende-ferrohornblende	$Ca_2(Fe^{2+},Mg)_4Al[Si_7AlO_{22}](OH)_2$
Tschermakite-ferrotschermakite	$Ca_2(Fe^{2+},Mg)_3Al_2[Si_6Al_2O_{22}](OH)_2$
Edenite-ferroedenite	$(Na, K)Ca_2(Fe^{2+},Mg)_5[Si_7AlO_{22}](OH)_2$
Pargasite-ferropargasite	$(Na, K)Ca_2(Fe^{2+},Mg)_4[AlSi_6AlO_{22}](OH)_2$
Magnesiohastingsite-hastingsite	$(Na, K)Ca_2(Fe^{2+},Mg)_4Fe^{3+}[Si_6Al_2O_{22}](OH)_2$

With the exception of tschermakite, hornblende is common in ultramafic to acid plutonic rocks and less common in hypabyssal rocks and lavas. They also occur in metamorphic rocks of greenschist to lower granulite facies.

Igneous

Hornblende is common in igneous rocks, particularly intermediate varieties such as diorite, granodiorite, andesite, and dacite, and can also be present in granitic/rhyolitic, syenitic, and phonolitic rocks. It may be a late phase in solidifying mafic rocks (e.g. gabbro, norite). Oxyhornblende is restricted to volcanic rocks. Hornblende may replace pyroxene with fine aggregates of tremolite–actinolite, known as uralite. Hornblende alters to biotite, chlorite, and other Fe–Mg silicates.

Metamorphic

Hornblende is common in medium- and high-grade metamorphosed mafic rocks (e.g. amphibolite and hornblende gneiss), and may occur in marble, skarn, and other metacarbonate rocks (typically as tremolite–actinolite). In general, Na and Al increase with increasing metamorphic grade. Conversion of Mg-hornblende to oxyhornblende occurs at ~800 °C.

Figure 1.32

A Hornblende in gabbro from Adamello (Italy). Green–brown hornblende with plagioclase (colorless). The green, blue–green to light brown color reflects composition; greater Fe results in darker colors. The upper crystal shows twinning along {100}. Plane-polarized light, 10× magnification, field of view = 2 mm.

B Hornblende in amphibolite from Alpe Arami (Switzerland). Note the vibrant colors at the higher end of the birefringence spectrum for hornblende. Cross-polarized light, 2× magnification, field of view = 7 mm.

C Pargasite in skarn from Elba (Italy). Pargasite showing typical blue–green colors (crystal in center). Cross-polarized light, 2× magnification, field of view = 7 mm.

D Oxyhornblende in andesite from unknown location (Ecuador). This euhedral phenocryst of oxyhornblende shows typical strong red–brown coloration associated with oxidation (Fe^{3+}). Plane-polarized light, 2× magnification, field of view = 7 mm.

Kaersutite–Ferrokaersutite

$(Na,K)Ca_2(Fe^{2+},Fe^{3+},Mg,Al)_4(Ti,Fe^{3+})[Si_6Al_2O_{22}](O,OH,F)_2$

Optical Properties

Color in thin-section:	Yellow–red or green–brown
Pleochroism:	Distinctive shades of yellow–green, green, blue–green, yellow–brown, brown, red–brown (composition-dependent)
Birefringence:	0.028–0.047 (may be distorted by mineral color)
Relief:	Moderate
Cleavage:	Perfect {110}, parting on {100} and {001}, typical amphibole cleavage intersection 56°
Crystal system/optic sign:	Monoclinic / (–)
Other features:	Simple and lamellar {100} twins are common. Hornblendes are distinguished from most other minerals by their elongate, prismatic shape, and kaersutite shows distinctive red–brown colors and pleochroism; typical amphibole cleavage and inclined extinction distinguishes it from titanian augite.

Paragenesis

Kaersutite occurs in alkaline igneous rocks, including trachybasalts, trachyandesite, trachytes, and alkali rhyolites, as well as in syenites and lamphrophyre and camptonite dikes. In more silica-rich rocks, it may occur as a groundmass phase. It often occurs as a reaction rim around olivine and titanian augite phenocrysts in trachybasalts; the kaersutite in these rocks is invariably surrounded by opaque margins of magnetite. It can be abundant in some monzonites and eclogites, and ferrokaersutite has been documented in spinel-bearing lherzolite xenoliths. Kaersutites typically have prismatic shape and Ti-rich varieties have strong reddish colors.

Igneous

Kaersutite occurs in basic, SiO_2-undersaturated rocks that have higher fO_2 than tholeiitic suites. Magmatic kaersutite is present in dikes and in gabbroic inclusions within dikes. Experimental results indicate that kaersutite in water-rich basalt and peridotite is stable at upper mantle P–T conditions (<80–100 km depths).

Kaersutite and alkali feldspar megacrysts often occur together in alkaline basalts and may or may not be in equilibrium with their host; when in equilibrium, kaersutite crystallizes before alkali feldspar and often has a cumulate texture. Hawaiian mantle-derived peridotite xenoliths contain kaersutite-bearing magma pockets (trapped trachybasaltic and trachyandesitic melts). Kaersutite is common in pyroxenite and websterite mantle xenoliths and occurs as antecrysts from earlier magmatic stages when associated with alkaline basalts. In plutonic rocks, they are inferred to have crystallized from evolved interstitial melts during slow intratelluric solidification.

Metamorphic

Some kaersutites are thought to be associated with metasomatic veins in both mantle and crustal sources.

Figure 1.33

A Kaersutite, Lugar Sill (Scotland). Note the characteristic reddish color and amphibole cleavage of these kaersutite phenocrysts (center). Strong colors are due to Ti. Plane-polarized light, 2× magnification, field of view = 7 mm.

B Kaersutite, Lugar Sill (Scotland). Same image as (A) but with crossed polarizers. Kaersutite can be twinned and have high birefringence (e.g. lower, central grain in image) but the latter is often muted by their natural color. Crossed-polarized light, 2× magnification, field of view = 7 mm.

C Kaersutite in trachybasalt sill near Komňa (Czech Republic). The elongate and partially resorbed kaersutite phenocryst shows a distinctive opacite reaction rim. Such "micro-oxide" rims are generated when kaersutite is transported from a high P–T to a lower P–T environment. Plane-polarized light, 2× magnification, field of view = 4 mm.

Image courtesy of L. Krmíček

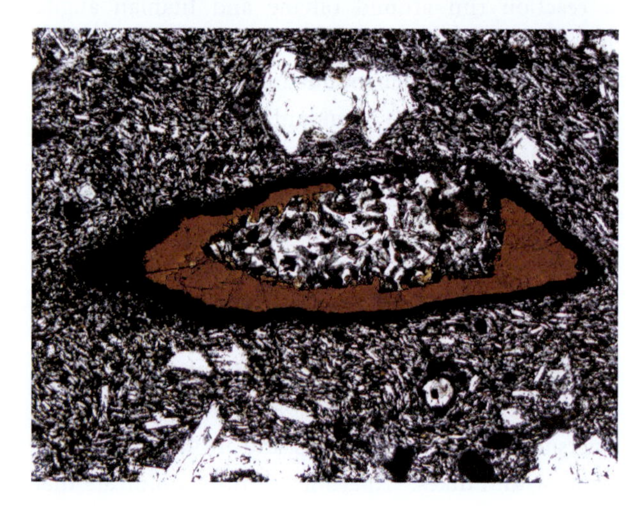

Richterite–Ferrorichterite $(Na,K)CaNa(Mg,Al,Fe^{2+},Mn)_5[Si_8O_{22}](OH)_2$
Katophorite $(Na,K)CaNa(Mg,Fe^{2+})_4[Si_7AlO_{22}](OH)_2$

Optical Properties

Color in thin-section:	(Ferro)richterite is colorless to pale yellow, violet; (Magnesio)katophorite is yellow, reddish-brown, bluish-green.
Pleochroism:	Variable; pale yellow, oranges, and reds, sometimes violets and blues
Birefringence:	0.015–0.029 (richterite/Fe-richterite); 0.007–0.021 (katophorite)
Relief:	Moderate
Cleavage:	Perfect {110}, parting on {100} and {001}, typical amphibole cleavage intersection 56°
Crystal system/optic sign:	Monoclinic / (–)
Other features:	Simple and lamellar {100} twins are common. Richterite is recognized by its distinctive pleochroism and katophorite by its yellow, reddish-yellow, and brownish absorption colors.

Paragenesis

Richterites are associated with metamorphosed carbonate rocks (e.g. contact-metamorphosed limestones and skarns) and occur as hydrothermal alteration products. Ferrorichterites occur in pantellerites and in the late stages of granitic intrusions. K-rich magnesiorichterite has an extensive stability field and occurs in alkaline to peralkaline basalts, lamprophyres, and mica-peridotites. Katophorite is more rare, but when it occurs is associated with arfvedsonite and aegirine in alkaline rocks such as phonolite and trachyte.

Figure 1.34 Richterite in glimmerite from the Siilinjärvi carbonatite complex (Finland). (*Left*) (plane-polarized light) Euhedral, lozenge-shaped richterite crystals with typical 120° cleavage. The two crystals have cavities and numerous apatite inclusions. The groundmass is formed of phlogopite (red–orange color). (*Right*) The same image but with crossed polarizers. Richterite has medium (yellow–orange), phlogopite high (pink), and apatite low (gray) birefringence. Both images at 10× magnification, field of view = 2 mm.

Tremolite–Ferroactinolite $Ca_2(Mg, Fe^{2+})_5[Si_8O_{22}](OH)_2$

Optical Properties

Color in thin-section:	Colorless (tremolite) to deep green (ferroactinolite)
Pleochroism:	Tremolite, none; ferroactinolite, pale yellowing green to green, blue–green, and dark green (stronger with higher Fe)
Birefringence:	0.022–0.027 (tremolite); 0.017–0.032 (ferroactinolite)
Relief:	Moderate to high
Cleavage:	Good on {110}, parting on {100}
Crystal system/optic sign:	Monoclinic/ (–)
Other features:	Simple lamellar twins are common and an acicular or fibrous form are diagnostic. Actinolite is distinguished by its green color and pleochroism; ferroactinolite (rare) is dark green to black. Pale color distinguishes tremolite and actinolite from hornblendes, though colored varieties can be more difficult to distinguish. Mg varieties are distinguished from cummingtonite–grunerite by negative optic sign and Fe-rich varieties by their lower birefringence; orthoamphiboles are easily distinguished by their straight extinction in all [001] zone sections.

Paragenesis

The tremolite–actinolite–ferroactinolite series forms a solid solution in which tremolite contains Mg > 90%, actinolite 90% > Mg > 50%, and ferroactinolite Mg < 50%. Ferroactinolite is uncommon, while tremolite and actinolite are associated with low-grade metamorphism.

Igneous

This series does not form primary igneous minerals.

Metamorphic

Tremolite and actinolite are typically found in low-grade contact and regionally metamorphosed rocks. Tremolite is common in metamorphosed impure dolomitic limestones and occurs with calcite, dolomite, forsterite, garnet, diopside, wollastonite, talc, and epidote. Tremolite forms by the reaction:

$$\underset{\text{dolomite}}{5CaMg(CO_3)_2} + \underset{\text{quartz}}{8SiO_2} + \underset{\text{fluid}}{H_2O} \rightarrow \underset{\text{tremolite}}{Ca_2Mg_5Si_8O_{22}(OH)_2} + \underset{\text{calcite}}{3CaCO_3} + \underset{\text{fluid}}{7CO_2}$$

At higher metamorphic grades, tremolite reacts with calcite and quartz to form diopside, or with dolomite to give forsterite and calcite.

Actinolite occurs as a retrograde metamorphic product of basic rocks in association with carbonates, antigorite, talc, epidote, and chlorite. During prograde metamorphism, actinolite gives way to hornblende via the reaction:

$$\text{actinolite} + \text{epidote} + \text{chlorite} + \text{quartz} \rightarrow \text{hornblende} + H_2O$$

Ferroactinolite is limited to metamorphosed ironstones, where it may be associated with calcite, dolomite, and cummingtonite. The series may alter to chlorite, talc, and carbonate. It may also form uralite, the common alteration product of pyroxenes and hornblendes.

Figure 1.35

A Tremolite in altered peridotite from Elba island (Italy). Clusters of colorless, acicular crystals are typical of tremolite. Plane-polarized light, 10× magnification, field of view = 7 mm.

B Tremolite in altered peridotite from Elba island (Italy). Same image as (A) but with crossed polarizers showing tremolite's normal birefringence. Cross-polarized light, 10× magnification, field of view = 7 mm.

C Actinolite in schist from Merano (Italy). Elongate actinolite crystals with typical yellow to green colors. Plane-polarized light, 2× magnification, field of view = 7 mm.

D Actinolite in schist from Merano (Italy). Same image as (C) but with crossed polarizers. Elongate actinolite crystals showing typical birefringence and twinning. Cross-polarized light, 2× magnification, field of view = 7 mm.

Aegirine $(Na,Fe^{3+})Si_2O_6$
Aegirine–Augite $(Na,Ca)(Fe^{3+},Fe^{2+},Mg)Si_2O_6$

Optical Properties

Color in thin-section:	Pale to dark green or yellow–green; darker with higher Fe
Pleochroism:	Distinctive emerald green and bright greens, brown–green
Birefringence:	0.030–0.060
Relief:	High
Cleavage:	Good {110}, parting on {100} with near 90° intersection
Crystal system/optic sign:	Monoclinic / Aegirine (–), Aegirine–augite (+) or (–)
Other features:	Aegirine and aegirine–augite are distinguished from other pyroxenes by diagnostic pleochroism; aegirine may form pointed terminations or blocky prisms and is optically negative, whereas aegirine–augite typically has a stubby prismatic form and can be either positive or negative. Simple and lamellar twins are common and may resemble some alkali amphiboles, but the pyroxene cleavage is diagnostic in basal and near-basal sections.

Paragenesis

Aegirine and aegirine–augite are typical clinopyroxenes of alkali-rich magmas and many peralkaline rocks $(Na_2O + K_2O > Al_2O_3)$ and are therefore common in alkali granites, quartz syenites, syenites, and nepheline syenites and their associated pegmatites and veins. They are present in ultra-alkaline basic rocks such as phonolites and are also associated with carbonatite melts and their associated alkali-rich silicate magmas. Aegirine–augites are found in regionally metamorphosed rocks, such as glaucophane- and riebeckite-bearing schists. Aegirine occurs in some quartzose rocks and granulites. It commonly alters to uralite (fine-grained amphibole) or chlorite.

Igneous

In alkaline igneous rocks, sodic pyroxene typically occurs with potassium feldspar, a sodium-rich amphibole and either a feldspathoid or quartz. In these rocks, sodium-rich pyroxenes are associated with arfvedsonitic amphiboles and magnesioriebeckite. Aegirine often forms late in the crystallization sequence and although it commonly mantles earlier augitic or sodium-rich pyroxenes, it is not a reaction product between these earlier phases and residual liquids.

Metamorphic

Na-rich metasomatism associated with carbonatite intrusions may produce sodic pyroxenes via "fenitization" reactions. In alpine metamorphic belts, the occurrence of aegirine and aegirine–augite is restricted to blueschists and related rocks.

Figure 1.36 Aegirine in fine-grained syenite from Lovozaro (Russia). Note the distinctive "grass" green pleochroism and sharp crystal terminations. The crystals sit in a feldspar-rich groundmass. Plane-polarized light, 2× magnification, field of view = 7 mm.

Augite–Ferroaugite $(Ca,Na)(Mg,Fe,Ti)(Si,Al)_2O_6$

Optical Properties

Color in thin-section:	Pale brown, green–brown, purplish-brown, green
Pleochroism:	Pale green, brown, yellow, violet; stronger in Ti-rich varieties
Birefringence:	0.018–0.033
Relief:	High
Cleavage:	Good {110}, partings {100} {010} with near 90° intersection
Crystal system/optic sign:	Monoclinic / (+)
Other features:	Due to their solid solution, it is not always possible to distinguish between the clinopyroxenes. Augite generally has lower birefringence; Ti-rich augites often have violet pleochroism and typical sector zoning. Orthopyroxene has straight extinction, whereas clinopyroxene has inclined extinction.

Paragenesis

There is complete solid solution between the calcic clinopyroxenes that form the diopside–hedenbergite–augite series. Augitic pyroxenes occur in a wide variety of igneous rocks, particularly in gabbros, dolerites, and basalts. They occur less frequently in ultrabasic and intermediate rocks. Augites can be present in granulites, charnockites, and other high-grade metamorphic rocks. Iron-rich augites occur in highly differentiated ferro-gabbros, iron-rich dolerites and their pegmatites, syenites, acid volcanic glasses, and metamorphosed iron formations. Calcium-poor augites are typical of quickly chilled basic basalts and andesites. Titanian augites are typical of basic alkaline rocks, such as nepheline dolerite. Augite commonly alters to a greenish-blue (uralitic) amphibole, either as a single crystal or as an aggregate of small prismatic crystals.

Igneous

Augite occurs in basic igneous rocks, ultramafic rocks, and ultrabasic nodules, and as megacrysts in basic rocks. A wide and continuous range of composition occurs in augite from layered basic intrusions, changing from diopsidic to augitic. Calcium-rich augite of the tholeiitic rock series show strong iron enrichment coupled initially with a decrease, and later with an increase, in calcium content. Until the later stage of magmatic fractionation, augite is associated with orthopyroxene or pigeonite. Augite of the alkaline series show a more limited iron enrichment and a fractionation trend approximately parallel to the diopside–hedenbergite join. Exsolution of a Ca-poor phase (orthopyroxene/pigeonite) within the Ca-rich augitic host reflects the pyroxene miscibility gap (see *pigeonite*). Mg-rich augites exsolve orthopyroxene lamellae, while Fe-rich augites exsolve lamellae of pigeonite which may later invert to orthopyroxene; the first set of lamellae form approximately parallel to (001) and a second set form parallel to the (100) plane of the augite host. In tholeiitic lavas, augites are commonly sub-calcic in composition due to rapid cooling from high temperatures.

Metamorphic

Alumina-rich augite is associated with spinel in metamorphosed limestones. It also occurs at igneous–carbonate contacts associated with garnet, vesuvianite, scapolite, epidote, and pargasitic amphiboles, as well as in eclogite xenoliths in nephelinite and kimberlitic pipes. Titanian augite is a relatively common constituent of alkaline dikes such as camptonites.

Figure 1.37

A Augite in cumulate from the Rum complex (Scotland). Pale pink–green augite poikilitically enclosing plagioclase and chromite. Plane-polarized light, 10× magnification, field of view = 2 mm.

B Cumulate augite, Rum complex (Scotland). Same view as (A) but with crossed polarizers. Note the uniform birefringence of the clinopyroxene and typical polysynthetic twinning of plagioclase. Cross-polarized light, 10× magnification, field of view = 2 mm.

C Augite in leucite-bearing tephrite from Vesuvius (Italy). The large euhedral phenocryst of augite has typical high birefringence and sector (hour glass) zoning. Cross-polarized light, 10× magnification, field of view = 2 mm.

D Augite exsolving orthopyroxene from the Bushveld complex (South Africa). A single augite crystal (blue birefringence) is host to lamellae of exsolved orthopyroxene (gray). Cross-polarized light, 10× magnification, field of view = 2 mm.

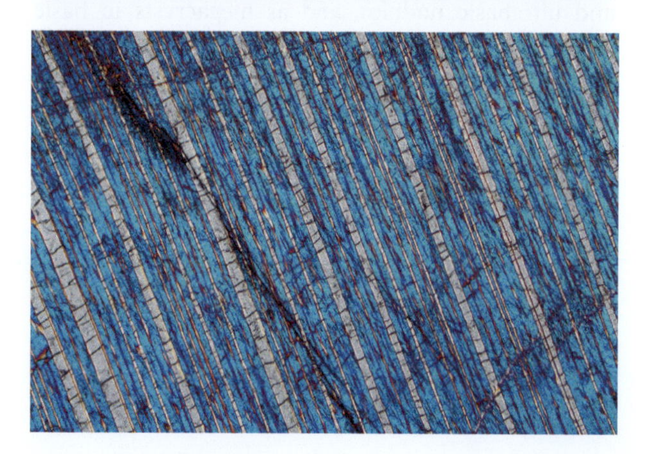

Diopside–Hedenbergite $Ca(Mg,Fe)Si_2O_6$

Optical Properties

Color in thin-section:	Colorless to brownish-green
Pleochroism:	Weak in pale green, green–brown, blue–green, and yellow–green
Birefringence:	0.024–0.034
Relief:	High
Cleavage:	Good {110}, partings {100} {010} with near 90° intersection
Crystal system/optic sign:	Monoclinic / (+)
Other features:	Due to their solid solution, it is not always possible to distinguish between the clinopyroxenes. Augite generally has lower birefringence; orthopyroxene has straight extinction, whereas clinopyroxene is inclined.

Paragenesis

There is complete solid solution between the clinopyroxenes that form the augite–diopside–hedenbergite series. Diopside and hedenbergite are most common in metamorphosed calcareous rocks and characteristic of contact-metamorphosed calcium-rich sediments. Diopside may occur in ultramafic and mafic igneous rocks and in ultramafic nodules associated with alkali olivine basalts and kimberlites. Lamellar twinning parallel to (100) occurs in diopside of deformed basic and metamorphic rocks. Diopside is often colorless or pale green, whereas hedenbergite is usually brownish-green. Ca-pyroxenes often alter to fine-grained, light-colored amphibole (uralite), serpentine, biotite, chlorite, carbonates and/or clay.

Igneous

Cr-rich diopside is common in some ultramafic and mafic igneous rocks. Lamellae of orthopyroxene may be present in igneous diopside; in Cr-rich specimens, in addition to the broader orthopyroxene lamellae, narrow lamellae of spinel form parallel to (010). Diopside also occurs in more strongly alkaline rocks (e.g. wyomingite). Hedenbergite is a constituent of nepheline- and quartz-bearing syenites, in which it is usually associated with fayalite.

Metamorphic

Although often preceded by tremolite or forsterite, diopside occurs relatively early during the metamorphism of siliceous dolomites:

$$\underset{\text{dolomite}}{CaMg(CO_3)_2} + \underset{\text{quartz}}{SiO_2} \longrightarrow \underset{\text{diopside}}{CaMgSi_2O_6}$$

It is commonly associated with tremolite–actinolite, grossular garnet, epidote, wollastonite, fosterite, calcite, and dolomite. In regionally metamorphosed calcium-rich sediments and basic igneous rocks of upper amphibolite facies, compositions along the diopside–hedenbergite solid solution occur via the reaction:

$$\underset{\text{hornblende}}{Ca_2(Mg,Fe^{2+})_3Al_4Si_6O_{22}(OH)_2} + \underset{\text{calcite}}{3CaCO_3} + \underset{\text{quartz}}{4SiO_2} \longrightarrow$$

$$\underset{\text{diopside}}{3Ca(Mg,Fe^{2+})Si_2O_6} + \underset{\text{anorthite}}{2CaAl_2Si_2O_8} + 3CO_2 + H_2O$$

Hedenbergite occurs in contact-metamorphosed iron-rich sediments. In regionally metamorphosed ironstones and eulysites, it is associated with grunerite and fayalite. All members of the diopside–hedenbergite series occur in skarns, marbles, and other metamorphosed carbonate rocks; in skarns they are associated with humite minerals, monticellite, vesuvianite, scapolite, garnet, and bustamite.

Figure 1.38

A Diopside in lherzolite from Finero mafic complex, Finero (Italy). Note the green pleochroism of diopside; olivine (colorless) and chromite (black) are also present. Plane-polarized light, 2× magnification, field of view = 7 mm.

B Hedenbergite in skarn from Elba island (Italy). Nice euhedral crystals with compositional zoning from light brown to green (upper-right). Plane-polarized light, 2× magnification, field of view = 7 mm.

C Green diopside in marble from Mergozzo (Italy). Note distinctive green color, granular habit in a carbonate host. Plane-polarized light, 2× magnification, field of view = 7 mm.

D Uralite alteration in tonalite from Cerro Colorado (Panama). Note amphibole (green with typical 120° cleavage) mantles a pale brown core of clinopyroxene. Other minerals include colorless quartz, black magnetite, and red iron oxide (probably hematite). Plane-polarized light, 10× magnification, field of view = 2 mm.

Jadeite $NaAlSi_2O_6$

Optical Properties

Color in thin-section:	Colorless
Pleochroism:	None
Birefringence:	0.006–0.021
Relief:	Moderately high
Cleavage:	Good {110}, sometimes parting {100}, with near 90° intersection
Crystal system/optic sign:	Monoclinic / (+)
Other features:	Stubby to elongate crystals with four- to eight-sided cross-section; can be granular or aggregates of acicular to fibrous grains. High Fe content may cause anomalous interference colors and lower birefringence. With anomalous interference colors it resembles zoisite, but the latter has straight extinction. Simple and lamellar twins are sometimes present.

Paragenesis

Jadeite is restricted to high-pressure and moderate-temperature metamorphic rocks, and is common in Alpine-type metamorphic terranes where it is usually associated with albite. It is relatively common in metagreywackes and related rocks associated with regional metamorphism. In blueschist facies greywacke it coexists with quartz and in metabasalt with omphacite or sodic pyroxene. Jadeite may be replaced by amphibole or less commonly by analcime or nepheline.

Igneous

There are no known occurrences of igneous jadeite.

Metamorphic

Jadeite is formed by the reactions:

nepheline + albite = 2 jadeite

albite = jadeite + quartz

Such reactions result in a volume reduction and an increase in density from ~2.6 to ~3.3 g/cm^3. Pseudomorphs of feldspar and relicts of lawsonite may persist in jadeite via the reaction:

$$NaCaAl_3Si_5O_{16} + 2H_2O \longrightarrow NaAlSi_2O_6 + CaAl_2Si_2O_7(OH)_2H_2O + SiO_2$$

plagioclase *jadeite* *lawsonite* *quartz*

With increasing metamorphic grade, the omphacitic pyroxene component increases at the expense of the jadeitic component. Gabbros under conditions of regional metamorphism have jadeite–pumpellyite-rich assemblages and regionally metamorphosed mafic lavas may have jadeite-bearing amygdales. Under blueschist facies conditions, jadeite and analcite can occur in veins and pods.

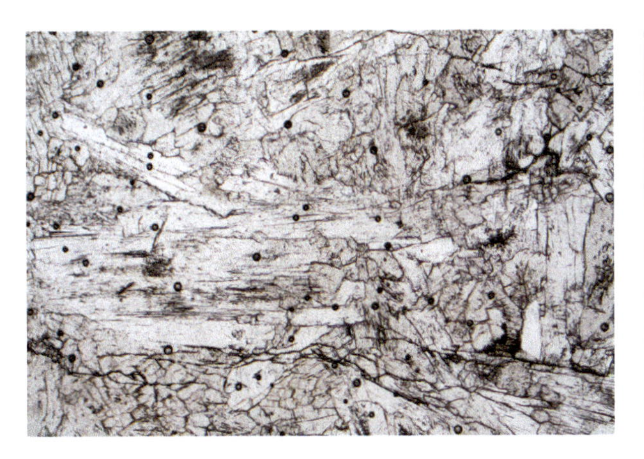

Figure 1.39 Jadeite from Syros (Greece). Prismatic jadeite has no pleochroism (plane-polarized light, *left*) and high birefringence (cross-polarized light, *right*). Small, high-relief "black dots" are trapped air bubbles. Both images: 2× magnification, field of view = 7 mm.

Omphacite $(Ca,Na)(Mg,Fe,Al)Si_2O_6$

Optical Properties

Color in thin-section:	Colorless to pale green
Pleochroism:	Weak; pale green to pale blue–green
Birefringence:	0.012–0.028
Relief:	High
Cleavage:	Good {110}, parting {100}, with near 90° intersection
Crystal system/optic sign:	Monoclinic / (+)
Other features:	May form stubby crystals but anhedral grains are more common. Omphacite has darker color and higher birefringence than jadeite and the color of aegirine and aegirine–augite is distinct; twins (simple, lamellar) and exsolution lamellae of augitic pyroxene are common; in blueschist it may be difficult to differentiate omphacite from calcic amphibole, but acicular grains with pronounced cleavage traces across the grain are diagnostic of omphacite.

Paragenesis

Omphacite is a high-pressure mineral reflecting lower-crust and upper-mantle conditions. It represents a solid solution between jadeite (20–80%) and augite (80–20%) but with a slightly different mineral structure, and has components of sodium, calcium, and aluminum-rich clinopyroxene. It typically occurs in eclogite, eclogite xenoliths in kimberlites, and high-pressure metamorphic/igneous basaltic/gabbroic rocks, and is commonly associated with garnet. Associated minerals in eclogite include garnet, quartz or coesite, rutile, kyanite, phengite, and lawsonite. It is less commonly found in blueschist, metagraywacke, and related rocks of alpine metamorphic belts.

Igneous

The chemical composition of many eclogites is comparable to those of basic igneous rocks. Overlap in the mineral stability fields of igneous and high-grade metamorphic rocks suggests that omphacite may be an igneous product and the cognate cumulus omphacite + garnet layers in the Breaksea orthogneiss preserve igneous textures and mineral compositions.

Metamorphic

The formation of omphacite due to the recrystallization of pre-existing igneous rocks under high P–T conditions follows the reaction:

$$\underbrace{3CaAl_2Si_2O_8 + 2NaAlSi_3O_8}_{labradorite} + \underbrace{3Mg_2SiO_4}_{olivine} + \underbrace{nCaMgSi_2O_6}_{diopside} \rightarrow$$

$$\underbrace{2NaAlSi_2O_6 + nCaMgSi_2O_6}_{omphacite} + \underbrace{3CaMg_2Al_2Si_3O_{12}}_{garnet} + \underbrace{2SiO_2}_{quartz}$$

During retrograde metamorphism, alteration of omphacite can produce a border of hornblende and fibers of amphibole within the pyroxene, or a symplectic intergrowth of diopside and plagioclase. In mantle-derived eclogite, omphacite may break down to an aluminum-rich augite intergrown with the host omphacite + quartz, possibly via the reaction:

$$2Ca_{0.5}\square_{0.5}AlSi_2O_6 \rightarrow CaAl_2SiO_6 + 3SiO_2$$

Figure 1.40

A Omphacite in basite layers from the Breaksea orthogneiss (New Zealand). Blue–green omphacite occurs in cumulate layers with pink garnet (upper corners) and plagioclase + K-feldspar + quartz (colorless). A few, smaller deep red crystals of rutile are also present. Plane-polarized light, 3× magnification, field of view = 5 mm.

Image courtesy of G. Clark & T. Chapman

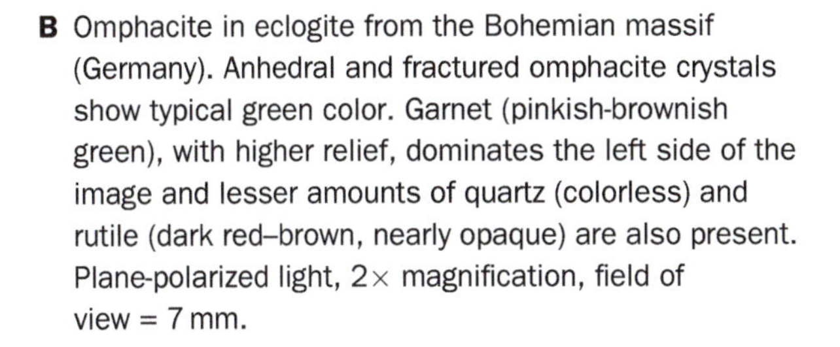

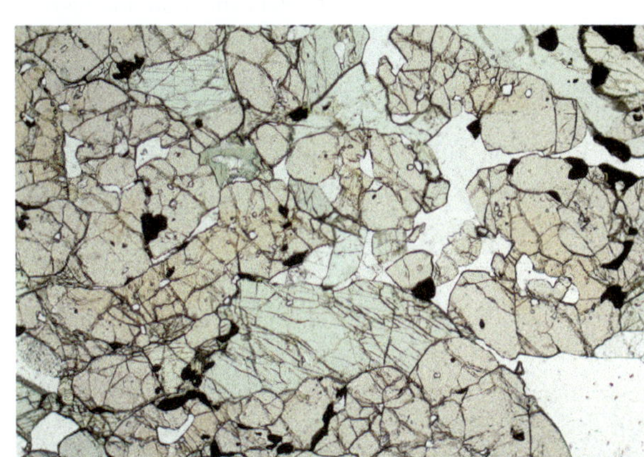

B Omphacite in eclogite from the Bohemian massif (Germany). Anhedral and fractured omphacite crystals show typical green color. Garnet (pinkish-brownish green), with higher relief, dominates the left side of the image and lesser amounts of quartz (colorless) and rutile (dark red–brown, nearly opaque) are also present. Plane-polarized light, 2× magnification, field of view = 7 mm.

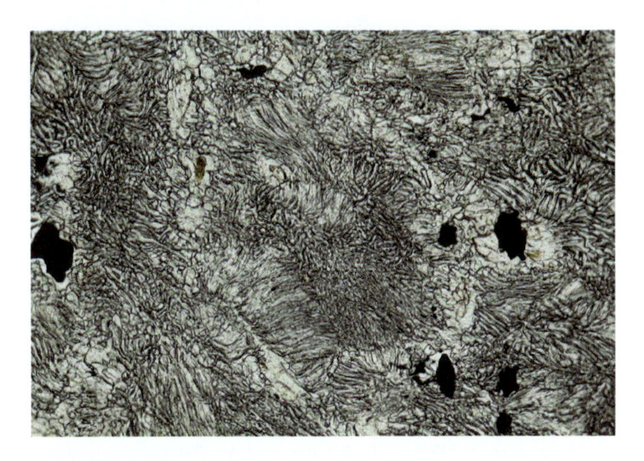

C Omphacite in eclogite from Sardinia (Italy). This is a fantastic retrograde association of symplectic diopside and albite after omphacite. Black magnetite is also present. Plane-polarized light, 10× magnification, field of view = 2 mm.

D Omphacite in eclogite from Sardinia (Italy). Same image as (C) but with crossed polarizers. Vermicular intergrowth of diopside (high birefringence) and albite (low colors) nicely visible. Cross-polarized light, 10× magnification, field of view = 2 mm.

Pigeonite $(Mg,Fe^{2+},Ca)Si_2O_6$

Optical Properties **THE PYROXENES**

Color in thin-section:	Colorless or pale pink–green (Mg-rich); red/green (Fe-rich)
Pleochroism:	Absent or pale brownish-green, pale yellowish-green
Birefringence:	0.023–0.029
Relief:	High
Cleavage:	Good {110}, partings {100} {010} with near 90° intersection
Crystal system/optic sign:	Monoclinic / (+)
Other features:	Prismatic crystals with square or octagonal cross-section, exsolution lamellae common, inclined extinction and higher birefringence distinguishes it from orthopyroxene.

Paragenesis

Pigeonite is the monoclinic high-temperature, low-Ca clinopyroxene. Higher temperature allows more Ca within the pigeonite crystal lattice and most pigeonite has Mg:Fe ratios between 70:30 and 30:70. Slow cooling allows Ca to exsolve along planes parallel to (001), producing lamellae of Ca-rich augite in the Ca-poor pigeonite host. The monoclinic–orthorhombic structural inversion is temperature-dependent and decreases from ~1,100 °C for Mg-rich pigeonite to ~950 °C for Fe-rich pigeonite. In rapidly quenched rocks, augite exsolution lamellae in pigeonite do have time to form, but lamellae can be small (<300 nm wide), so chemical homogeneity cannot be assumed.

Igneous

Pigeonite may be present in shallow layered mafic intrusions, but most slower-cooling plutonic occurrences invert to orthopyroxene during cooling. Inverted pigeonite is a common cumulus phase in basic plutonic rocks with tholeiitic affinities. The change from enstatite as an early cumulate phase to pigeonite in later differentiates usually occurs at Fe = 30–35%. In higher-temperature basaltic melts, magnesium-rich liquids are above the orthorhombic inversion and pigeonite is absent, whereas it is a characteristic constituent of more evolved volcanic rocks. In magmatic liquids with increasing fractionation and iron enrichment, pigeonite is the normal low-calcium phase and pigeonite phenocrysts (or microphenocrysts) may coexist with augite or form groundmass grains. Alteration products include serpentine, talc, chlorite, and fine-grained amphibole.

Metamorphic

High-temperature metamorphism of iron formations may produce pigeonite; however, it mostly converts to orthopyroxene upon cooling.

Figure 1.41 Pigeonite in gabbro from Mutoko (Zimbabwe). Note the exsolution lamellae (Ca-rich augite with bright blue birefringence) in the pigeonite host. Cross-polarized light, 2× magnification, field of view = 7 mm.

Enstatite–Ferrosilite $(Mg,Fe^{2+})[SiO_3]$

Optical Properties **THE PYROXENES**

Color in thin-section: Colorless or pale pink–green (Mg-rich); reddish/greenish colors (Fe-rich)

Pleochroism: None (enstatite) or weak to strong pink, violet, purple–brown, reddish-brown, yellow, yellow–green/brown, pale brown, pale green or brown, greenish brown (ferrosilite)

Birefringence: 0.016–0.020

Relief: Medium to high

Cleavage: Good {210}, partings {100} {010} with near 90° intersection

Crystal system/optic sign: Orthorhombic / Mg- and Fe-rich end-members (+) and intermediate compositions (–)

Other features: Twinning is common, straight extinction, near-90° cleavage, weak pink–green pleochroism, and weak to moderate birefringence.

Paragenesis

The orthopyroxenes represent a solid solution between Mg (enstatite) and Fe (ferrosilite). Mg-rich enstatite is common in ultramafic and mafic plutonic rocks, while ferrosilite is more typical in diorites, syenites, or granites. In metamorphic rocks it is diagnostic of the granulite facies. A high-temperature, high-pressure $(Mg,Fe)SiO_3$ phase with an ilmenite-type structure has been synthesized (P = 24–27 GPa, T = 1,100–1,400 °C) and may be an important constituent of the Earth's lower mantle. Mg-rich varieties alter to serpentine and talc, generating pseudomorphs when the original structure is preserved. Alteration to a fine-grained pale amphibole (uralite) is also common. The optic sign reflects composition, with end-members (En_{100-88} and En_{2-0}) optically positive and intermediate compositions (En_{88-12}) optically negative.

Igneous

Enstatite is present in the early differentiates of many layered intrusions and in norites reflects crystallization from basic magma contaminated by argillaceous material. Ferrosilite occurs in some acid rocks such as adamellites and granodiorites. Compositional zoning is rare in plutonic and metamorphic orthopyroxene. In volcanic rocks, enstatite phenocrysts occur in olivine tholeiites and tholeiitic andesites and are commonly associated with olivine, diopside, and spinel; more iron-rich orthopyroxene (up to about Fe_{50}) occurs in some dacites and rhyolites. In alkaline lavas, orthopyroxene megacrysts are generally Al-rich. In some lavas, orthopyroxene and magnetite form reaction rims around olivine. Compositional zoning of orthopyroxene is common in volcanic rocks, with Mg-rich cores giving way to Fe-rich margins; this "normal" zoning is typical, but reversed zoning may also occur. Exsolution lamellae are also common.

Metamorphic

Orthopyroxene is characteristic of the charnockite rock series and typical of granulite facies. In argillaceous hornfels orthopyroxene is derived from the breakdown of biotite:

$$\text{Bt} + \text{Qz} + \text{Pl}\ (+\text{Grt}) \longrightarrow \text{Opx} + \text{Crd} + \text{Liq}$$

At higher temperatures, it is associated with cordierite via reactions such as:

$$\text{Bt} + \text{Qz} = \text{Opx} + \text{Crd} + \text{Kfs} + \text{Liq}$$

Ferrosilite is associated with fayalite, hedenbergite, grunerite, and almandine/spessartine garnet in eulysite, a regionally metamorphosed iron-rich sediment.

Figure 1.42

A Enstatite in norite from an unknown locality (Germany). The crystal in the center of the image is mantled by later clinopyroxene (augite). Plane-polarized light, 2× magnification, field of view = 7 mm.

B Enstatite in norite from an unknown locality (Germany). Same view as in (A) but with crossed polarizers. Note the higher birefringence of augitic clinopyroxene. Cross-polarized light, 2× magnification, field of view = 7 mm.

C Enstatite in ignimbrite from Vulsini volcano (Italy). Note the pronounced twinning along the basal section of this euhedral phenocryst. Cross-polarized light, 10× magnification, field of view = 2 mm.

D Enstatite–ferrosilite in mafic granulite from Hartmannsdorf (Germany). Note its beautiful pink–green pleochroism and granular texture. Plagioclase and magnetite also present. Plane-polarized light, 10× magnification, field of view = 2 mm.

Sapphirine $(Mg,Fe,Al)_8(Al,Si)_6O_{20}$

Optical Properties

Color in thin-section:	Light blue, blue–gray, green, greenish-gray
Pleochroism:	Colorless to shades of blue and green
Birefringence:	0.005–0.009
Relief:	High
Cleavage:	Moderate {010}, poor {100} and {001}
Crystal system/optic sign:	Monoclinic / generally (–)
Other features:	Distinctive color, pleochroism, and associated minerals make sapphirine easy to identify. It is biaxial (–) but anomalous (+) has been reported; it can be distinguished from corundum and from kyanite by poorer cleavage and lower birefringence.

Paragenesis

Sapphirine, despite being considered a rare and uncommon rock-forming mineral, is increasingly recognized in deep crustal, Mg-rich and SiO_2-poor rocks associated with high-temperature and ultra-high-temperature amphibolite–granulite facies. Sapphirine is often associated with numerous other minerals, such as orthopyroxene, cordierite, spinel, gedrite, phlogopite, kyanite, sillimanite, chlorite, talc, kornerupine, mullite, corundum, Ca-plagioclase, and garnet. It is also found in Al-rich xenoliths from kimberlites, in peridotites, and layered igneous intrusions, as well as thermally metamorphosed rocks (e.g. calc-silicate skarns and hornfels). Sapphirine may occur as a primary magmatic mineral in silica-poor rocks and in peraluminous pegmatites, but its presence in igneous rocks is otherwise rare. Sapphirine alters to a mixture of corundum and biotite (± talc).

At high pressure, the upper stability of sapphirine is defined by the reaction:

$$\underset{\textit{sapphirine}}{3Mg_2Al_4SiO_{10}} \leftrightarrow \underset{\textit{pyrope}}{Mg_3Al_2Si_3O_{12}} + \underset{\textit{spinel}}{2MgAl_2O_4} + \underset{\textit{corundum}}{2Al_2O_3}$$

During water-present, retrograde metamorphism, sapphirine is unstable and breaks down:

$$\underset{\textit{sapphirine}}{2Mg_2Al_4SiO_{10}} + 4H_2O \leftrightarrow \underset{\textit{chlorite}}{Mg_4Al_2(Si_2Al_2)O_{10}(OH)_8} + \underset{\textit{corundum}}{2Al_2O_3}$$

Disequilibrium associations, such as relict spinel inclusions in sapphirine or the symplectic intergrowth of sapphirine being replaced by cordierite, document changing temperature and pressure conditions. Such textures may be explained by reactions such as:

$$\underset{\textit{spinel}}{3MgAl_2O_4} + SiO_2 \leftrightarrow \underset{\textit{sapphirine}}{Mg_2Al_2(Al_2Si)O_{10}}$$

or

spinel + cordierite ↔ sapphirine + orthopyroxene

Figure 1.43

A Sapphirine in granulite from Val Codera (Italy). Prismatic sapphirine crystals with distinctive strong blue–green pleochroism, surrounded by cordierite (colorless). Orthopyroxene (pink–brown) also present. Plane-polarized light, 2× magnification, field of view = 7 mm.

B Sapphirine in granulite from Val Codera (Italy). Same image as (A) but with crossed polarizers. Sapphirine (low-interference colors partially masked by its natural color) is enclosed by cordierite (gray birefringence and complex twins; lower-right); orthopyroxene has first-order orange interference colors. Cross-polarized light, 2× magnification, field of view = 7 mm.

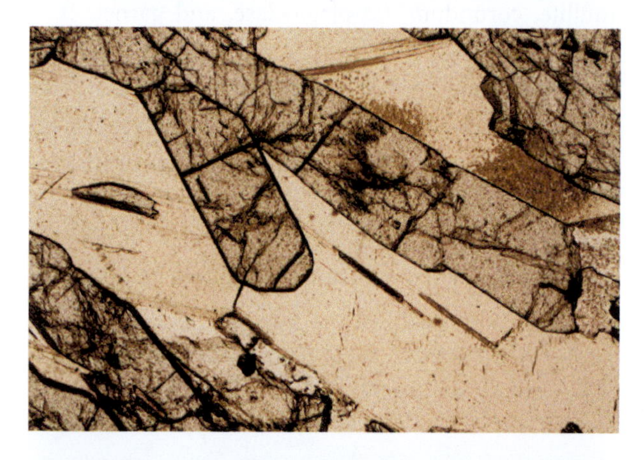

C Sapphirine in schist from Fiskernæs (Greenland). Euhedral sapphirine crystals (high relief, cleaved, and weakly colored) associated with muscovite (low relief, colorless). Plane-polarized light, 10× magnification, field of view = 2 mm.

D Sapphirine in schist from Fiskernæs (Greenland). Same image as (C) but with crossed polarizers. Sapphirine crystals (gray) show low-interference colors, whereas muscovite has high birefringence. Cross-polarized light, 10× magnification, field of view = 2 mm.

Wollastonite $CaSiO_3$

Optical Properties

Color in thin-section:	Colorless
Pleochroism:	None
Birefringence:	0.013–0.014
Relief:	Moderate to moderately high
Cleavage:	Perfect {100}, good {001} and {102}
Crystal system/optic sign:	Triclinic / (–)
Other features:	Commonly occurs as bladed or fibrous crystal masses or more rarely as twinned acicular single crystals. Its three cleavages when seen are diagnostic; it is distinguished from tremolite and pectolite by its weaker birefringence. Diopside has higher relief.

Paragenesis

Wollastonite is a pyroxenoid because it has a modified pyroxene structure. It mostly occurs in metamorphosed or metasomatized impure limestone and may exist in a fairly pure form or substitute appreciable amounts of Fe and Mn for Ca. It often forms from the reaction of calcite and quartz via the reaction:

$$CaCO_3 + SiO_2 \leftrightarrow CaSiO_3 + CO_2$$

During the early stages of progressive metamorphism of siliceous dolomites, talc, tremolite, diopside, forsterite, wollastonite, periclase, monticellite form; the formation of wollastonite approximates the highest grade of contact metamorphism (e.g. associated with intrusion of granite). Wollastonite can occur in regionally metamorphosed amphibolite and granulite facies rocks of appropriate composition. Wollastonite can also occur in alkaline igneous rocks, such as in the Alnö alkaline complex in Sweden, the ijolite alkaline rocks of East Africa, and in wollastonite phonolites.

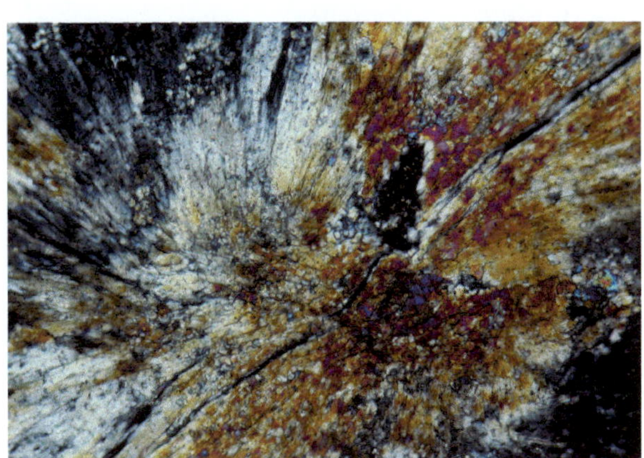

Figure 1.44 Wollastonite in marble from Elba island (Italy). Wollastonite occurs as a fibrous, radiating aggregate filling the entire image (*left*, plane-polarized light). In crossed polarizers (*right*) its birefringence varies from low to high first-order colors. Both images: 10× magnification, field of view = 2 mm.

APPLICATION 1.2

Imaging Minerals in Thin-Section

Thin-section imaging is used to enhance what is visible or even reveal what is invisible in a petrographic thin-section. Imaging methods are used, for example, to map minerals or the compositions of minerals, to show structures and fabrics, or alteration. A variety of imaging methods are used today, and many of them utilize an X-ray energy source (e.g. the secondary electron microscope [SEM] or the electron microprobe [EMP]); these may be combined with different detectors to provide different types of images (e.g. energy-dispersive X-ray spectroscopy [EDS], backscattered electron [BSE], or cathodoluminescence [CL] detectors). Such methods allow us to map mineral compositions, textures, and structures.

Energy-dispersive spectroscopy is capable of elemental analysis and chemical characterization, and it can be used to estimate the relative abundance of different elements/minerals present. The spectra obtained are semi-unique, and by cross-referencing these to mineral standards compositions can be estimated. Automated systems such as QEMSCAN can collect and map spectral data in a high-density grid pattern using a stepping interval of less than 50 μm. The collected spectral data are used to produce "false color" images such as the one shown here. This QEMSCAN image of peridotite highlights the mineral phases present and has identified olivine (light green), orthopyroxene (purple–blue), magnesian clinopyroxene (dark green), and chrome spinel (dark pink). Important textures are also highlighted, such as the infilling of orthopyroxene along olivine grain boundaries and cracks, as well as the entrainment of Cr-spinel within orthopyroxene. The mineral compositional data also defines the olivine as forsterite and the orthopyroxene as enstatite. This image was made using a 1 μm stepping interval. Field of view = 1 cm.

Image courtesy of J. Omma (Rocktype)

Background
Olivine Fo90-100
Orthopyroxene En90-100
Mg-Clinopyroxene
Cr-Spinel

Backscattered electron (BSE) image of the Agoult eucrite from Morocco. This BSE image of a polished thin-section shows an annealed granulitic texture of monoclinic tridymite (black) within plagioclase (dark gray), pyroxene (lighter grays) and scattered oxides; the white regions reflect secondary gold coating. 170× magnification, field of view = 700 μm.

Image courtesy of H. Ono

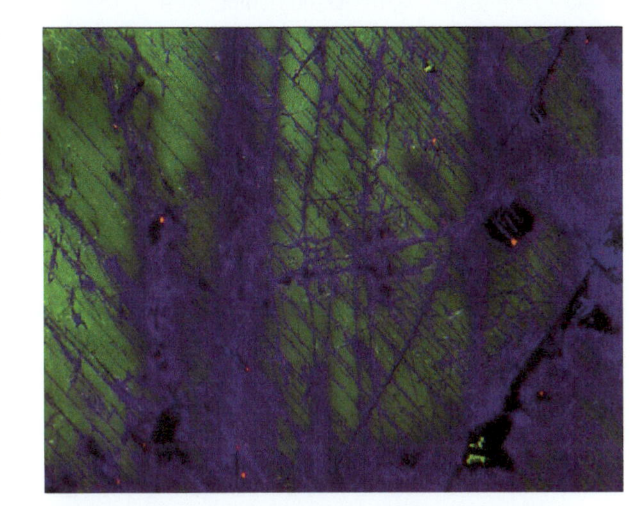

Cathodoluminescence (CL) images are useful for visualizing internal structures that are not discernible with a regular petrographic microscope. In this CL image of albite from Spruce Pine (United States), primary mineral textures and secondary alteration features are both visible. Primary feldspar is green (CL activated by Mn^{2+}) and secondary alteration of albite is purple–blue (CL activated by REE^{3+}). Field of view = 1.5 mm.

Image courtesy of J. Götze

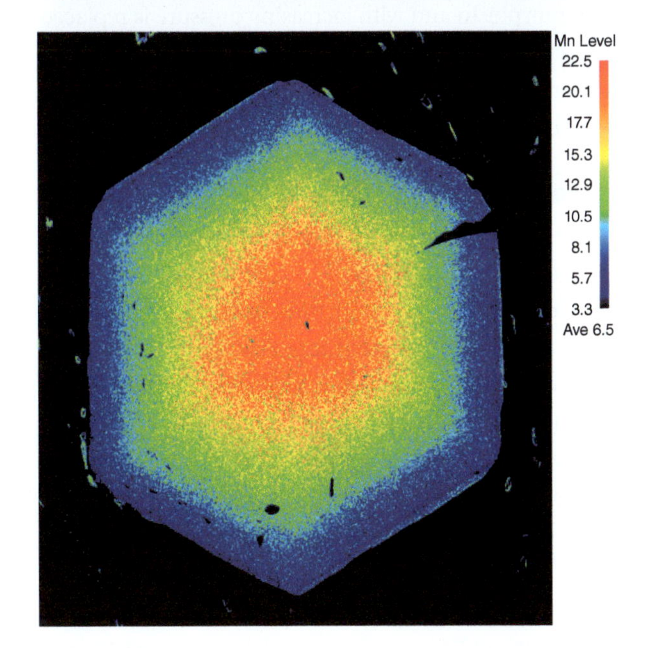

Electron microprobe (EMP) analysis is a powerful method for obtaining element maps of minerals in thin-section. In this false color image, Mn in garnet shows clear compositional zoning. The decrease in Mn from core to rim documents growth during increasing temperature and is interpreted to reflect prograde metamorphism. Field of view = 2.3 mm across.

Image courtesy of P. Manzotti

1.5 Disilicates, Orthosilicates, and Ring Silicates

Andalusite Al_2SiO_5

Optical Properties **THE ALUMINOSILICATES**

Color in thin-section:	Colorless to pale pink, with a patchy color distribution
Pleochroism:	Weak rose-pink to greenish-yellow
Birefringence:	0.009–0.012
Relief:	High
Cleavage:	Good on {110} but only visible in large crystals
Crystal system/optic sign:	Orthorhombic / (–)
Other features:	Andalusite has a square cross-section, straight extinction, and is length-fast, whereas sillimanite is length-slow and kyanite has oblique extinction and higher birefringence. The variety chiastolite is recognized by its cruciform pattern of inclusions. Andalusite is commonly altered to sericite.

Paragenesis

Andalusite is the high-temperature, low-pressure polymorph of Al_2SiO_5. It is characteristic of metamorphosed argillaceous rocks and commonly associated with cordierite. Andalusite forms during contact metamorphism (e.g. hornfels facies) along with muscovite, biotite, almandine garnet, cordierite, and sillimanite, kyanite, or staurolite. It also forms during regional metamorphism and with increasing pressure is replaced by kyanite. It is a relatively common accessory mineral in peraluminous granites. The pink and red varieties contain Fe, whereas the green varieties contain Mn.

Igneous

The formation of andalusite in felsic peraluminous igneous rocks is strongly influenced by (1) pressure and temperature, and (2) the amount of water and other components present (e.g. Al_2O_3, Be, B, Li, P, F). The latter influences the position of the water-saturated granite solidus. Magmatic andalusite is highly susceptible to subsolidus reactions, resulting in it being partially or totally replaced by sericite.

Metamorphic

The major mineralogical distinction between lower- and higher-pressure metapelites is the presence of cordierite (which occur in the former) and andalusite. These two minerals may be introduced at 450–500 °C. Under low-temperature metamorphic conditions, andalusite forms via the breakdown of chlorite in the presence of muscovite and quartz through the reaction:

$$\text{chlorite} + \text{muscovite} + \text{quartz} \rightarrow \text{andalusite} + \text{cordierite} + \text{biotite} + H_2O$$

At slightly higher temperatures, if no chlorite is available, andalusite is produced through the reaction:

$$\text{cordierite} + \text{muscovite} + \text{quartz} \rightarrow \text{andalusite} + \text{biotite} + H_2O$$

During prograde metamorphism (through amphibolite–hornfels facies conditions), andalusite is replaced by sillimanite at higher temperature and by kyanite at higher pressure. The coexistence of andalusite and sillimanite is commonly observed and attributed to the sluggish kinetics of the andalusite → sillimanite transition; the metastable coexistence of these two phases occurs until higher temperature is reached. In most metamorphic aureoles, andalusite forms oval porphyroblasts that are commonly rich in carbonaceous inclusions (e.g. chiastolite). Chiastolite forms in the early stages of contact metamorphism by the selective attachment of impurities at the corners of the crystal.

Figure 1.45

A Andalusite in granitic aplite from Elba island (Italy). Note brighter sericite alteration along porphyroblast margins. Cross-polarized light, 10× magnification, field of view = 2 mm.

B Andalusite in schist from the Adamello massif (Italy). The porphyroblasts have rectangular sections (parallel to the c-axis) and show compositional zoning. Pink to red colors are due to Fe, whereas green colors reflect Mn. Plane-polarized light, 10× magnification, field of view = 2 mm.

C Chiastolite in graphitic schist from Karakorum (India). The graphite inclusions in the porphyroblast define the classic cruciform arrangement. Plane-polarized light, 2× magnification, field of view = 7 mm.

D Andalusite in hornfels schist from Valmalenco (Italy). The andalusite porphyroblasts (extinct gray colors) occur in a muscovite + quartz-rich matrix and also contain numerous inclusions of quartz (gray) and mica (bright). Cross-polarized light, 10× magnification, field of view = 2 mm.

Kyanite Al_2SiO_5

Optical Properties

THE ALUMINOSILICATES

Color in thin-section: Colorless to pale blue

Pleochroism: None

Birefringence: 0.012–0.016

Relief: High

Cleavage: Perfect on {100} and good on {010}

Crystal system/optic sign: Triclinic / (–)

Other features: Kyanite is distinguished from other Al_2SiO_5 polymorphs by its high relief, distinctive cleavage, and birefringence. Hypersthene shows similar relief but has straight extinction in longitudinal sections, while kyanite always shows oblique extinction. Crystallographic intergrowth with staurolite is common.

Paragenesis

Kyanite is the high-pressure polymorph of Al_2SiO_5 and characteristic of medium- and high-pressure regionally metamorphosed pelitic rocks. Kyanite is a metamorphic index mineral in metapelites, since during prograde metamorphism it forms after staurolite and before sillimanite; it is commonly replaced by sillimanite. Kyanite is a relatively common mineral in high-pressure rocks known as whiteschists (kyanite–talc–quartz) and in ultra-high-pressure coesite-bearing whiteschists. Kyanite is also an accessory mineral in eclogite and amphibolite, and is found in granitic pegmatite.

Igneous

In rare cases, kyanite can form in igneous rocks, as do the other Al_2SiO_5 polymorphs. Kyanite can be found in peraluminous leucogranites and pegmatites, where it may crystallize directly from the melt or from melt segregation processes in the final stages of melt solidification.

Metamorphic

During prograde metamorphism of Al-rich pelites, kyanite forms via the breakdown of staurolite at 600–620 °C:

$$\text{staurolite} + \text{chlorite} + \text{muscovite} + \text{quartz} \longrightarrow \text{kyanite} + \text{biotite} + H_2O$$

The production of kyanite continues through the reaction:

$$\text{chlorite} + \text{muscovite} + \text{quartz} \longrightarrow \text{kyanite} + \text{biotite} + H_2O$$

During the first stages of high-pressure/low-temperature metamorphism (lower blueschist facies conditions), the breakdown of carpholite forms chloritoid, chlorite, talc, and kyanite. At higher pressures (eclogite facies) chloritoid and talc break down to form kyanite through the reaction:

$$\text{talc} + \text{chloritoid} \ (+ \text{muscovite}) \longrightarrow \text{biotite} + \text{kyanite} + \text{quartz} + H_2O$$

Ultra-high-pressure metapelites (whiteschists) are characterized by talc + phengite + pyrope + kyanite + silica polymorph (quartz or coesite, depending on the conditions of metamorphism). Kyanite can also develop from a mafic protolith at amphibolite and eclogite facies.

Figure 1.46

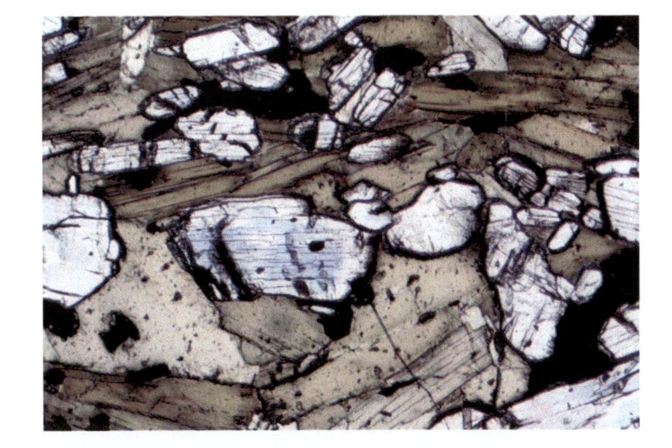

A Kyanite in gneiss from Sudbury (England). The kyanite crystals have pale blue cores and are surrounded by biotite. Plane-polarized light, 10× magnification, field of view = 2 mm.

B Kyanite in gneiss from Sudbury (England). This kyanite porphyroblast shows perfect {100} cleavage. Plane-polarized light, 2× magnification, field of view = 7 mm.

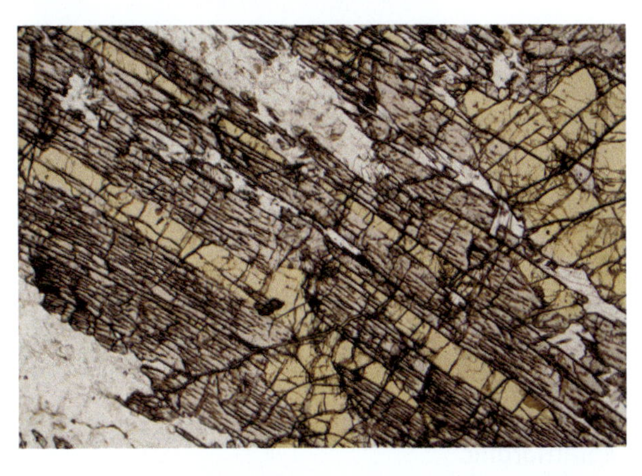

C Kyanite in schist from Alpe Sponda (Switzerland). Kyanite (pale brown) with notable cleavage intergrown with staurolite (gold–yellow). Plane-polarized light, 2× magnification, field of view = 7 mm.

D Kyanite in gneiss from Sudbury (England). The large twinned kyanite porphyroblast (center) is associated with biotite (second-order colors) and quartz (gray birefringence). Cross-polarized light, 2× magnification, field of view = 7 mm.

Sillimanite Al_2SiO_5

Optical Properties **THE ALUMINOSILICATES**

Color in thin-section:	Colorless; fibrolite can be yellow–brown
Pleochroism:	None
Birefringence:	0.018–0.022; basal section shows low first-order interference colors
Relief:	High
Cleavage:	Perfect {010} visible in larger crystals, diagonal to the pseudotetragonal basal section
Crystal system/optic sign:	Orthorhombic / (+)
Other features:	Sillimanite has straight extinction parallel to the cleavage trace (seen in larger crystals). It might be confused with other Al_2SiO_5 polymorphs, but relative to andalusite it is length-slow and has higher birefringence. Relative to kyanite, sillimanite may have similar (low) birefringence but has lower relief.

Paragenesis

The Al_2SiO_5 polymorphs develop in alumina-rich pelites under different conditions of temperature and pressure. Sillimanite is the high-temperature polymorph of Al_2SiO_5 and forms acicular to fibrous crystals. It is stable above ~550 °C and is an index mineral in metamorphic rocks, as are andalusite and kyanite.

Igneous

The presence of sillimanite in igneous rocks is uncommon. However, due to the overlap in the stability fields of sillimanite and the felsic melt solidus, sillimanite can occur with/without andalusite in some peraluminous granites. Magmatic sillimanite tends to invert to andalusite (andalusite pseudomorphs after sillimanite) during slow subsolidus cooling. Sillimanite can also occur in high-temperature metasedimentary xenoliths, present as a residual product in aluminous rocks after partial melting.

Metamorphic

Sillimanite can develop in (1) thermally metamorphosed pelitic rock (hornfels) around intrusions, and (2) high-grade regionally metamorphosed rocks. In thermally metamorphosed pelitic rock, sillimanite commonly derives from the breakdown of biotite or from earlier-formed andalusite. In regionally metamorphosed rocks sillimanite can form from the breakdown of muscovite or quartz by the following reactions:

muscovite + plagioclase + quartz ⟶ hydrous melt (monzogranitic)
⟶ alkali feldspar + garnet + sillimanite

staurolite + muscovite + quartz ⟶ sillimanite + biotite + garnet + H_2O

It is possible to find sillimanite (which occurs over a wide range of pressures and temperatures) and other Al_2SiO_5 polymorphs in the same rock. At higher temperatures both andalusite and kyanite invert to sillimanite. This can also happen via retrograde metamorphic reactions, where kyanite and sillimanite may coexist due to the sluggish kinetics associated with the transition of kyanite to sillimanite.

Figure 1.47

A Sillimanite in granulite from Le Serre (Italy). Prismatic crystals of sillimanite (mostly colorless with high relief) are oriented from upper-left to lower-right. Fractures perpendicular to crystal long axes are distinctive and typical of sillimanite. It is present with biotite (lower relief in browns) and colorless quartz (upper-right). Plane-polarized light, 2× magnification, field of view = 7 mm.

B Sillimanite in granulite from Le Serre (Italy). Same image as (A) but with crossed polarizers showing typical birefringence of sillimanite. Cross-polarized light, 2× magnification, field of view = 7 mm.

C Sillimanite in granulite facies metapelite from Le Serre (Italy). Basal sections with low birefringence show typical pseudotetragonal outlines and diagonal cleavage. These sillimanite crystals are enclosed in a large crystal of biotite. Cross-polarized light, 10× magnification, field of view = 2 mm.

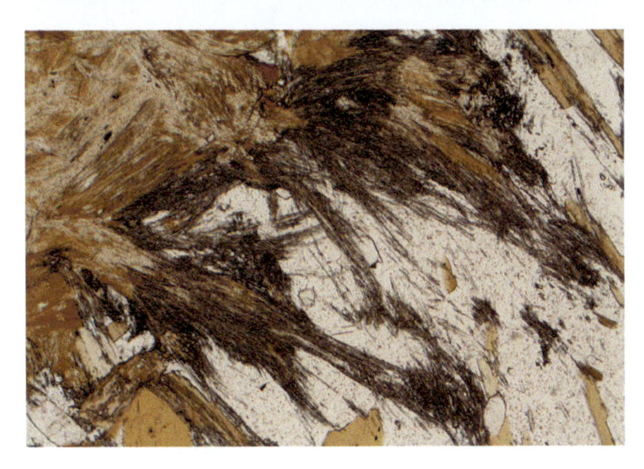

D Sillimanite in schist from Ivrea Verbano (Italy). These yellow–brown, hair-like crystals are known as *fibrolite.* This aggregate form of sillimanite is common. Plane-polarized light, 10× magnification, field of view = 2 mm.

Beryl $Be_3Al_2[Si_6O_{18}]$

Optical Properties

Color in thin-section:	Colorless
Pleochroism:	Rare; if present, pale blue to pale green
Birefringence:	0.004–0.009
Relief:	Low–moderate
Cleavage:	Poor on {0001}
Crystal system/optic sign:	Hexagonal / (–)
Other features:	Beryl may be mistaken for quartz but the latter has higher birefringence and is optically uniaxial (+). Apatite has higher relief, nepheline has a different paragenesis, and topaz is optically biaxial (+).

Paragenesis

Despite the low beryllium content of crustal rocks, it forms many minerals of which the most abundant and common is beryl. The occurrence of beryl rather than other Be-silicates in granitic rocks is due to the small size of the Be^{2+} ion, which is incompatible in all the most common silicate mineral structures. Beryl occurs principally in peraluminous granite and as a pneumatolytic mineral in pegmatites (especially in vugs and miarolitic cavities); it is also found in alkali rhyolites and in hydrothermal veins, and rarely occurs in nepheline syenites. Beryl can develop idiomorphic crystals from millimeter to meter size in granitic pegmatites, sometimes of gem quality. The green variety, emerald, is usually found in regionally metamorphosed mica schist, marbles, and gneisses. Beryl easily alters to clay minerals during hydrothermal alteration:

$$\underset{\textit{beryl}}{Be_3Al_2Si_6O_{18}} + 4H_2O = \underset{\textit{kaolinite}}{Al_4Si_4O_{10}(OH)_8} + 5SiO_2 + \underset{\textit{phenakite}}{2Be_2SiO_4}$$

Figure 1.48 Beryl in pegmatite from Trentino-Alto Adige (Italy). (*Left*) In plane-polarized light beryl has a cloudy appearance and moderate relief, while quartz and plagioclase are colorless; the three darker brown–green crystals are tourmaline. (*Right*) Under crossed polarizers the birefringence of beryl is typical. Both images: 2× magnification, field of view = 7 mm.

Chloritoid $(Fe^{2+},Mg,Mn)_2(Al,Fe^{3+})Al_3O_2[SiO_4]_2(OH)_4$

Optical Properties

Color in thin-section:	Colorless, green, blue–green
Pleochroism:	Strong, from pale green to blue and colorless to pale yellow
Birefringence:	0.005–0.022, masked by color of mineral, often shows anomalous blue colors
Relief:	High
Cleavage:	Perfect {001}, moderate {110}, parting {010}
Crystal system/optic sign:	Monoclinic / (+)
Other features:	Its green–blue color, strong pleochroism, and low-interference colors are diagnostic. Compared to chloritoids, chlorite has a lower relief and commonly shows anomalous interference colors. Biotite is optically negative. All other micas have higher birefringence. Chloritoids commonly show polysynthetic twins on (001) and zoning with hour glass structure.

Paragenesis

Chloritoid is very common in aluminous and iron-rich metapelite, under both intermediate-pressure Barrovian and high-pressure subduction-related metamorphic gradients. In high-Al pelite undergoing Barrovian metamorphism, at low metamorphic grade chloritoid is introduced by the reaction:

$$\text{chlorite} + \text{pyrophyllite} \longrightarrow \text{chloritoid} + \text{quartz} + H_2O$$

whereas at medium grade, chloritoid-bearing assemblages may also include staurolite due to the reaction:

$$\text{chloritoid} + \text{quartz} \longrightarrow \text{staurolite} + \text{garnet} + H_2O$$

At upper greenschist facies conditions garnet may join the chloritoid-bearing assemblage via the reaction:

$$\text{chloritoid} + \text{biotite} + H_2O \longrightarrow \text{garnet} + \text{chlorite}$$

The ongoing chloritoid-consuming, garnet- and staurolite-producing reactions lead to the complete disappearance of chloritoid from the metamorphic assemblage at medium grade (~550–600 °C) conditions.

Together with various amounts of talc, micas (phengite and paragonite), and sodic amphibole, Mg-rich chloritoid-bearing assemblages are typical for high-pressure metapelites. Along high-pressure gradients, chloritoid is introduced by the reaction:

$$\text{carpholite} \longrightarrow \text{chloritoid} + \text{quartz} + H_2O$$

Chloritoid is also reported in high-pressure (i.e. eclogite facies) mafic rocks such as Mg–Al-rich gabbro in the Western Alps.

Figure 1.49

A Chloritoid in garnet–micaschist from Île de Groix (France). Chloritoid crystals (green–blue) occur within the E–W foliation. Chloritoid crystals are associated with garnet (pale pink, subhedral, fractured), quartz, and muscovite (both colorless). Plane-polarized light, 2× magnification, field of view = 7 mm.

B Chloritoid in garnet–micaschist from Île de Groix (France). Same image as (A) but with crossed polarizers. Chloritoid has low birefringence, masked by its natural green color; muscovite has high birefringence, garnet is isotropic, and quartz (abundant in the left pressure-shadow around the garnet) has gray interference colors. Cross-polarized light, 2× magnification, field of view = 7 mm.

C Chloritoid in chloritoid–schist from Val di Susa (Italy). Euhedral chloritoid crystal with low-interference colors (near center of image) shows good {001} cleavage and polysynthetic twinning; it is surrounded by quartz (low birefringence) and muscovite (high birefringence). Cross-polarized light, 10× magnification, field of view = 2 mm.

D Chloritoid in chloritoid–schist from Calabria (Italy). Euhedral chloritoid with distinctive interpenetrating crystals surrounded by a deformed muscovite- and quartz-rich matrix. Muscovite has high (orange) and quartz has low (gray) birefringence. Cross-polarized light, 10× magnification, field of view = 2 mm.

Cordierite $(Na, K)_{0\text{-}1}(Mg, Fe, Mn, Li)_2[Al_4Si_5O_{18}] \cdot n(H_2O, CO_2)$

Optical Properties

Color in thin-section:	Colorless to pale blue
Pleochroism:	None
Birefringence:	0.008–0.018
Relief:	Low
Cleavage:	Moderate on {100}; poor along {001}, {010}
Twinning:	Common single, lamellar, and cyclic twinning along {110}, {310}
Crystal system/optic sign:	Orthorhombic / (–)
Other features:	Cordierite is distinguished from quartz by its negative optic sign and from plagioclase by the presence of cyclic twinning and/or pinite. If cyclic twinning is absent, it can be difficult to distinguish from plagioclase.

Paragenesis

Cordierite is the product of high-temperature, low-pressure metamorphism of metapelitic rocks and is typically found in contact-metamorphosed metasediments or high-grade amphibolite to granulite facies terrains. It can also occur in metasedimentary xenoliths within granites. Cordierite may occur in granites, gabbros, rhyolites, and andesites. As an accessory mineral it indicates peraluminous compositions. It commonly alters to pinite, a yellowish mixture of fine-grained sericitic mica.

Igneous

Cordierite occurs in many peraluminous felsic intrusive and extrusive rocks. It is found predominantly in rhyolites, granitoids, and pegmatites, but has also been reported in norites. There are several ways cordierite may be incorporated in a crystallizing magma: (1) as xenocrysts from metamorphic rocks; (2) as a peritectic phase that forms directly during partial melting of the source granitic melt; and (3) as magmatic cordierite. Igneous cordierite forms in association with rapidly decreasing temperature and pressure associated with ascending magmas, which favors its nucleation over other aluminosilicates.

Metamorphic

Cordierite forms together with andalusite in medium-grade, low-pressure metapelites from the destabilization of chlorite at 450–500 °C:

$$\text{chlorite} + \text{muscovite} + \text{quartz} \longrightarrow \text{cordierite} + \text{andalusite} + \text{biotite} + H_2O$$

In the early stages of metamorphism, cordierite commonly forms ovoid inclusion-rich poikiloblasts. At higher temperatures, such as in medium-grade hornfels, several muscovite, biotite, and andalusite/sillimanite-consuming reactions may produce cordierite in association with K-feldspar, garnet, spinel, or orthopyroxene. Cordierite is common in many granulite facies rocks, where it is associated with sapphirine, spinel, biotite, and feldspar. Cordierite can also form from the partial melting (pyrometamorphism) of arenaceous and argillaceous sediments, as in buchites. Finally, cordierite is a common peritectic product of melting reactions occurring in low-pressure migmatites together with sillimanite, garnet, orthopyroxene, feldspars, biotite, and sapphirine.

Figure 1.50

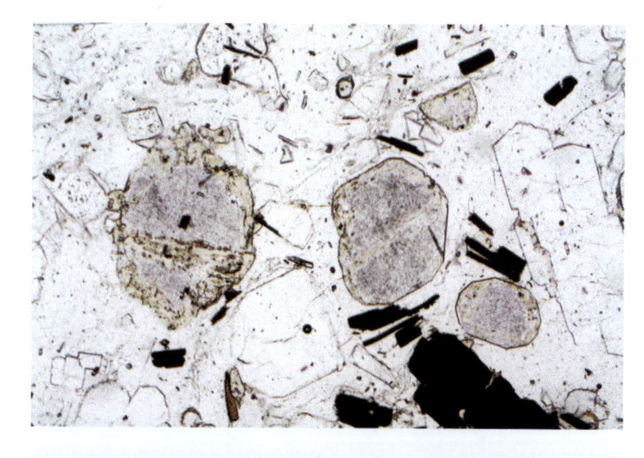

A Cordierite in rhyolite from Tuscany (Italy). The cordierite crystals (center) are rounded and have a "cloudy" appearance due to a high concentration of microinclusions. The larger crystals show pinite alteration (yellow) along their margins. The cordierite crystals, together with plagioclase phenocrysts (similar in size to cordierite but colorless) and biotite (very dark brown) are set in a colorless, low-relief, glassy groundmass. Plane-polarized light, 2× magnification, field of view = 7 mm.

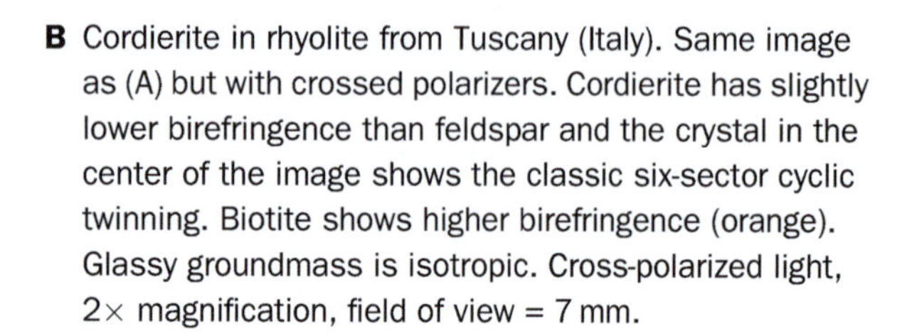

B Cordierite in rhyolite from Tuscany (Italy). Same image as (A) but with crossed polarizers. Cordierite has slightly lower birefringence than feldspar and the crystal in the center of the image shows the classic six-sector cyclic twinning. Biotite shows higher birefringence (orange). Glassy groundmass is isotropic. Cross-polarized light, 2× magnification, field of view = 7 mm.

C Cordierite in thermally metamorphosed shale from Adamello mass (Italy). Ovoid cordierite crystals are aligned and have a "cloudy" appearance due to numerous minute inclusions. Cordierite is surrounded by a fine-grained, biotite-rich matrix. Plane-polarized light, 2× magnification, field of view = 7 mm.

D Cordierite–andalusite hornfels from the Skiddaw metamorphic aureole, Cumbria (England). Cordierite crystal with inclusions defining chiastolite cross (upper-right corner) and cyclic twinning (left of center) associated with andalusite can be seen. The fine-grained matrix is rich in micas and quartz. Cross-polarized light, 2× magnification, field of view = 7 mm.

Allanite $Ca(REE)^{3+}(Al_2Fe^{2+})(Si_2O_7)(SiO_4)O(OH)$

Optical Properties

Color in thin-section:	Brownish-yellow or reddish-brown
Pleochroism:	Red–brown
Birefringence:	0.013–0.036; colors may be masked by crystal color; when metamict can be isotropic
Relief:	High
Cleavage:	Imperfect on {001}, poor on {100} and {110}
Crystal system/optic sign:	Monoclinic/ (–) (+)
Other features:	Often forms rounded to irregular-shaped, compositionally zoned grains rimmed by epidote; commonly metamict due to high REE content and pleochroic halos are not uncommon.

Paragenesis

The allanite family is one of the main REE-rich accessory minerals and represents the most diverse phase of the epidote group. It occurs in a wide range of intrusive rocks (e.g. granites, granodiorites, monzonites, syenites, and pegmatites), including gabbros and diorites. Allanite is commonly metamictic due to α-particle emission by radioactive constituents; this reduces its stability and makes it susceptible to alteration/hydration, resulting in crystal expansion and the formation of cracks that radiate into adjacent minerals. Allanite is less common in low-grade (or retrograde) meta-igneous and metasedimentary rocks like gneisses, amphibolites, and schists. In rocks from high- or ultra-high-pressure environments like eclogites or granulites it is uncommon but stable. It represents one of the main reservoirs of REEs in subduction zones. It can also be present in contact metamorphic zones as a primary or metasomatic phase in limestone skarn and carbonate or quartz veins, along with diopside and other clinopyroxenes.

Figure 1.51

A Allanite in granite from Ampar (Brazil). Euhedral allanite shows concentric zoning and its strong natural red-brown color results in anomalous birefringence under crossed polarizers. It occurs with biotite (uniform reddish-brown) and alkali feldspar (colorless). Plane-polarized light, 10× magnification, field of view = 2 mm.

B Allanite in granite from Ampar (Brazil). This allanite crystal is metamict, replaced by hydrous minerals, and due to the volume expansion resulting from alteration cracks radiate into adjacent minerals. In crossed polarizers it would be isotropic. Biotite (dark brown–green), quartz (clear), and alkali feldspar are also present. Plane-polarized light, 10× magnification, field of view = 2 mm.

Clinozoisite $Ca_2Al_3(Si_2O_7)(SiO_4)O(OH)$

Optical Properties

Color in thin-section:	Colorless
Pleochroism:	Usually non-pleochroic
Birefringence:	0.004–0.015
Relief:	High
Cleavage:	Perfect on {001}
Crystal system/optic sign:	Monoclinic / (+)
Other features:	Clinozoisite has oblique extinction which distinguishes it from zoisite. Vesuvianite and melilite are optically uniaxial; lawsonite and pumpellyite have higher birefringence.

Paragenesis

Clinozoisite is the monoclinic dimorph of zoisite and, along with epidote, occurs mainly as a hydrothermal alteration product of plagioclase in a wide range of plutonic and volcanic rocks. Both are also found in low-grade metamorphic rocks, such as greenschist, calcareous phyllite, and blueschist. During retrograde metamorphism epidote and zoisite are formed due to loss of the plagioclase anorthite component (a saussuritization processes). Anorthite is unstable at high pressures and the following reactions are common:

$$\text{anortite} + H_2O \longrightarrow \text{zoisite (or clinozoisite)} + \text{kyanite} + \text{quartz}$$

$$\text{albite} + \text{anortite} + H_2O \longrightarrow \text{zoisite (or clinozoisite)} + \text{paragonite} + \text{quartz}$$

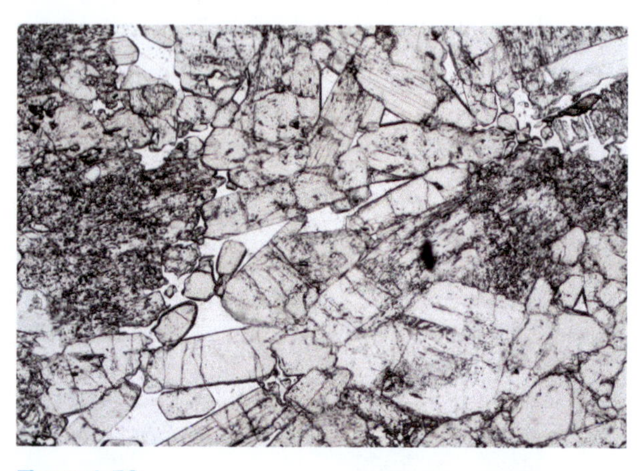

Figure 1.52

A Clinozoisite in retrogressed eclogite from Liguria (Italy). Clinozoisite has high relief and in some cases a prismatic cleavage. To the left and right of the image, two inclusion-rich epidote crystals are present. Plane-polarized light, 10× magnification, field of view = 2 mm.

B Clinozoisite in retrogressed eclogite from Liguria (Italy). Same image as (A) but with crossed polarizers. Clinozoisite shows characteristic low birefringence; some crystals are near extinction. Epidote is also present and shows high-interference colors (orange–violet, left and right). Cross-polarized light, 10× magnification, field of view = 2 mm.

Epidote $Ca_2Al_2Fe^{3+}[Si_2O_7][Si_2O_4]O(OH)$

Optical Properties — **THE EPIDOTES**

Color in thin-section:	Colorless to pale yellowish-green
Pleochroism:	Slightly pleochroic from colorless to pale yellow and pale green
Birefringence:	0.015–0.051
Relief:	High
Cleavage:	Perfect on {001} and clear on {100} with an intersection angle of ~115°
Crystal system/optic sign:	Monoclinic / (–)
Other features:	Epidote is easily recognized due to its intense and anomalous interference colors. It is distinguished from clinopyroxene by its negative optic sign and characteristic pleochroism.

Paragenesis

The clinozoisite–epidote series minerals occur in a wide variety of parageneses. They are primary minerals in some acid igneous rocks, are common in regional metamorphic rocks from greenschist to epidote–amphibolite facies, and occur in contact metamorphic environments. Clinozoisite–epidote composition ranges from $Ca_2Al_3Si_3O_{12}(OH)$ to $Ca_2Fe^{3+}Al_2Si_3O_{12}(OH)$. The Al $\leftrightarrow$ Fe^{3+} substitution affects the optical characteristics (refractive indices, birefringence, and pleochroism) and Ca,Fe compositions have the typical pistachio-green color. The clinozoisite–epidote solid solution may result in strong compositional zoning, which can be continuous, change abruptly, or be reversed.

Igneous

Epidote occurs as a primary accessory mineral in intermediate igneous rocks such as the tonalite–trondhjemite–granodiorite series and in monzogranites. According to experimental studies, epidote is stable above the wet granite solidus near 0.5 GPa at 680 °C. However, the presence of epidote and zoisite in igneous rocks is commonly linked to hydrothermal alteration (saussuritization) of plagioclase.

Metamorphic

Epidote occurs in a wide variety of metamorphic rocks, from low-grade to high-pressure and ultra-high-pressure conditions. The transition from lower-grade (prehnite–pumpellyite facies) rocks to the lower greenschist facies is marked by the following reaction:

$$\underset{\textit{pumpellyite}}{2Ca_4Fe^{2+}Fe^{3+}Al_4Si_6O_{21}(OH)_7} + 0.5O_2 \rightarrow \underset{\textit{epidote}}{4Ca_2Fe^{3+}Al_2Si_3O_{12}(OH)} + 5H_2O$$

At greenschist facies, the typical mineral association is epidote, chlorite, albite, quartz, and carbonate. Near amphibolite facies conditions, epidote is still present despite being partially consumed by reactions such as:

epidote + chlorite + quartz → Al-rich amphibole + anorthite

At the transition to blueschist facies, epidote occurs via reactions such as:

actinolite + chlorite + albite → epidote + glaucophane + quartz

Like other epidote group minerals, epidote can be stable at high-pressure or even ultra-high-pressure conditions, but its occurrence is restricted by the availability of H_2O.

Figure 1.53

A Epidote in retrogressed blueschist from As Sifah (Oman). Epidote crystals in this sample are euhedral, have high relief, are yellow–brown in color, and are associated with cholorite (yellow–green to green) and inclusion-rich albite (colorless, in center). Plane-polarized light, 10× magnification, field of view = 2 mm.

B Epidote in retrogressed blueschist from As Sifah (Oman). Same image as (A) but with crossed polarizers. Epidote has high birefringence that can vary within the crystal. Chlorite (anomalous dark green–brown) and albite (twinned with gray–white birefringence) are also present. Cross-polarized light, 10× magnification, field of view = 2 mm.

C Epidote in a gneiss from an unknown locality in the Alps (Italy). The large, rectangular, epidote crystal shows perfect intersecting {001} and {100} cleavage. Cross-polarized light, 10× magnification, field of view = 2 mm.

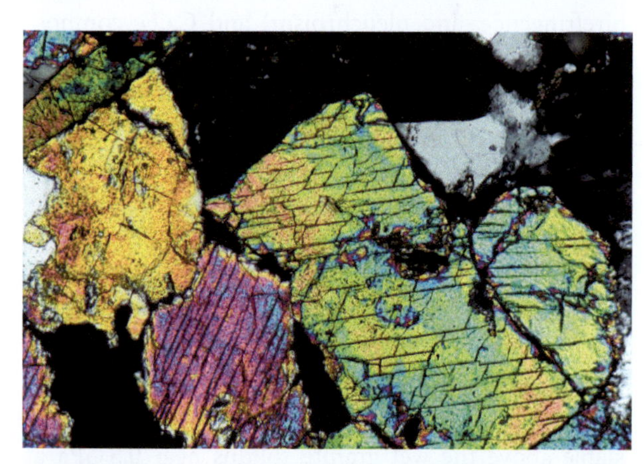

D Epidote in a greenschist from Trentino-Alto Adige (Italy). Epidote crystals have high (red to blue) interference colors and are associated with muscovite (yellow–azure interference colors) and quartz (granoblastic texture). Cross-polarized light, 10× magnification, field of view = 2 mm.

Piemontite $Ca_2Al_2Mn^{3+}(Si_2O_7)(SiO_4)O(OH)$

Optical Properties

Color in thin-section:	Violet or pink
Pleochroism:	Diagnostic, ranging from lemon yellow to amethystine and to carmine red or magenta
Birefringence:	0.025–0.073; birefringence is masked by its strong natural colors
Relief:	High
Cleavage:	Perfect on {001}
Crystal system/optic sign:	Monoclinic / (+)
Other features:	Piemontite is easily distinguished from other epidote group minerals by its distinctive pleochroism. Lamellar twinning on {100} can be present. The Mn-bearing zoisite (thulite) has a smaller optic axial angle and straight extinction.

Paragenesis

Piemontite has a strong relationship and dependence on Mn content, and can be found from low to high metamorphic grades in diverse assemblages, even in rare skarn Mn ore-related deposits. A typical protolith that leads to its formation is the Mn-rich chert formed at ocean bottoms. At low metamorphic grade, piemontite occurs as an accessory phase accompanied by chlorite, actinolite, magnetite, pumpellyite, stilpnomelane, calcite, albite, and Mn varieties of allanite. The highest-grade occurrence reported for piemontite-bearing rocks is in jasper–manganese ore deposits affected by granulite facies metamorphism. It is accompanied mainly by tremolite, diopside, quartz, garnet, and sillimanite.

Figure 1.54

A Piemontite in Mn ore from Aosta valley (Italy). These prismatic piemontite crystals have colors ranging from purple to yellow–orange. They are associated with plagioclase (colorless). Plane-polarized light, 2× magnification, field of view = 7 mm.

B Piemontite in Mn ore from Aosta valley (Italy). Same image as (A) but with crossed polarizers. Piemontite birefringence is masked by its strong natural colors. Plagioclase shows twins and low (gray–white) birefringence. Cross-polarized light, 2× magnification, field of view = 7 mm.

Zoisite $Ca_2Al_3(Si_2O_7)(SiO_4)O(OH)$

Optical Properties **THE EPIDOTES**

Color:	Colorless; thulite has purple–pink to yellowish pleochroism
Pleochroism:	Weak
Birefringence:	0.003–0.008
Relief:	High
Cleavage:	Perfect {100} prismatic cleavage, poor {001} cleavage sometimes present
Crystal system/optic sign:	Orthorhombic / (+)
Other features:	Unlike clinozoisite, zoisite has straight extinction on prism edge or {100} cleavage, and can be distinguished from other epidote group minerals by its weaker birefringence (anomalous blue interference colors are characteristic).

Paragenesis

Zoisite is the orthorhombic dimorph of clinozoisite. It occurs mainly as a hydrothermal alteration product in basic igneous rocks associated with calcic plagioclase. Hydrothermal alteration of anorthite-rich plagioclase results in a fine-grained aggregate of zoisite, epidote–clinozoisite, sericite, and calcite (saussurite). Zoisite is a primary component in low-grade (prehnite–pumpellyite facies) metamorphic rocks; it is a common constituent of epidote–amphibolite facies rocks derived from calcareous shales and sandstones, and in mafic amphibolites. Above amphibolite facies, zoisite is unstable and is commonly replaced by calcium-rich plagioclase:

$$\underset{\text{zoisite}}{2Ca_2Al_3Si_3O_{12}OH} + CO_2 \leftrightarrow \underset{\text{anorthite}}{3CaAl_2Si_2O_8} + \underset{\text{calcite}}{CaCO_3} + H_2O$$

Zoisite is proposed to be the main H_2O source in ultra-high-pressure subduction environments and has been documented at thickened crustal conditions of about 7 GPa.

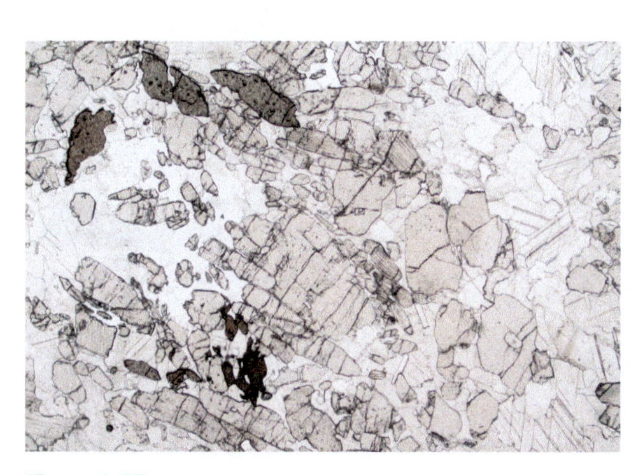

Figure 1.55

A Zoisite in calc-schist from Val Sondrio (Italy). Zoisite crystals with high relief occupy most of the image. Zoisite occurs with calcite and titanite (green–brown, high relief, upper-left). Plane-polarized light, 2× magnification, field of view = 7 mm.

B Zoisite in calc-schist from Val Sondrio (Italy). Same image as (A) but with crossed polarizers. Zoisite shows typical anomalous blue interference colors. Cross-polarized light, 2× magnification, field of view = 7 mm.

Almandine $Fe^{2+}{}_3Al_2(SiO_4)_3$

Optical Properties

THE GARNETS

Color in thin-section:	Colorless to pinkish red
Pleochroism:	Absent
Birefringence:	Isotropic
Relief:	High to very high
Cleavage:	None, but crystals are often fractured
Twinning:	None
Crystal system/optic sign:	Cubic
Other features:	Almandine, like other garnet group minerals, has a six- or eight-sided cross-section. High relief, pale pink color, and isotropic character with inclusions are distinctive of almandine; spinel group minerals are distinct from garnets as they are more colorful and have {111} cleavage.

Paragenesis

Almandine is one of the most common garnets and is found in Al-rich pelitic rocks, from phyllites to high-grade gneisses. It is also found in some igneous rocks such as Al-rich granites, granodiorites, and calc-alkaline volcanic rocks. Almandine commonly alters to chlorite.

Igneous

The presence of almandine in igneous rocks is not rare and may be magmatic or xenocrystic in origin. Garnets of magmatic origin may represent a primary magmatic phase in calc-alkalic granites, rhyolites, dacites, or a late crystallization phase in aplites and pegmatites. They may be derived from the reaction between magma and pelitic Al–Mn-rich xenoliths in the marginal facies of granitic intrusions, or from the reaction of early-formed phases and silicate melt, such as:

$$\text{liquid} + \text{biotite} \longrightarrow \text{garnet} + \text{muscovite}$$

$$\text{liquid} + \text{aluminosilicate} + \text{biotite} \longrightarrow \text{garnet}$$

Xenocrystic garnets may represent: (1) porphyroblasts from high-level pelitic country rocks, or (2) restitic minerals from refractory zones of partial melting.

Metamorphic

Almandine is a characteristic mineral during regional metamorphism of argillaceous sediment and basic igneous protoliths. It appears in the biotite zone associated with the reaction of chlorite:

$$\text{chlorite} + \text{quartz} \longrightarrow \text{garnet} + \text{Mg-chlorite} + H_2O$$

With increasing temperature, almandine may also be produced by the breakdown of mica or by the reaction between quartz and staurolite:

$$\text{staurolite} + \text{quartz} + \text{muscovite} \leftrightarrow \text{garnet} + \text{biotite} + \text{kyanite (or sillimanite)} + H_2O$$

The presence of almandine-rich garnet is common in high-pressure rocks (e.g. granulite, eclogite, blueschist). It can also occur in thermally metamorphosed pelitic rocks represented by the common hornfels assemblage almandine, cordierite, orthopyroxene, and plagioclase.

Figure 1.56

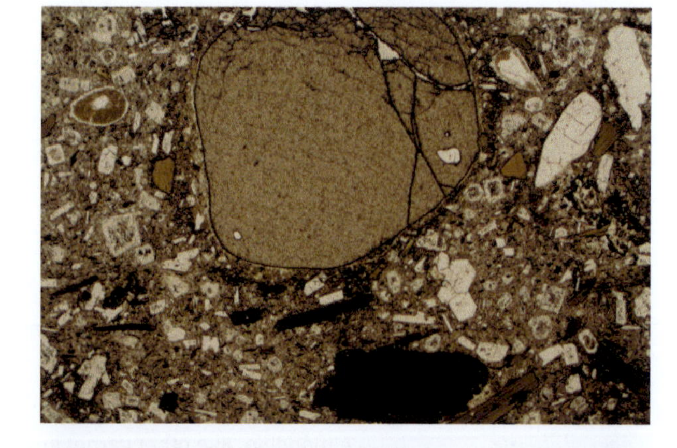

A Almandine in dacite from el Hoyazo de Níjar (Spain). Large, rounded almandine phenocryst (pale brown) set in a glassy plagioclase-rich groundmass. Some plagioclase crystals have sieve textures. The dark brown crystals are biotite. Plane-polarized light, 2× magnification, field of view = 7 mm.

B Almandine in garnet–micaschist from Val Lanterna (Italy). Rounded and fractured almandine porphyroblasts (pale pink color, high relief) in a matrix dominated by muscovite (colorless), biotite (brown), and lesser amounts of quartz (colorless) and opaques (black). Plane-polarized light, 2× magnification, field of view = 7 mm.

C Almandine in garnet–micaschist from Val Lanterna (Italy). Same image as (B) but with crossed polarizers. The almandine porphyroblasts are isotropic and have quartz inclusions. Muscovite has higher birefringence than biotite (low second order) and quartz (low first order). Cross-polarized light, 2× magnification, field of view = 7 mm.

D Almandine in eclogite from Liguria (Italy). On the right side of the image almandine (small, pale pink) crystal mantle omphacite (grass green). The eclogite is partially retrogressed, and between the almandine crystals pale blue–green secondary amphiboles have formed. Plane-polarized light, 2× magnification, field of view = 7 mm.

Andradite $Ca_3(Fe^{3+},Ti)_2(SiO_4)_3$

Optical Properties

Color in thin-section:	Yellowish to dark brown
Pleochroism:	None
Birefringence:	Isotropic but often anomalous birefringence, particularly in larger crystals
Relief:	High to very high
Cleavage:	None
Twinning:	Sector twins are not uncommon
Crystal system/optic sign:	Cubic
Other features:	Garnet of the ugrandite series (uvarovite, grossular, andradite) may show complex sector twins composed of 6, 12, or 24 pyramids with vertices meeting at the center of the crystal. Zoning is a common feature, especially in andradite of contact metamorphic skarn deposits.

Paragenesis

Andradite is a typical product of impure, thermally metamorphosed calcareous rocks and skarn deposits. It is commonly accompanied by magnetite and hedenbergite. In the latter, andradite formation is related to the introduction of Fe_2O_3 ($\pm$ SiO_2) in the system:

$$\underset{\text{calcite}}{3CaCO_3} + Fe_2O_3 + 3SiO_2 \rightarrow \underset{\text{andradite}}{Ca_3Fe_3Si_3O_{12}} + 3CO_2$$

If FeO is available, hedenbergite may form together with andradite. Uncommon varieties like demantoid and topazolite occur only in serpentinites and chlorite schists. Ti-rich andradite (melanite) is found in alkaline igneous rocks like nephelinites, phonolites, and nepheline syenite.

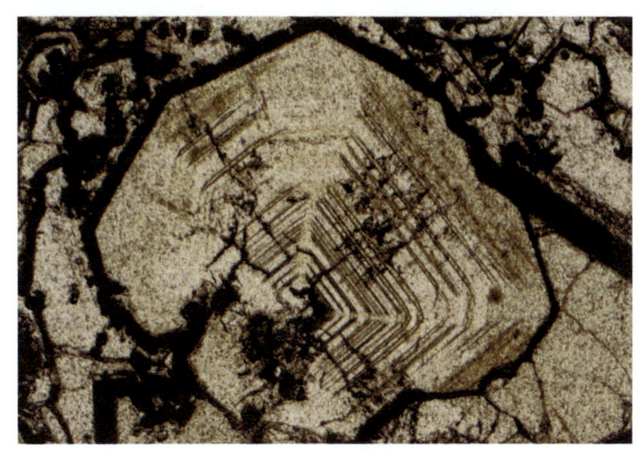

Figure 1.57

A Andradite in alkali syenite from the Khibiny complex (Russia). Large, anhedral andradite (brown crystal) with partially sericitized alkali feldspar. Plane-polarized light, 2× magnification, field of view = 7 mm.

B Andradite in skarn from the Haytor Mine (England). Partially idioblastic andradite crystal, with concentric zoning. Plane-polarized light, 10× magnification, field of view = 2 mm.

Grossular–Hydrogrossular $Ca_3Al_2(SiO_4)_3$-$Ca_3Al_2(SiO_4)_{3-x}(OH)_{4x}$

THE GARNETS

Optical Properties

Color in thin-section:	Colorless to pale green
Pleochroism:	Absent
Birefringence:	Isotropic; may be weakly anisotropic
Relief:	High to very high
Cleavage:	None, but crystals are often fractured
Twinning:	None
Crystal system/optic sign:	Cubic
Other features:	Form and color are diagnostic; garnets of the ugrandite series (uvarovite, grossular, andradite) may show complex sector twins.

Paragenesis

Grossular is stable at ~500 °C and 0.1 GPa, but above 850 °C it breaks down to wollastonite, gehlenite, and anorthite. Grossular is characteristic of regional and thermal metamorphism of impure calcareous rocks or calcareous shales and is found in abundance in skarns and metamorphosed basalts. It may be associated with vesuvianite and diopside during regional metamorphism. Less commonly it can be found in granitic pegmatites, serpentinites, and rodingites. Despite being a typical mineral phase associated with thermal metamorphism, there is no conclusive evidence that grossular has appreciable amounts of molecular water. Hydrogrossular is the hydrated (OH) phase, also known as katoite or hydrogarnet, and is chemically less resistant than grossular. It can be found in metamorphosed marls and in metasomatized gabbros, rodingites, and pyroxenites.

Figure 1.58

A Hydrogrossular crystals in a skarn from Giglio Island (Italy). Idiomorphic hydrogrossular (pale green and inclusion-rich) in association with small epidote crystals. A large hydrogrossular crystal, partially visible, is present in the lower-right corner. Plane-polarized light, 2× magnification, field of view = 7 mm.

B Hydrogrossular crystals in a skarn from Giglio Island (Italy). Same image as (A) but with crossed polarizers. Idiomorphic hydrogrossular with anomalous birefringence and a sector zoning. Epidote shows high-interference colors. Cross-polarized light, 2× magnification, field of view = 7 mm.

Pyrope $Mg_3Al_2(SiO_4)_3$

THE GARNETS

Optical Properties

Color in thin-section:	Colorless to pale pink
Pleochroism:	Absent
Birefringence:	Isotropic
Relief:	High to very high
Cleavage:	None, but crystals are often fractured
Twinning:	None
Crystal system/optic sign:	Cubic
Other features:	Typical high relief, isotropic nature and color may distinguish pyrope.

Paragenesis

Pyrope is a common aluminous phase in some ultrabasic igneous rocks, especially garnet–peridotite derived from the Earth's upper mantle. It also occurs in many orogenic peridotites and in peridotite xenoliths in kimberlites, where it is xenocrystic. It is a common phase in Mg, Al-rich metasediment of high metamorphic grade and in granulite and eclogite (almandine–pyrope) facies rocks.

Igneous

Pyrope-rich garnet is an essential aluminous mineral of garnet–peridotite and similar rocks of mantle origin. In these rocks, pyrope may have a fibrous kelyphitic corona, a structure that records the high- to low-pressure transition of garnet- to spinel-peridotite stability at around 20 kbar. Kelyphitic reactions can constrain exhumation processes that affected high-pressure mantle rocks and range between two extremes, dry and wet:

garnet + olivine $\longrightarrow$ orthopyroxene + spinel + clinopyroxene

garnet + olivine + H_2O $\longrightarrow$ orthopyroxene + spinel + amphibole

Metamorphic

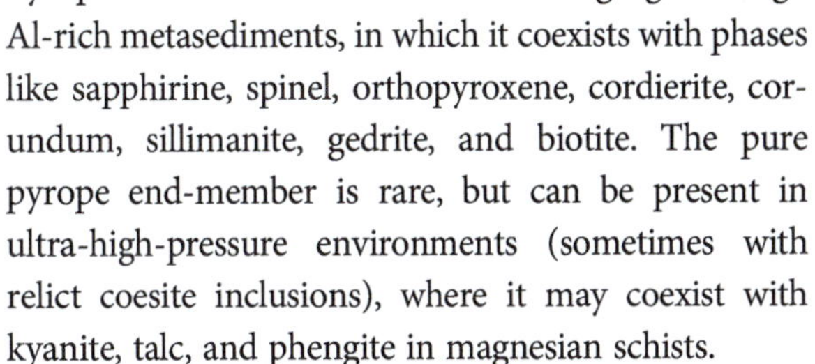

Pyrope is a characteristic mineral in high-grade Mg–Al-rich metasediments, in which it coexists with phases like sapphirine, spinel, orthopyroxene, cordierite, corundum, sillimanite, gedrite, and biotite. The pure pyrope end-member is rare, but can be present in ultra-high-pressure environments (sometimes with relict coesite inclusions), where it may coexist with kyanite, talc, and phengite in magnesian schists.

Figure 1.59 Pyrope in garnet–peridotite (lherzolite) from Como (Italy). Two large, subhedral pyrope crystals, surrounded by dark, fibrous kelyphitic rims due to retrogression under amphibolite facies conditions, in a matrix of partially serpentinized olivine crystals. (*Left*) Plane-polarized and (*right*) cross-polarized light, 2× magnification, field of view = 7 mm.

Spessartine $Mn^{2+}{}_3Al_2(SiO_4)_3$

THE GARNETS

Optical Properties

Color in thin-section:	Pale pink to pale brown; may show color zonation
Pleochroism:	Absent
Birefringence:	Isotropic; may be weakly anisotropic
Relief:	High to very high
Cleavage:	None, but crystals are often fractured
Twinning:	None
Crystal system/optic sign:	Cubic
Other features:	May have black oxide/hydroxide alteration.

Paragenesis

In the isomorphous garnet series, spessartine represents the Mn-rich end-member. It has a wide range of compositions and forms a solid solution series with almandine. It is stable between 410 °C and 900 °C at 0.05–0.3 GPa, and below 600 °C is a heavily hydrated phase. In igneous rocks, spessartine and compositions between spessartine and almandine are found in felsic igneous rocks such as granites and rhyolites. It can also occur in pegmatites and aplite intrusions, although in these rocks it is commonly present as the "spessartine component" in almandine garnets. Spessartine is less common than the other garnet varieties and is typically metasomatic (i.e. associated with manganesian skarns along with rhodonite, pyroxmangite, and tephorite). It also occurs in certain low-grade metamorphic phyllites.

High Mn content produces spessartine of a light orange color; this changes to reddish-brownish with increasing almandine (Fe) component. Sodic varieties tend to be reddish-orange. The alteration and surface oxidation of spessartine gives rise to a mixture of black manganese oxides and hydroxides. The geochemical association of yttrium with manganese can result in yttrian spessartines containing $>2\%$ Y_2O_3.

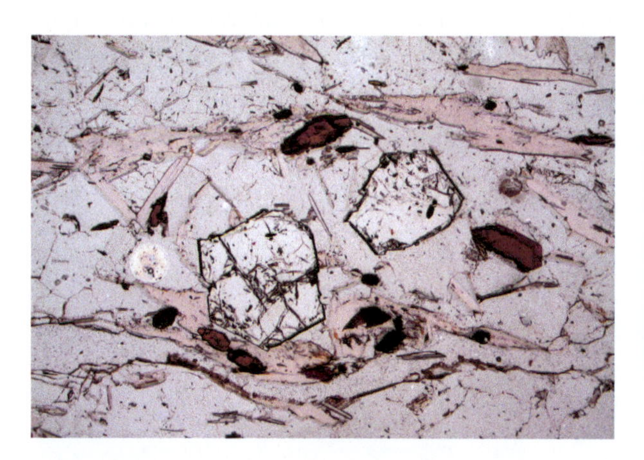

Figure 1.60 Spessartine in a schist from Aosta valley (Italy). Two colorless to pale pink spessartine crystals with high relief (center) surrounded by quartz (colorless), pink Mn-bearing muscovite (var. alurgite), and small red piemontite crystals (*left*, plane-polarized light). Garnet is isotropic, quartz has gray interference colors, and piemontite shows vibrant second-order colors (*right*, cross-polarized light). Both images: 2× magnification, field of view = 7 mm.

Uvarovite $Ca_3Cr_2(SiO_4)_3$

THE GARNETS

Optical Properties

Color in thin-section:	Green
Pleochroism:	Absent
Birefringence:	Isotropic; may be weakly anisotropic
Relief:	High to very high
Cleavage:	None, but crystals are often fractured
Twinning:	None
Crystal system/optic sign:	Cubic
Other features:	Garnet of the ugrandite series (uvarovite, grossular, andradite) may show complex pyramidal sector twins.

Paragenesis

Among the six common members of the garnet group, uvarovite is the most rare. Its presence is restricted to a few lithologies such as metamorphosed limestone, sulfide Cu–Ni–Co-related skarn, and in serpentinite often associated with chromite. In metasomatic environments the occurrence of uvarovite can be related to the introduction of Cr from basic or ultrabasic igneous rocks, or due to metasomatism of ultramafic rocks like lherzolites, harzburgites, or wehrlites. It also has been documented in podiform chromitites related to ophiolites, generally accompanied by calcite, titanite, rutile, and chromium clinochlore.

Figure 1.61

A Uvarovite crystals in a serpentinized chromitite from Skyros Island (Greece). Green uvarovite crystals grow as a corona around black chromite crystals. Uvarovite has high relief and a characteristic emerald green color. Chromite crystals are set in a serpentine-rich matrix. Plane-polarized light, 2× magnification, field of view = 7 mm.

Image courtesy of M. Bussolesi

B Uvarovite crystals in a serpentinized chromitite from Skyros Island (Greece). Higher-magnification image of (A) showing the green color of uvarovite and its corona texture. Plane-polarized light, 10× magnification, field of view = 2 mm.

Image courtesy of M. Bussolesi

Lawsonite $CaAl_2[Si_2O_7](OH)_2.H_2O$

Optical Properties

Color in thin-section:	Colorless to bluish-green
Pleochroism:	Usually non-pleochroic
Birefringence:	0.019–0.021
Relief:	High
Cleavage:	Very good on {010}, good on {100}, imperfect on {101}; cleavage intersection 67°
Crystal system/optic sign:	Orthorhombic / (+)
Other features:	Lawsonite is distinguished from clinozoisite/zoisite by the anomalous blue interference colors of the latter. Epidote is yellow–green; pumpellyite has oblique extinction; tremolite lacks 67° cleavage intersections.

Paragenesis

Lawsonite occurs in metamorphic environments in which fluid-overpressure is combined with low thermal gradients. Lawsonite is an index mineral for blueschist facies rocks and is associated with glaucophane, jadeite, aragonite, calcite, and pumpellyite. However, the existence of lawsonite-bearing eclogites in nature and experiments suggests it is also representative for eclogite facies. Lawsonite-bearing eclogites are predicted to be common in cold subduction zones at 45–300 km depths and pressures up to at least 10 GPa. Xenoliths of rare high-pressure lawsonite blueschists within rhyolitic tephra deposits have been reported and of ultra-high-pressure lawsonite-bearing eclogites from kimberlitic pipes. Under high-pressure conditions, lawsonite becomes unstable and can be replaced by paragonite + zoisite + quartz, according to the reactions:

4 lawsonite + 1 albite ↔ 1 paragonite + 2 zoisite + 2 quartz + 6 H_2O

4 lawsonite + 1 jadeite ↔ 1 paragonite + 2 zoisite + 1 quartz + 6 H_2O

Lawsonite is an important H_2O-bearing mineral, with up to 11.5 wt% H_2O in its structure. The H_2O contained in lawsonite may be subducted and released, triggering partial melting of the slab or mantle wedge, or can be incorporated in hydrous phases which transport H_2O to greater depth in the mantle.

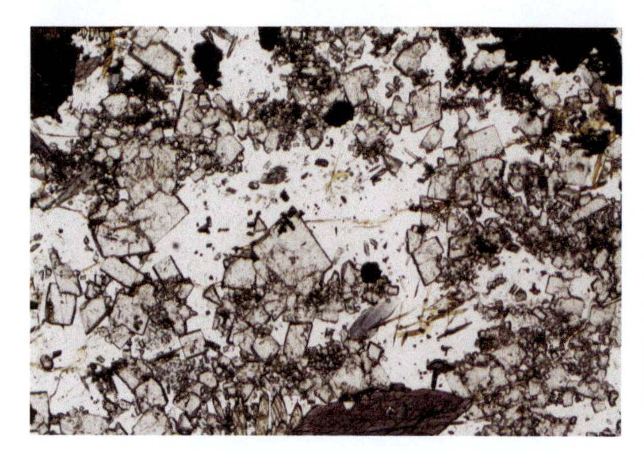

Figure 1.62 Lawsonite in blueschist from Milos (Greece). Euhedral lawsonite has square outlines, high relief, and is enclosed in a large colorless plagioclase crystal (*left*, plane-polarized light). Lawsonite shows moderate interference colors and plagioclase polysynthetic twinning (*right*, crossed polarizers). Both images: 10× magnification, field of view = 2 mm.

Gehlenite $Ca_2[Al_2SiO_7]$
Åkermanite $Ca_2[MgSi_2O_7]$

Optical Properties **THE MELILITES**

Color in thin-section:	Colorless to pale yellow
Pleochroism:	None
Birefringence:	0.008 (åkermanite); 0.011 (gehlenite)
Relief:	Moderate
Cleavage:	{001} moderate, {110} poor
Crystal system/optic sign:	Tetragonal; Gehlenite (–), melilite (+) or (–), åkermanite (+)
Other features:	The two end-members are distinguished by their optic sign, and low birefringence values indicate intermediate compositions (melilite). Their limited occurrence, low birefringence, and anomalous interference colors make them easily distinguishable.

Paragenesis

The name melilite is used to describe this mineral group that represents an isomorphic solid solution between the two end-members: gehlenite (Al end-member) and åkermanite (Mg end-member). The mineral name melilite is used to define compositions intermediate to gehlenite and åkermanite. The most common minerals of the melilite group are present in strongly silica-undersaturated rocks such as nephelinites, leucitites, carbonatites, and in rare plutonic rocks. Melilite minerals are also found in metamorphosed impure limestones in high-temperature contact aureoles around plutonic intrusions. Hydrothermal alteration of the melilite group minerals produces zeolite.

Igneous

Melilite is a common constituent in silica-undersaturated igneous rocks, both intrusive (melilitolites) and extrusive (melilitites). Melilite-bearing plutonic rocks are relatively common in alkaline volcanic provinces and melilite is associated with perovskite, olivine, haüyne, nepheline, and pyroxene. In extrusive melilite-bearing rocks, melilite is confined to the groundmass (phenocrysts are rare) and its composition is:

$$(Ca,Na)_2(Al,Mg,Fe^{2+})[(Al,Si)_2O_7]$$

Metamorphic

Melilite is a common mineral in high-temperature thermally metamorphosed impure limestones and dolomites (sanidinite facies); in these rocks, melilite forms from the reaction between diopside and calcite:

$$\underset{\text{diopside}}{CaMgSi_2O_6} + \underset{\text{calcite}}{CaCO_3} \leftrightarrow \underset{\text{åkermanite}}{Ca_2MgSi_2O_7} + CO_2$$

As the temperature rises, åkermanite reacts to form merwinite:

$$\underset{\text{åkermanite}}{Ca_2MgSi_2O_7} + CO_2 \leftrightarrow \underset{\text{merwinite}}{Ca_3MgSi_2O_8} + CO_2$$

Figure 1.63

A Melilite in melilitolite from San Venanzo (Italy). The image is dominated by a large "dusty" melilite crystal enclosing other phases such as leucite, olivine, nepheline, and magnetite. Plane-polarized light, 2× magnification, field of view = 7 mm.

B Melilite in melilitolite from San Venanzo (Italy). Same image as (A) but with crossed polarizers. Melilite shows variable, anomalous blue birefringence; olivine has higher birefringence (pink to blue). Nepheline (rectangular with gray interference colors) and leucite are also present. Cross-polarized light, 2× magnification, field of view = 7 mm.

C Melilite in a skarn from Brad (Romania). Idioblastic melilite crystals have rectangular outlines and pale brown colors. The high-relief, brown crystal (bottom) is garnet (var. melanite) and the black crystals are magnetite. The thin-section is full of voids (smooth, uniform regions between melilite crystals). Plane-polarized light, 10× magnification, field of view = 2 mm.

D Melilite in a skarn from Brad (Romania). Same image as (C) but with crossed polarizers. Idioblastic melilite crystals, with rectangular outline and low-interference colors. The garnet crystal is isotropic. Cross-polarized light, 10× magnification, field of view = 2 mm.

Forsterite–Fayalite $(Fe,Mg)_2SiO_4$

THE OLIVINES

Optical Properties

Color in thin-section:	Colorless or pale green, yellow with increasing Fe
Pleochroism:	None or pale yellow/orange–yellow/reddish-brown Fe-rich varieties
Birefringence:	0.035–0.052
Relief:	High
Cleavage:	Good {010}, {100}, but moderate {010} and poor {100} in Fe-rich varieties
Crystal system/optic sign:	Orthorhombic / Forsterite (+), fayalite (–)
Other features:	Although the birefringence and relief of some pyroxenes may appear similar, olivine may be distinguished by its poor cleavage.

Paragenesis

Olivine is an important rock-forming mineral group with an unusual yellow–green to deep green color. The forsterite (Mg-rich) and fayalite (Fe-rich) end-members form a solid solution series; forsterite, strictly speaking, has no more than 10% iron substituting for magnesium, and fayalite denotes species with no more than 10% magnesium substituting for iron. Forsteritic olivine is never found in equilibrium with quartz. Fayalitic olivine can coexist in equilibrium with quartz in iron-rich rocks. Forsterite is believed to be the most abundant constituent of the Earth's upper mantle. Fayalite occurs in high-temperature metamorphic rocks, lunar basalts, and some meteorites.

Igneous

Forsterite is a characteristic mineral of ultramafic and basic magmatic rocks (e.g. peridotite, dunite, picrite, gabbro, alkali olivine basalt, basanite). Olivine in gabbro and basalt is typically forsteritic (Fo_{80}) to intermediate (Fo_{40}) in composition. Early-formed Mg-rich crystals often have increasingly Fe-rich rims. In mafic volcanic rocks, olivine is often present as two generations (e.g. phenocrysts and groundmass).

Fayalite occurs in both alkaline and acid plutonic and hypabyssal rocks, typically in association with iron-rich amphibole and pyroxene. It also occurs in small amounts in many acid and alkaline volcanic rocks such as rhyolite, trachyte, and phonolite.

Olivine is typically associated with calcic plagioclase, magnesium-rich pyroxenes, and iron–titanium oxides such as magnetite and ilmenite. This mineralogical association is diagnostic of the relatively high temperature of crystallization associated with mafic rock types.

Metamorphic

Olivine often forms during prograde metamorphism of serpentinite by reactions such as:

$$\text{serpentine} + \text{diopside} \longrightarrow \text{forsterite} + \text{tremolite}$$

$$\text{serpentine} \longrightarrow \text{forsterite} + \text{talc}$$

Olivine also forms during the metamorphism of dolomite (e.g. forsterite marble) under anhydrous conditions via the reaction dolomite + SiO_2 = forsterite + calcite + CO_2. Under more hydrous conditions olivine is formed by the reaction:

$$\text{tremolite} + \text{dolomite} \longrightarrow \text{forsterite} + \text{calcite} + CO_2 + H_2O$$

Figure 1.64

A Olivine troctolite from Flakstadøya (Norway) showing interlocking grain boundaries typical of plutonic rocks. Forsteritic olivine (center) sits in a groundmass of calcic plagioclase and shows typical serpentinization along fractures. Cross-polarized light, 2× magnification, field of view = 7 mm.

B Olivine basalt from Etna volcano (Italy). Perfect hexagonal crystal shape for forsteritic olivine through (001) section with typical high-interference colors and poor cleavage. Cross-polarized light, 10× magnification, field of view = 2 mm.

C Olivine marble from Ivrea verbano (Italy). Anhedral forsteritic olivine (rounded, high relief) is enclosed in a host of calcite and oxides. Plane-polarized light, 2× magnification, field of view = 7 mm.

D Olivine marble from Ivrea verbano (Italy) showing the same image as (C) but with crossed polarizers. Note that the range of birefringence from yellow to blue associated with forsteritic olivine depends on the orientation of the grain. Calcite twins with very high interference colors are visible. 2× magnification, field of view = 7 mm.

Monticellite $CaMgSiO_4$

Optical Properties

Color in thin-section:	Colorless
Pleochroism:	None
Birefringence:	0.014–0.017
Relief:	Moderate
Cleavage:	Poor {010}
Crystal system/optic sign:	Orthorhombic / (–)
Other features:	Monticellite has lower birefringence than Mg,Fe-olivine; diopside has poorer cleavage and negative optic sign.

Paragenesis

Calcium-bearing olivine group minerals are represented by a solid solution between monticellite ($CaMgSiO_4$) and kirschsteinite ($CaFeSiO_4$). The latter has been synthesized but does not occur in nature. Monticellite, on the other hand, occurs in two distinct parageneses: (1) It is developed during progressive metamorphism of siliceous and magnesian limestone, and (2) it forms in skarns at metasomatized contacts between carbonates and acid, alkaline, and basic intrusions. Less frequently, monticellite can occur as a constituent of ultrabasic rocks such as kimberlites, alnöites, and carbonatites. When manganese replaces magnesium in monticellite, it is called glaucochroite. Glaucochroite has been reported from ore deposits.

Figure 1.65

A Monticellite in a plutonic carbonatite (sövite) from the Oka complex, Quebec (Canada). Monticellite (central field of view) has very high relief and is fractured. It occurs with calcite and two small, rounded apatite crystals (center right). Plane-polarized light, 2× magnification, field of view = 7 mm.

B Monticellite in a plutonic carbonatite (sövite) from the Oka complex, Quebec (Canada). Same image as (A) but with crossed polarizers. Monticellite shows high first-order interference colors. Typical calcite twins and two small, round anomalous blue apatite crystals (center-right edge of the field of view). Cross-polarized light, 2× magnification, field of view = 7 mm.

BOX 1.4 Olivine Alteration

Olivine is highly susceptible to weathering, hydrothermal alteration, and metasomatic alteration. Alteration products are often a complex mixture of fine-grained minerals that cannot be distinguished by optical means (cryptocrystalline). Some characteristic forms of olivine alteration include:

Serpentine: Serpentine alteration is common in plutonic and metamorphic rocks and occurs along fractures. It is the result of hydrothermal or deuteric processes and may completely convert olivine to serpentine with a characteristic "mesh" structure following the olivine fracture pattern.

Iddingsite and bowlingite: Iddingsite and bowlingite alteration are characteristic of oxidation, deuteric alteration, or weathering. Iddingsite appears as a reddish-brown replacement of olivine along fractures and/or rims, or as a complete replacement. Iddingsite may appear optically homogeneous, with high relief and high-interference colors. Bowlingite is a fine-grained aggregate containing smectite, goethite, chlorite, calcite, silica, and/or talc.

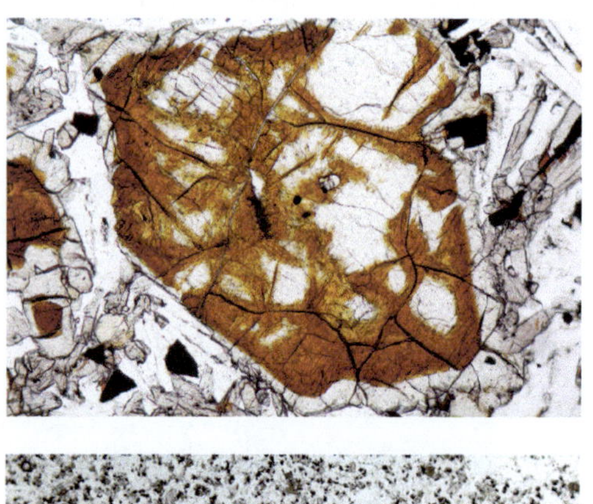

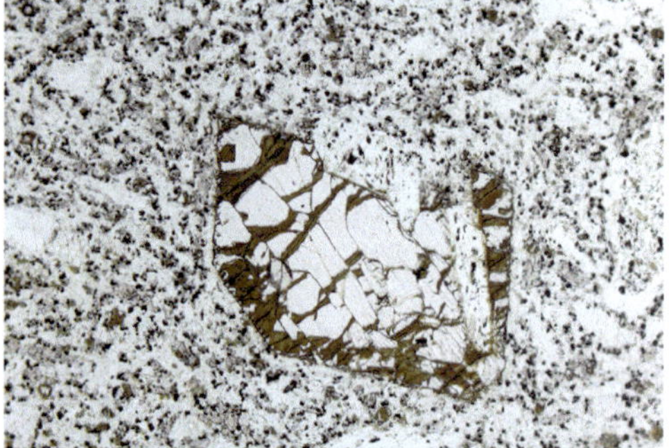

Upper-left: olivine partially altered to red iddingsite; plane-polarized light, 10× magnification, field of view = 2 mm. *Upper-right*: altered olivine with remnant fresh olivine (bright colors) rimmed by serpentine (gray colors) + magnetite (black) veins; cross-polarized light, 10× magnification, field of view = 2 mm. *Lower-left* (plane polarizers) and *lower-right* (crossed polarizers) show an euhedral olivine phenocryst (center). Note partial bowlingite alteration of olivine (greenish in plane light); 2× magnification, field of view = 7 mm.

Pumpellyite $Ca_2Al_2(Al,Fe^{3+},Fe^{2+},Mg)_1[Si_2(O,OH)_7][SiO_4](OH,O)_3$

Optical Properties

Color in thin-section:	Colorless, green, brown, or yellows
Pleochroism:	Typically yellow to pale green
Birefringence:	0.010–0.020 (increases with Fe content); may be anomalously blue or yellowish-brown
Relief:	High
Cleavage:	Good {001}, moderate {100}
Crystal system/optic sign:	Monoclinic / (+); Fe-rich varieties can be optically negative
Other features:	Typically radiating aggregates; Fe-rich varieties are more darkly colored; the blue–green absorption of Fe-rich varieties distinguish them from epidote, from zoisite by inclined extinction on (010), from lawsonite by poorer cleavage and pleochroism, and from clinochlore by the lower birefringence of the latter; weakly colored varieties look similar to clinozoisite but the latter has lower birefringence; pumpellyite with anomalous birefringence may be confused with chlorite; the "oak leaf" habit associated with vein-related pumpellyite is diagnostic.

Paragenesis

Pumpellyite occurs in low-grade metamorphosed igneous and sedimentary rocks; it is not a primary igneous mineral. Compositional variation is due to the substitution of Fe^{3+} for Al, and Fe^{2+}, Fe^{3+}, and Al for Mg, and tends to be more Fe-rich at lower P–T. Minor Na and/or K may substitute for Ca.

Pumpellyite is commonly associated with zeolites and chlorite group minerals and is also common in glaucophane schist and related low-temperature/high-pressure assemblages of alpine metamorphic belts. At blueschist facies, pumpellyite occurs with glaucophane, sodic amphiboles, aegirine, epidote, and lawsonite – it may be related to retrograde metamorphism as pumpellyite is seen replacing garnet, lawsonite, and/or epidote.

As metamorphic grade increases and the prehnite–pumpellyite facies goes to greenschist facies, pumpellyite may also occur with epidote via the reaction:

$$\text{pumpellyite} \leftrightarrow \text{zoisite} + \text{grossular} + \text{chlorite} + \text{quartz} + H_2O$$

Pumpellyite-producing reactions occur at metamorphic pressures between 50 and 10 MPa and temperatures between 50 and 150 °C. In greenschist facies assemblages, it occurs with actinolite, chlorite, and epidote. In metamorphosed or hydrothermally altered mafic igneous rocks (e.g. basalt and gabbro) it often fills vesicles.

Figure 1.66

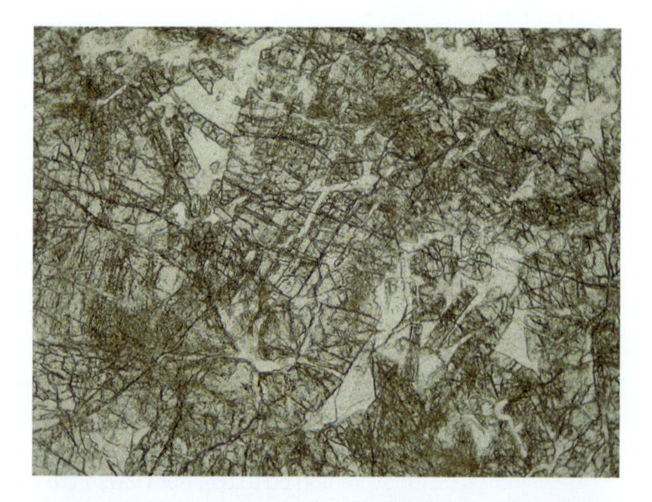

A Pumpellyite in altered gabbro from Livorno (Italy) shows typical pale green color and high relief. Interstitial prehnite fills the low-relief, colorless regions. Plane-polarized light, 2× magnification, field of view = 7 mm.

B Pumpellyite in altered gabbro from Livorno (Italy) shows blue birefringence and prehnite second-order yellows. Crossed-polarized light, 2× magnification, field of view = 7 mm.

C Pumpellyite in glaucophane schist, Tiburon Peninsula, California (United States). Pumpellyite shows diagnostic "oak leaf" shape (center) and second-order birefringence. Cross-polarized light, 44× magnification, field of view = 2.5 mm.

From MacKenzie & Guilford (1980), with permission.

Staurolite $(Fe^{2+}Mg)_2(AlFe^{3+})_9O_6[SiO_4]_4(O,OH)_2$

Optical Properties

Color in thin-section:	Pale honey-yellow/tan
Pleochroism:	Colorless to pale yellow, yellow–brown, or golden yellow to reddish-brown
Birefringence:	0.011–0.014
Relief:	High
Cleavage:	Poor {010}, brittle with subconchoidal fracture
Crystal system/optic sign:	Monoclinic(pseudoorthorhombic) / (+)
Other features:	Prismatic and elongate crystals, sector zoning, and/or penetrating cruciform twins, yellow pleochroism, and straight extinction are distinctive. May resemble brown tourmaline but the latter is uniaxial.

Paragenesis

Staurolite has no igneous occurrence and is common in medium-grade regionally metamorphosed pelitic schists. It generally appears before kyanite and defines the staurolite regional metamorphic zone. It is replaced at higher grade by kyanite and almandine (less commonly by sillimanite and almandine) via the reaction:

$$\text{staurolite} + \text{muscovite} + \text{quartz} \longrightarrow \text{almandine} + Al_2SiO_5 + \text{biotite} + H_2O$$

The staurolite–almandine–garnet–kyanite assemblage is typical of regional metamorphism, while the staurolite–cordierite–andalusite/sillimanite assemblage is typical of lower-pressure Buchan-type metamorphism. Inclusions in staurolite porphyroblasts provide evidence of crystallization during late to post-deformational metamorphism. Staurolite formation also occurs in some aluminous quartzofeldspathic schists and it can be present in metabauxites, associated with kyanite, corundum, and magnetite.

The similarities between the staurolite and kyanite mineral structures makes it easy for kyanite to nucleate on staurolite and this explains why the intergrowth of these two minerals often occurs in nature. Staurolite is also commonly associated with andalusite, kyanite, garnet, cordierite, chloritoid, muscovite, and biotite. The color intensity of staurolite is controlled by Ti content. Zoning, including sector zoning, is common and inclusions can be abundant (forming a sponge-like or chiastolitic arrangement). During retrograde metamorphism from the staurolite zone, staurolite can be mantled by or partially replaced by chlorite; it also alters to fine-grained white mica (sericite).

Figure 1.67

A Staurolite in micaschist from Sardinia (Italy). Staurolite shows diagnostic honey-yellow colors and cruciform, penetrative twins; the small green crystals (center-left) are tourmaline and are set in a muscovite–quartz-rich matrix. Plane-polarized light, 2× magnification, field of view = 7 mm.

B Staurolite in micaschist from Sardinia (Italy). Staurolite crystals with abundant quartz inclusions. The central crystal also has a cruciform twin. Muscovite (blue birefringence) and quartz (grays) are also present. Cross-polarized light, 10× magnification, field of view = 2 mm.

C Staurolite in schist from Sponda (Switzerland). Staurolite (tan color, perpendicular fractures) intergrown with kyanite (brown–pink, distinct cleavage). Plane-polarized light, 10× magnification, field of view = 2 mm.

D Staurolite in schist from Sponda (Switzerland). Same image as (C) but with crossed polarizers. Staurolite shows yellow birefringence, while kyanite shows variable from low to first-order orange interference colors. Muscovite (blue birefringence) is also present. Cross-polarized light, 10× magnification, field of view = 2 mm.

Titanite $CaTi(SiO_4)(O,OH,F)$

Optical Properties

Color in thin-section:	Brown to dark brown, but can be colorless or yellow
Pleochroism:	Colored varieties may be moderately pleochroic from pale yellow to brownish-yellow and orange–brown
Birefringence:	0.100–0.192
Relief:	Extremely high
Cleavage:	Good {110}
Twinning:	Single twins on {100}; occasional lamellar twinning
Crystal system/optic sign:	Monoclinic / (+)
Other features:	Typical brown color and rhombic cross-sections are diagnostic. Twins are common; thorium content can cause pleochroic halos; zoned crystals have a higher refractive index along their margins. Rutile, zircon, and cassiterite are uniaxial.

Paragenesis

Titanite is a dominant titanium-bearing mineral and therefore is a widespread accessory mineral in magmatic and metamorphic rocks. It is a common late magmatic phase in acidic to foid-bearing plutonic rocks (granite, granodiorite, tonalite, syenite, and nepheline syenite), but absent in gabbro/basalt where titanium is incorporated into pyroxene (titanaugite) and ilmenite. It is rare in volcanic rocks. Titanite is a common mineral in low-temperature Alpine-type hydrothermal veins and in pegmatites. It also occurs in low- to medium-grade metamorphic rocks rich in ferromagnesian minerals, in amphibolite, in marble and calc-silicate rocks, and in skarns.

Figure 1.68

A Titanite in titanite ore from Kukisvumchorr (Russia). Titanite crystals with well-developed rhombic cross-sections show characteristic colors (tan–brown). The large crystal has a simple lamellar twin (note the different colors). The groundmass is feldspar-rich (colorless), with minor secondary chlorite (green). Plane-polarized light, 2× magnification, field of view = 7 mm.

B Titanite in gneiss from Val Martello (Italy). Large euhedral titanite crystal (brown) with a simple twin along its centerline and numerous cleavage planes; it is adjacent to a large quartz crystal (colorless). Note titanite's high relief compared to the surrounding minerals. Other phases are hard to see but include epidote ("dusty" crystals) and chlorite (pale green). Plane-polarized light, 10× magnification, field of view = 2 mm.

Topaz $Al_2(SiO_4)(OH,F)_2$

Optical Properties

Color in thin-section:	Colorless
Pleochroism:	None (but may be present in thick sections)
Birefringence:	0.008–0.011
Relief:	Moderate
Cleavage:	Perfect {001}
Twinning:	None
Crystal system/optic sign:	Orthorhombic / (+)
Other features:	Topaz forms short columns but may occur in granular form and may be rich in fluid inclusions; may be confused with quartz and feldspar (they have similar birefringence), but topaz has higher relief. Beryl is optically uniaxial (–); andalusite has a patchy pinkish color and is biaxial (–).

Paragenesis

Topaz occurs mainly in acid igneous rocks such as granites and granite pegmatites, and can be a primary phase in some topaz rhyolites. Its formation is linked to the presence of fluorine-bearing vapor derived from the late-stage crystallization of igneous rocks. It is often associated with quartz, cassiterite, tourmaline, fluorite, and beryl. It is a common constituent of rocks subjected to pneumatolytic processes, called greisen. Although rare, it is found in some skarn and metabauxite deposits. Topaz may alter to clay minerals such as kaolinite and sericite during the latest stage of magmatic solidification via the reaction:

$$\underset{\textit{topaz}}{2Al_2SiO_4(OH,F)_2} + 2H_2O + 2SiO_2 \longrightarrow \underset{\textit{kalonite}}{Al_4Si_4O_{10}(OH,F)_8}$$

Figure 1.69

A Topaz in greisen from Saxony (Germany). The large, prismatic topaz crystal shows typical fractures perpendicular to its long axis; it is surrounded by quartz (white–gray birefringence); muscovite is also present (high birefringence, lower-right). Cross-polarized light, 2× magnification, field of view = 7 mm.

B Topaz in greisen from Saxony (Germany). Short prismatic topaz crystals surrounded by quartz. Note the similar birefringence of quartz and topaz on the right side of image. Cross-polarized light, 10× magnification, field of view = 2 mm.

Tourmaline $(Na,Ca)(Mg,Fe,Mn,Li,Al)_3Al_6Si_6O_{18}(BO_3)_3(OH,F)_4$

Optical Properties

Color in thin-section:	Highly variable; colorless, blue, green, yellow
Pleochroism:	Variable, normally strong except on basal sections
Birefringence:	0.025–0.035
Relief:	Moderate
Cleavage:	Poor on {110} and {101}
Twinning:	Rare
Crystal system/optic sign:	Trigonal / (–)
Other features:	Short columnar crystals, triangular cross-sections, and distinctive pleochroism are diagnostic; may form needle-shaped, spherulitic, or acicular crystals; compositional zoning is common.

Paragenesis

Tourmaline typically occurs in granites and granitic pegmatites. Fe^{2+}-bearing tourmaline (schorl) is the most common. During plutonic solidification, late-stage fluids concentrate elements such as Li and F, resulting in the formation of Li-rich tourmaline (elbaites). During the pneumatolytic stage, B-rich fluids can cause intense substitution processes or "tourmalinization," with the replacement of magmatic minerals by tourmaline. If tourmalinization goes to completion, a tourmaline–quartz rock results. In pneumatolytic veins, tourmaline occurs with topaz, spodumene, cassiterite, fluorite, and apatite. Tourmaline is common in metamorphic rocks, where it may form microlites and is a product of boron metasomatism. In metamorphic or metasomatic assemblages, Mg-rich tourmaline (dravite) may occur.

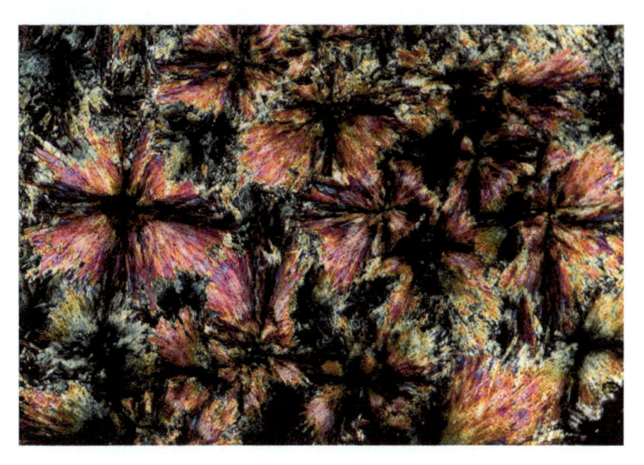

Figure 1.70

A Tourmaline in granite from Gavorrano (Italy). Semi-triangular cross-sections and compositional zoning are diagnostic. Tourmaline crystals are surrounded by colorless quartz. Plane-polarized light, 2× magnification, field of view = 7 mm.

B Tourmalinized granite from Luxulyan (England). Boron-rich fluids and element substitution generated this "luxullianite." The radiating aggregate of fibrous tourmaline (the "tourmaline sun") is typical. Cross-polarized light, 2× magnification, field of view = 7 mm.

Zircon $ZrSiO_4$

Optical Properties

Color in thin-section:	Colorless or pale brown but hard to see in small grains
Pleochroism:	Weak
Birefringence:	0.042–0.065
Relief:	Very high
Cleavage:	Imperfect {110}, poor {111}
Crystal system/optic sign:	Tetragonal / (+)
Other features:	High relief, bright birefringence, and radioactive halos are distinctive; rutile is darker and has higher birefringence; xenotime has higher birefringence.

Paragenesis

Zircon occurs in many igneous, metamorphic, and sedimentary rocks, and is generally identified microscopically. It can incorporate U, Th, and Hf and is therefore widely used in geochronology, reconstruction of magmatic history, and studies of crustal genesis. Due to its hardness and poor cleavage, it can survive more than one cycle of weathering and sedimentation, and is consequently an important mineral for sediment provenance investigations. Unless metamict, zircon is highly stable and does not weather or alter to other minerals.

Igneous

Zircon occurs in plutonic rocks as small crystals enclosed in later minerals, as large well-developed crystals in granite pegmatites (particularly in nepheline syenites), and in volcanic rocks as inclusions in other minerals or as discrete crystals (e.g. in crystal tuffs). The radioactive elements in zircon can produce pleochroic halos in host minerals such as biotite, amphibole, or other colored silicates. The size and morphological character of zircon may be similar within a single granitic body, may be rounded due to magmatic resorption, or corroded due to metasomatism. Magmatic growth zoning can sometimes be seen in thin-section.

Metamorphic

Metamorphic zircon commonly forms at amphibolite facies and higher grades, where it occurs as neoblastic or recrystallized grains. Less commonly, zircon of hydrothermal origin has been documented to nucleate on existing crystals in the presence of a fluid at greenschist facies. The latter can produce thin (5–30 μm) rims that if large enough can be analyzed and may date the time of rim overgrowth.

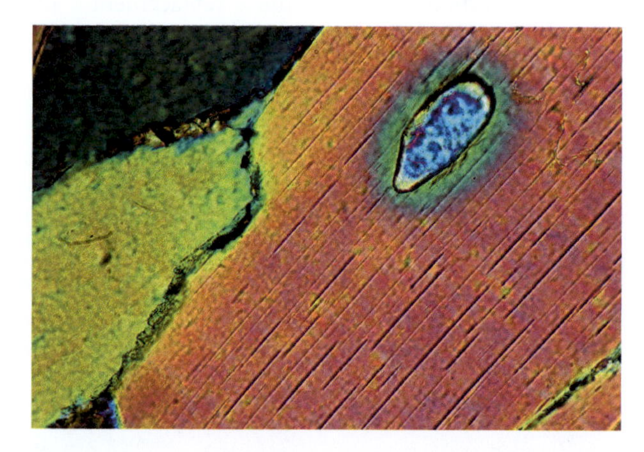

Figure 1.71 Zircon in granite from Gavorrano (Italy). Zircon with bright blue birefringence is enclosed in biotite (pink–orange). Radiation damage from zircon has resulted in a pleochroic halo (blue–green) in the biotite. Cross-polarized light, 20× magnification, field of view = 1 mm.

APPLICATION 1.3

Thermobarometry

Thermobarometry combines temperature (thermometry) and pressure (barometry). These are used to determine the absolute temperature and pressure of an equilibrium mineral assemblage in an igneous or metamorphic rock. Their successful application relies on the chemical equilibrium between minerals or the chemical composition of individual minerals, and reflects the fact that mineral pairs/assemblages vary their compositions as a function of temperature and pressure.

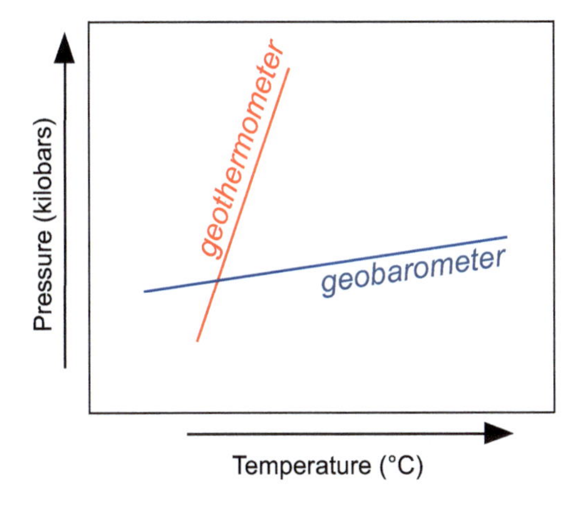

Reactions used as thermobarometers include:

- univariant equilibria
- solvus equilibria
- multivariant (or net transfer) equilibria
- exchange reactions

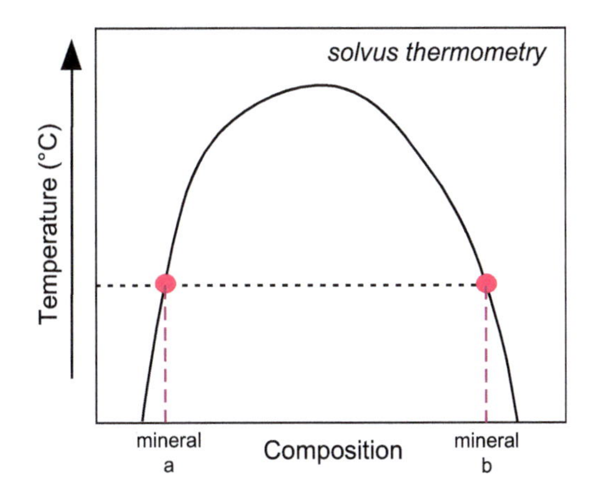

Geothermometers use reactions that depend strongly on temperature (independent of pressure). Solvus equilibria are based on mineral systems which exsolve two phases from a homogeneous single phase, such as the Na–K exchange in alkali feldspar or the Ca–Mg exchange in carbonates or pyroxenes. Another widely applied geothermometer is based on the Fe–Mg exchange between coexisting garnet and biotite. The relative proportions of Fe and Mg in garnet and biotite change with increasing temperature and measuring Fe and Mg in garnet and biotite provides the Fe–Mg distribution between them, allowing us to determine their temperature of crystallization.

Other geothermometers:

- Ti in biotite
- Ti in coesite
- Ti in zircon
- Zr in rutile
- Fe–Mg exchange between olivine and melt/glass inclusions
- Fe–Mg exchange between garnet–biotite and garnet–amphibole
- Mg–Fe exchange between pigeonite and augite
- Mg–Ca in coexisting orthopyroxene and clinopyroxene
- Coexisting Fe–Ti oxides

Geobarometers use reactions that depend strongly on pressure (independent of temperature). For example, the net transfer reaction of Fe^{2+}–Ca^{2+} in garnet is pressure-sensitive, regardless of temperature.

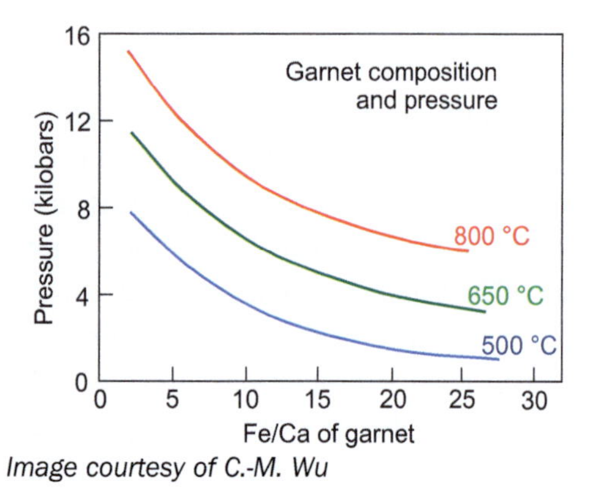

Image courtesy of C.-M. Wu

Successful geobarometers often have a large molar volume contrast between mineral phases. For example, quartz has a lower volume under higher pressure and this results in a density increase (as the mass does not change); therefore quartz polymorphs are sensitive indicators of pressure. Feldspar combined with a denser mineral like pyroxene or garnet makes a good barometer.

Some other geobarometers:

- garnet–pyroxene
- garnet–quartz (silica)–plagioclase (GASP)
- garnet–plagioclase–muscovite–biotite (GPMB)
- garnet–plagioclase–hornblende–quartz (GPHQ)
- hornblende

Uncertainties and assumptions The data for geothermometers and geobarometers are derived from both experimental and theoretical investigations; the former includes laboratory studies of artificial mineral assemblages grown at known temperatures and pressures, and their chemical equilibrium is measured directly. The latter are derived from the calibration of natural systems. The errors associated with geothermometry and geobarometry derive from inappropriate standard selection, correction factors, and analytical imprecision – these are typically ±50 °C, ±1 kbar. Additional sources of error are related to whether or not:

- equilibrium conditions have been met (using field and textural evidence);
- correct mineral compositions are used, especially when minerals are compositionally zoned;
- the oxidation state is known/constant;
- the activity of water is known/constant; and
- whether there are compositional deviations from the calibration dataset used.

1.6 Nonsilicates

Aragonite $CaCO_3$

THE CARBONATES

Optical Properties

Color in thin-section:	Colorless
Pleochroism:	Absent
Birefringence:	0.155–0.156
Relief:	Relief varies depending on the orientation of the section
Cleavage:	Imperfect prismatic cleavage on {010}
Crystal system/optic sign:	Orthorhombic / (–)
Other features:	Occurs mostly in radiating acicular aggregates, lacking the perfect rhombohedral cleavage of calcite or dolomite, and has higher refractive indices than calcite.

Paragenesis

Aragonite is less common than calcite and is the high-pressure $CaCO_3$ polymorph. Under most conditions it is metastable and readily converts to calcite. Some metastable aragonite crystals may form directly from ocean water and it is possible to find it in shallow marine deposits outside its stability field. Dissolution of aragonite occurs below ~2,000 m in ocean water (the aragonite compensation depth). Aragonite is a relatively common precipitate in geothermal areas (calcareous sinter). In volcanic rocks aragonite occurs as a secondary mineral, often in association with zeolites in cavities (e.g. amygdales). Aragonite is a widespread metamorphic mineral under high-pressure and low-temperature metamorphic conditions (e.g. blueschist facies), where it is stable at 300 °C and 6–10 kbar. It rarely survives uplift/exhumation processes as it typically inverts to calcite during exhumation.

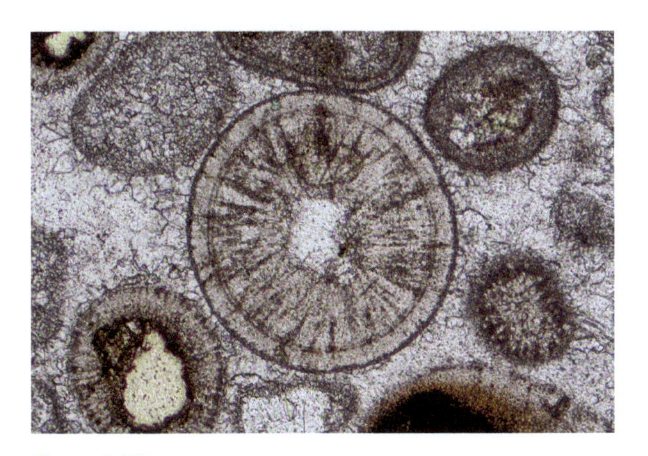

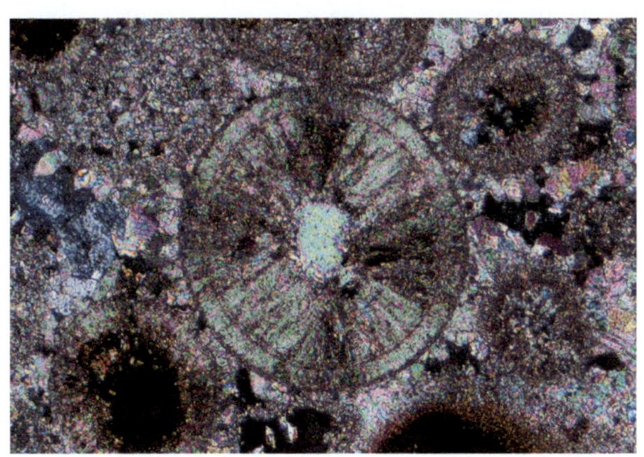

Figure 1.72

A Aragonite in a calcareous sinter deposit from Carlsbad (Germany). In the center of the photo, a perfect spherule is made of radial, fibrous aragonite crystals (light brown); these have nucleated on a carbonate fragment. Other spherules with less visible internal structure can be seen. Plane-polarized light, 10× magnification, field of view = 2 mm.

B Aragonite in a calcareous sinter deposit from Carlsbad (Germany). Same view as (A) but with crossed polarizers. Note the extremely high interference colors of aragonite fibers. Cross-polarized light, 10× magnification, field of view = 2 mm.

Calcite $CaCO_3$

THE CARBONATES

Optical Properties

Color in thin-section:	Colorless
Pleochroism:	Absent
Birefringence:	0.172–0.190
Relief:	Moderate with extreme variation; the crystal is said to "twinkle" during rotation
Cleavage:	Perfect rhombohedral on $\{10\bar{1}1\}$
Crystal system/optic sign:	Trigonal / (–)
Other features:	Perfect cleavage, typical twinning, and extremely high birefringence are diagnostic.

Paragenesis

Calcite is one of the most common nonsilicate minerals. It may be found in sedimentary rocks (occurring as carbonate shell material, as fine precipitates, and as clastic material), and in magmatic and metamorphic rocks. In igneous rocks it is mostly of secondary origin, but it can be a major rock-forming mineral in alkaline magmas (e.g. carbonatites). During metamorphism of pure calcitic limestone, calcite recrystallizes to a mosaic arrangement known as the "saccaroid" texture.

Igneous

Calcite occurs in some rare alkaline igneous rocks, like carbonatites, lamprophyres, kimberlites, and in some nepheline syenites. Calcite is usually the most abundant component in carbonatites (at least 50%) and is also a product of the late stages of hydrothermal metamorphism associated with magmatism when it fills veins or cavities.

Metamorphic

Calcite is a relatively common mineral in many sedimentary rocks that have been thermally or regionally metamorphosed. During metamorphism, calcite may react with other materials present in the original sediment to form various mineral assemblages. One important reaction in metamorphic rocks at high temperature (of considerable interest as a geothermometer) is that between calcite and quartz:

$$\underset{\text{calcite}}{CaCO_3} + \underset{\text{quartz}}{SiO_2} = \underset{\text{wollastonite}}{CaSiO_3} + CO_2$$

The formation of wollastonite is associated with the disappearance of the calcite–quartz assemblage due to the release of carbon dioxide. In thermally or regionally metamorphosed impure limestones, calcite may be found in association with calc-silicate minerals such as diopside, tremolite, vesuvianite, and grossular garnet, and also with forsterite. It is frequently associated with zeolites, epidotes, and quartz, and appears as a major component in marble.

Figure 1.73

A Calcite in carbonatite from Alnö (Sweden). Calcite with typical interference colors and twinning seen on the left side of the image. A large phlogopite (lower-right) shows purple and orange birefringence. Oriented N–S left of center, fractured monticellite (Ca-olivine) crystals with orange–gray interference colors are visible. Cross-polarized light, 2× magnification, field of view = 7 mm.

B Calcite amygdales in an altered basalt from San Venanzo (Italy). This rock has undergone strong hydrothermal alteration, resulting in carbonate pseudomorphs after olivine and the deposition of calcite in the volcanic vesicles to create calcite amygdales. Cross-polarized light, 2× magnification, field of view = 7 mm.

C Calcite in marble from Carrara (Italy). Small, polygonal calcite crystals, with mosaic (saccaroid) texture are apparent. Note the variable relief due to different orientations of some crystals. Plane-polarized light, 2× magnification, field of view = 7 mm.

D Calcite in marble from Carrara (Italy). Same image as (C) but with crossed polarizers. Note the high-interference colors; pale pinks are fourth order. Cross-polarized light, 2× magnification, field of view = 7 mm.

Dolomite $CaMg(CO_3)_2$

Optical Properties | **THE CARBONATES**

Color in thin-section:	Colorless
Pleochroism:	Absent
Birefringence:	0.179–0.185 (extremely high, higher than calcite)
Relief:	Low to moderate (varies with optic orientation)
Cleavage:	Perfect $\{10\bar{1}1\}$
Crystal system/optic sign:	Trigonal / (–)
Other features:	Dolomite can be distinguished from calcite by its extremely high birefringence and twinning parallel to the short axis of the rhomb.

Paragenesis

Dolomite is often associated with sedimentary evaporite deposits. In limestone, dolomite is the result of metasomatic alteration shortly after deposition or consolidation of limestone when Mg-rich solutions enter the rock. Dolomite can crystallize as a primary mineral in carbonatitic magmas. It is associated with ophiolite massifs, serpentines, and other ultramafic rocks, but is most common in contact or regionally metamorphosed Mg-rich limestones, where it may recrystallize to form a dolomitic marble. At a higher grade (~800 °C), dolomite may break down:

$$\underset{\text{dolomite}}{CaMg(CO_3)_2} = \underset{\text{calcite}}{CaCO_3} + \underset{\text{periclase}}{MgO} + CO_2$$

During prograde metamorphism of siliceous dolomite-bearing limestones, dolomite can give rise to numerous chemical reactions, with the formation of talc, tremolite, forsterite, and periclase. The typical sequence is talc → tremolite → diopside and/or forsterite → wollastonite.

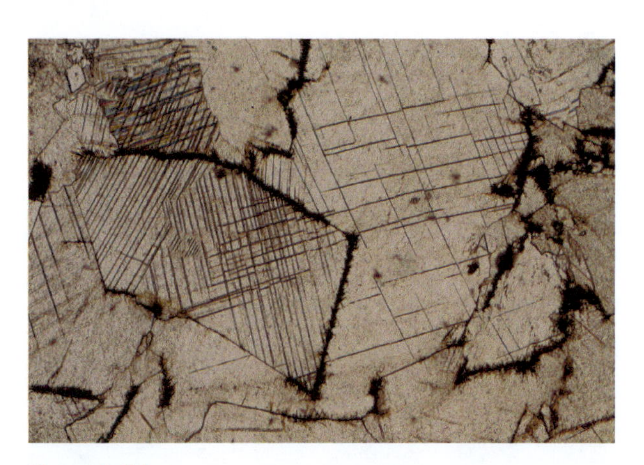

Figure 1.74

A Dolomite in marble from Adamello massif (Italy). Hypidiomorphic dolomite crystal with typical rhombohedral cleavage. Note the two sets of twins intersecting at a high angle. The dark rims around some dolomite crystals are iron oxide. Plane-polarized light, 10× magnification, field of view = 2 mm.

B Dolomite in marble from Adamello massif (Italy). Same thin-section as (A) but with crossed polarizers. Dolomite has very high-interference colors, higher than calcite. Cross-polarized light, 10× magnification, field of view = 2 mm.

Magnesite $MgCO_3$

Optical Properties

Color in thin-section:	Colorless
Pleochroism:	Absent
Birefringence:	0.190–0.219
Relief:	Moderate
Cleavage:	Perfect $\{10\bar{1}1\}$
Crystal system/optic sign:	Trigonal / (–)
Other features:	Magnesite is distinguished from dolomite and calcite by a lack of twinning and higher birefringence. In volcanic rocks, it is a common alteration product of olivine.

Paragenesis

Magnesite is a common alteration product of various magnesium-rich igneous and metamorphic rocks. In dunites and hartzburgites it typically occurs as fracture-fill. Peridotites and ultramafic rocks are commonly transformed to serpentinites during low- to medium-grade metamorphism by the infiltration of aqueous fluids. If CO_2 is available, the infiltration of CO_2-rich fluids can produce magnesite through the reaction:

$$\underset{\text{Mg-olivine}}{4Mg_2SiO_4} + 4H_2O + 2CO_2 = \underset{\text{serpentine}}{2Mg_3SiO_2O_5(OH)_4} + \underset{\text{magnesite}}{2MgCO_3}$$

Magnesite is a common product of olivine alteration in many volcanic rocks and commonly pseudomorphs after olivine. It also occurs as porphyroblasts in talc and chlorite schist.

Figure 1.75

A Olivine in basalt from Pergamon (Turkey). Partially skeletal olivine phenocryst pseudomorphed by magnesite (with minor serpentine). Cross-polarized light, 2× magnification, field of view = 7 mm.

B Magnesite as an alteration product in peridotite from Tuscany (Italy). Note the high-interference colors and multiple-veined growth of magnesite. Cross-polarized light, 2× magnification, field of view = 7 mm.

Siderite $FeCO_3$

Optical Properties

Color in thin-section:	Colorless to pale brown or pale yellow
Pleochroism:	Absent
Birefringence:	0.207–0.242
Relief:	Moderate
Cleavage:	Perfect on $\{10\bar{1}1\}$
Crystal system/optic sign:	Trigonal / (–)
Other features:	Extreme birefringence is diagnostic; titanite and cassiterite are optically (+).

Paragenesis

Siderite is a common mineral in banded iron formations and in oolitic ironstones. It occurs as veins in ore bodies, as the hydrothermal alteration product of other carbonates, and also forms from the interaction of Fe-rich fluids and limestone.

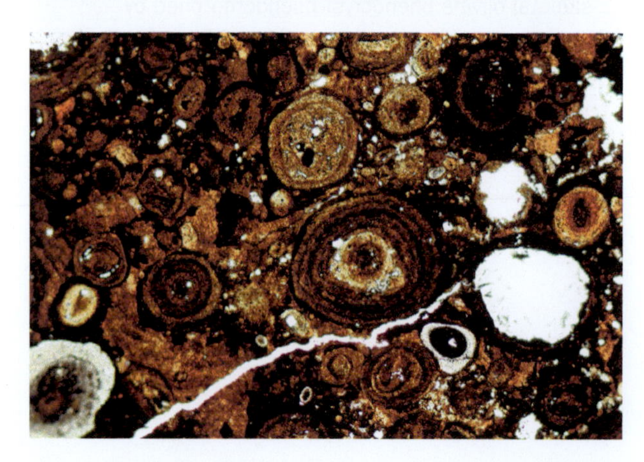

Figure 1.76

A Siderite in calcite–siderite vein from South Tyrol (Italy). Brownish siderite has distinct relief and color from calcite (center). Siderite may show Fe-oxide exsolution (the dark material in the crystal cleavage). Plane-polarized light, 2× magnification, field of view = 7 mm.

B Siderite in calcite–siderite vein from South Tyrol (Italy). Same image as (A) but with crossed polarizers. Note the higher interference colors of siderite crystals with respect to calcite. Cross-polarized light, 2× magnification, field of view = 7 mm.

C Siderite-rich ooids in an oolitic ironstone from an unknown locality (Germany). Rounded siderite-rich ooids with laminar, concentric growth. Plane-polarized light, 10× magnification, field of view = 2 mm.

Fluorite CaF_2

Optical Properties

Color in thin-section:	Colorless, pale green, or pale violet
Pleochroism:	None
Birefringence:	Isotropic
Relief:	Medium
Cleavage:	Perfect {111}
Crystal system/optic sign:	Isometric (cubic)
Other features:	Commonly occurs as interpenetrating octahedra or cubes; isotropic character and violet colors may be diagnostic.

Paragenesis

Fluorite is a halide and typically occurs as a late-stage accessory mineral in granitic systems (e.g. greisen or pegmatites) and is often associated with alkaline rocks. As a hydrothermal mineral, fluorite is associated with lead, zinc, and other metal ores. It may also occur as a hydrothermal product in limestone. Fluorite has been found in geodes of probable hydrothermal origin, and is associated with calcite, barite, and sphalerite. It has wide color variation, partly due to rare earth ions and/or oxygen complexing. Fluorite is sometimes the cementing material in sandstone. Violet fluorite grains are fairly common as a detrital mineral in sands derived from acid igneous rocks and hydrothermal deposits. Pseudomorphs of fluorite by quartz, clay, Mn minerals, and carbonate are common.

Igneous

Fluorite is often the result of late-stage (mainly hydrothermal) crystallization and is common in granitic pegmatites. Fluid inclusion studies indicate crystallization temperatures of 450–550 °C. Fluorite can be present in the drusy cavities of blocks ejected from volcanoes and as a volcanic sublimate – in these igneous occurrences associated minerals include cassiterite, topaz, apatite, and lepidolite. Minor amounts of Si, Al, and Mg are often present as inclusions or impurities. Most chemical substitutions are based on the replacement of Ca by Sr or Y + Ce.

Metamorphic

Fluorite may be associated with hydrothermal minerals not known to be directly related to any igneous body such as calcite, pyrite, and apatite. Fluid inclusion studies of hydrothermal fluorite from the North Pennines, (Pb, Zn, Ba)-fluorite deposits indicate very late crystallization at temperatures of 92–220 °C; the ore-forming fluid was probably a concentrated (Na, Ca, K)-chloride brine. The major chemical factors governing the formation of economic deposits are the dilution of mineralizing fluids by groundwater and the change in temperature gradient on contact with excess groundwater. Flow rates of 0.5–1.0 cm/s require 1,000 years to form a typical vein.

Figure 1.77

A Fluorite in limestone from Derbyshire (England). The red to purple colors and the cleavage shown are typical, as are the interpenetrant cubes. Plane-polarized light, 2× magnification, field of view = 7 mm.

B Fluorite in granite from Cornwall (England). This violet fluorite crystallized late in the solidification of the St. Austell granite complex. Plane-polarized light, 10× magnification, field of view = 2 mm.

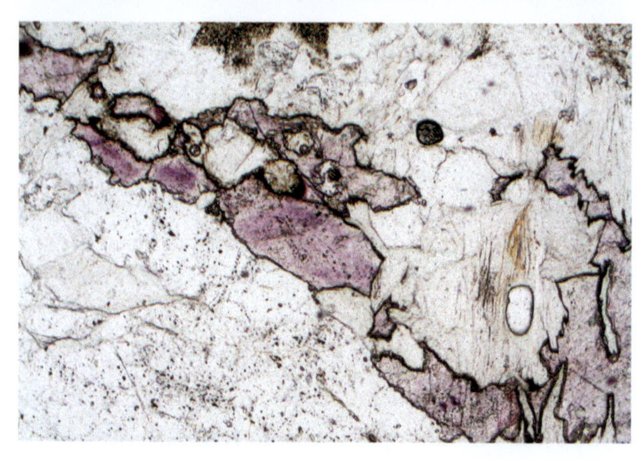

C Fluorite in granite from Cornwall (England). Same image as (B) but with crossed polarizers. Note the isotropic nature of fluorite. Cross-polarized light, 10× magnification, field of view = 2 mm.

D Fluorite in nepheline syenite from Norra Kärr (Sweden). Large fluorite crystal (shades of purple) with numerous fluid inclusions, surrounded by arfvedsonite amphiboles (green–blue) and feldspar (colorless). Plane-polarized light, 10× magnification, field of view = 2 mm.

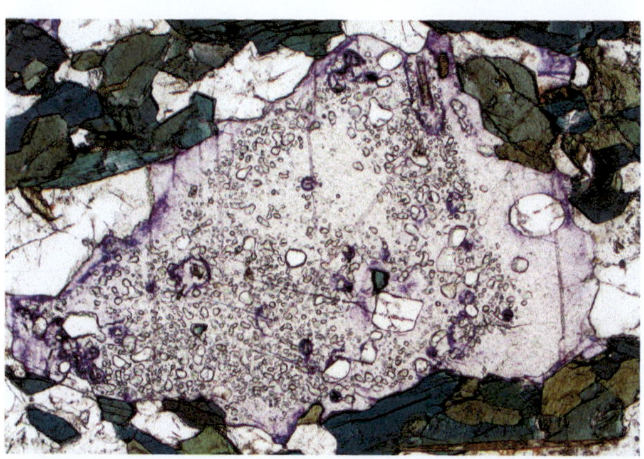

Brucite $Mg(OH)_2$

Optical Properties

Color in thin-section:	Colorless
Pleochroism:	Absent
Birefringence:	0.014–0.020
Relief:	Medium
Cleavage:	Perfect basal {0001}
Crystal system/optic sign:	Trigonal / (+)
Other features:	Brucite can be distinguished from muscovite and talc by their higher birefringence and negative optic sign, and from colorless chlorite by its lower birefringence.

Paragenesis

Brucite is one of the numerous hydrous minerals that occurs in serpentinites, where it forms due to the low-temperature hydrothermal alteration of olivine in Mg-rich ultramafic rocks. In serpentinites, brucite is commonly associated with other Mg-bearing silicate minerals such as chrysotile, lizardite, and talc. It contains ~31 wt% water and its dehydration during subduction is regarded as an important mechanism for the transfer of fluids into the mantle. Brucite is also a common alteration product of periclase in contact-metamorphosed dolomites.

Figure 1.78

A Brucite in dolomitic marble from Jakobsberg (Sweden). Micaceous, colorless brucite crystals (across the central part of the image) associated with higher-relief dolomite crystals. Plane-polarized light, 2× magnification, field of view = 7 mm.

B Brucite in dolomitic marble from Jakobsberg (Sweden). Same image as (A) but with crossed polarizers. The variation of brucite birefringence can be seen (yellows to blues). Dolomite shows its typical, very high-interference colors. Cross-polarized light, 2× magnification, field of view = 7 mm.

Cassiterite SnO_2

Optical Properties

Color in thin-section:	Commonly yellow, reddish or brown
Pleochroism:	Variable from very weak to strong; yellow, brown, or red
Birefringence:	0.096–0.098, but often masked by the color of the mineral
Relief:	High
Cleavage:	Poor on (100) and (110), and hardly observable under the microscope (visible only as faint traces)
Crystal system/optic sign:	Tetragonal / (+)
Other features:	Occurs as short to long columns or rounded, granular crystals. Variation in iron content leads to strong color zoning in plane and crossed polarizers. Twinning on (011) produces the "knee twin." Cassiterite is less colored than rutile, which has stronger pleochroism and higher birefringence. Allanite has lower birefringence and melanite garnet is only weakly birefringent.

Paragenesis

Cassiterite is a relatively common accessory mineral in acid igneous rocks such as granites, microgranites, and quartz porphyries. In association with high-temperature granite pegmatites and late-stage veins, fluid inclusion studies indicate formational temperatures of 300–600 °C. In high-temperature hydrothermal veins, greisen, stockworks, and skarn deposits it may be found with tourmaline and topaz, and it is commonly associated with other oxides such as wolframite, columbite, tantalite, scheelite, and hematite. Cassiterite also occurs in association with massive sulfide deposits, along with minerals such as arsenopyrite, molybdenite, and pyrrhotite. Due to the mineral's high density, it may also form placer deposits of economic importance. Containing c. 78% tin, it is the primary source of tin metal for the world.

Figure 1.79

A Cassiterite from granite in Cornwall (England). Cassiterite shows compositional zoning due to variation in iron content and the classic "knee twin." The blueish-green mineral associated with cassiterite is tourmaline. Plane-polarized light, 2× magnification, field of view = 7 mm.

B Cassiterite from granite in Cornwall (England). Same image as (A) but with crossed polarizers. Strong interference colors are typical. Cross-polarized light, 2× magnification, field of view = 7 mm.

Temperature and Oxygen Fugacity Using Fe–Ti Oxides

The Fe–Ti oxides are sensitive to temperature and oxidation state, which makes them useful for estimating the temperature (°C) and oxygen fugacity (fO_2) of magmatic systems. There are several solid solutions within the $FeO–TiO_2–Fe_2O_3$ system. The *titanomagnetite series* reflects the solid solution between magnetite and ulvospinel, while the *titanohematite series* reflects the solid solution between hematite and ilmentite; the ferrous-ferric pseudobrookite solid solution is limited since ferrous pseudobrookite is only stable at temperatures above 1,135 °C.

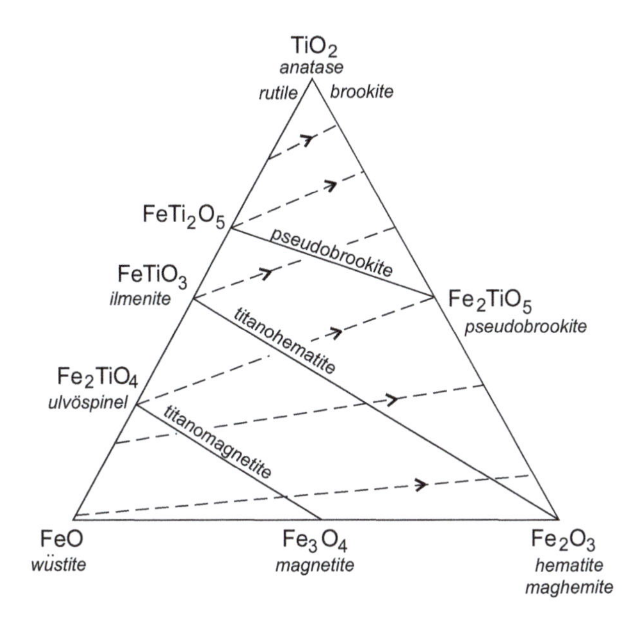

Ternary solid solution in the $FeO–TiO_2–Fe_2O_3$ system. The solid lines indicate existing solid solutions, while the dashed lines show increasing $fO_2$for constant Fe:Ti ratios.

After McElhinny (1979), *with permission from Cambridge University Press*

In the magnetite–ulvospinel series there is complete solid solution at temperatures greater than 600 °C. In the hematite-ilmenite series, complete solid solution occurs above 1,050 °C. At lower temperatures, both solid solutions are more restricted; in the magnetite–ulvospinel series two phases exsolve, while in the hematite–ilmenite series intermediate compositions are represented by the intergrowth of end-member compositions. Ionic replacement in both series takes the form:

$$2Fe^{3+} \leftrightarrow Fe^{2+} + Ti^{4+}$$

Natural titanomagnetites are often displaced toward higher oxidation compositions (the ilmenite–hematite series) during igneous cooling from 1,000 to 600 °C (i.e. deuteric conditions). In plutonic rocks, the following Fe–Ti mineral pairs occur and indicate solid solution compositions of increasing fO_2:

ulvospinel-rich magnetite + ilmenite

ulvospinel-poor magnetite + ilmenite

ulvospinel-poor magnetite + hematite

hematite + rutile

Fe–Ti oxide mineral pairs can be used to determine the temperature and fO_2 of magmatic systems. Ghiorso and Evans (2008) provided a geothermobarometer calibrated at 800–1,300 °C and ±3 log10 units of the nickel–nickel oxide (NNO) oxygen buffer accurate to ±30 °C and within one order of magnitude fO_2 (atm). Given the various $FeO–TiO_2–Fe_2O_3$ solid solutions, coexisting Fe–Ti minerals are common and once identified via reflecting light microscopy can be targeted for compositional analysis. The ilmenite and ulvospinel components are then used to determine T(°C) and fO_2 from the following figure.

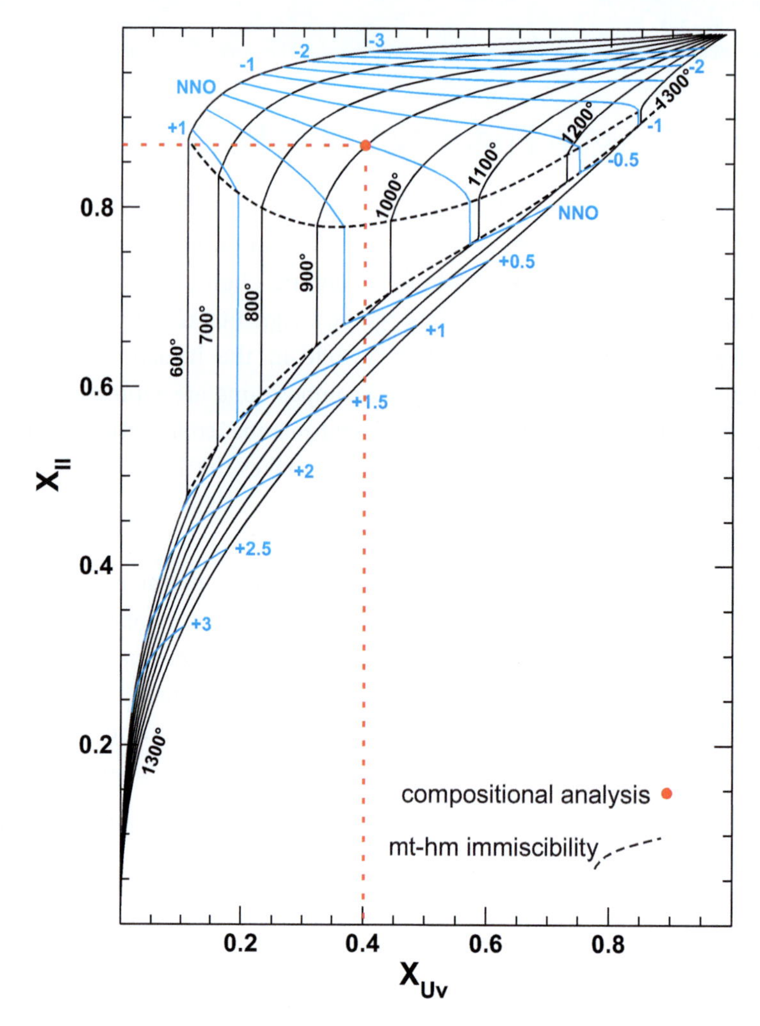

Fe–Ti two-oxide geothermometer and oxygen barometer (after Ghiorso & Evans, 2008). Lines of equal temperature (isotherms, °C) in black and lines of relative oxidation state (isopleths, displacement of log10 fO_2units from NNO) in blue. The dashed curves enclose the compositional miscibility gap. The red dot is a hypothetical chemical analysis of a coexisting mineral pair yielding T = 900 °C and fO_2= 0 (NNO).

After Ghiorso et al. (2008), with permission.

Corundum Al_2O_3

Optical Properties

Color in thin-section:	Colorless to weakly colored
Pleochroism:	Some colored varieties may present a pleochroism with blue shades
Birefringence:	0.008
Relief:	High
Cleavage:	None; parting on {0001} and on {11$\bar{2}$1} intersecting at an angle of 94°
Crystal system/optic sign:	Trigonal / (–)
Other features:	Corundum is rarely euhedral and typically occurs as small, rounded crystals. Low to moderate birefringence and simple twin lamellae are diagnostic. Apatite has lower relief and lower birefringence, vesuvianite has anomalous interference colors, tourmaline has higher birefringence and strong pleochroism; sapphirine is biaxial.

Paragenesis

Corundum occurs in silica-poor rocks such as nepheline syenites and other undersaturated alkali igneous rocks. Corundum may occur in contact metamorphic aureoles of thermally altered aluminous shales and in aluminous xenoliths found in high-temperature basic igneous plutonic and hypabyssal rocks. Red corundum is known as ruby and blue as sapphire. Corundum alters to fine-grained aggregates of muscovite, margarite, andalusite, sillimanite, kyanite, diaspore, gibbsite, or clay.

Igneous

Corundum may crystallize from undersaturated fractionated felsic melts at high pressure, as well as from syenitic melts of crustal to upper mantle origin. Corundum occurs as veins in amphibolite associated with dunite; it is observed in pegmatites and other rocks associated with nepheline syenites; it is also found in unusual dike rocks such as plumasites (oligoclase–corundum rock) and corundum-bearing plagioclasites – possibly derived by the desilication of an acid igneous rock in contact with a basic host rock, or possibly of hydrothermal origin. Many granitic and andesitic rocks are corundum normative, although they do not contain modal corundum; this may be due to amphibole fractionation. A small increase in Al_2O_3 (2%) in a magma has been shown experimentally to produce a large increase in the liquidus temperature. This implies that most granitic magmas do not have enough alumina to crystallize corundum.

Metamorphic

In metamorphic rocks, corundum is found in recrystallized limestones and marbles and is also formed from the metamorphic recrystallization of aluminous rocks, and by contact thermal processes. It often forms at high pressure and appears in xenoliths in rocks formed deep in the crust. Corundum is also found in metamorphic silica-poor hornfels, as well as in thermally and regionally metamorphosed bauxite reservoirs.

Figure 1.80

A Corundum in syenite–pegmatite from the southern Ural Mountains (Russia). Anhedral corundum with cleavage (center) with rounded inclusion of potassium feldspar (gray). Cross-polarized light, 10× magnification, field of view = 2 mm.

Image courtesy of E. Sorokina

B Corundum in anorthosite (kyshtymite) from the southern Ural Mountains (Russia). Euhedral corundum (sapphire) crystal with oscillatory compositional zoning; blue coloration is Fe-rich. The fine-grained matrix is predominantly plagioclase. Plane-polarized light, 10× magnification, field of view = 2 mm.

Image courtesy of M. Filina

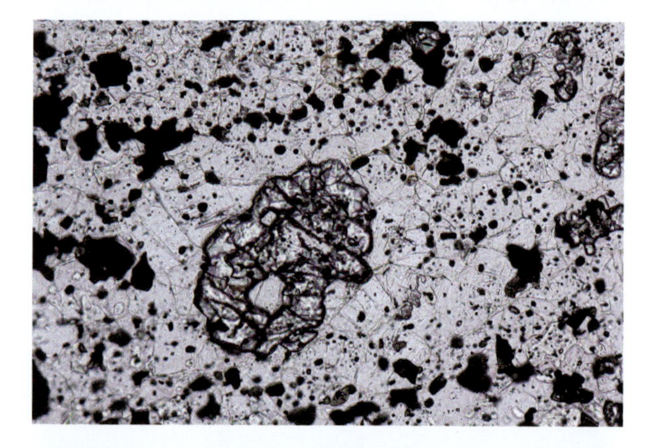

C Corundum in hornfels from the Tuscan Archipelago (Italy). Small, rounded corundum crystal (high relief in center of image) in association with plagioclase (colorless) and small magnetite crystals (black). Plane-polarized light, 20× magnification, field of view = 1 mm.

D Corundum in hornfels from the Tuscan Archipelago (Italy). Same image as (C) but with crossed polarizers. Corundum shows yellow interference colors and plagioclase first-order grays. Cross-polarized light, 20× magnification, field of view = 1 mm.

Perovskite $CaTiO_3$

Optical Properties

Color in thin-section:	Colorless to dark brown, pale yellow, yellowish-orange
Pleochroism:	Weak
Birefringence:	Weak
Relief:	High
Cleavage:	Good {100}, good {010}, good {001}
Crystal system/optic sign:	Orthorhombic (pseudocubic) / (+)
Other features:	Penetrative lamellar twinning on {101} and {121}. Small crystals can appear isotropic while larger specimens have extremely high relief, brownish color, weak birefringence, and complex lamellar twinning. Rutile has lower birefringence.

Paragenesis

The principal occurrence of perovskite is silica-undersaturated rocks because it cannot coexist with quartz, enstatite, albite, or sanidine; at temperatures of 600–800 °C at 100 MPa it reacts with these minerals to form titanite and other silicates. The perovskite mineral structure reflects dense cubic packing of the type ABO_3, where the A site is usually an alkaline earth or rare earth element, and the B site is filled with a transition metal element. With increasing pressure (mantle depths of 400–1,000 km), $CaSiO_3$, $FeSiO_3$, and $MgSiO_3$ adopt the perovskite structure.

Igneous

Perovskite associated with basic and ultrabasic alkaline igneous rocks is often deuteric and often occurs with melilite, leucite, or nepheline. It may occur as discrete crystals in the groundmass or as reaction rims around magnesian ilmenite in kimberlites. In perovskite there is considerable substitution of Ca with rare earth and alkali elements, and of Ti with Nb or Ta.

Metamorphic

In metamorphic rocks it may be associated with contact-metamorphosed impure limestones of the Ce- or Nb-bearing varieties.

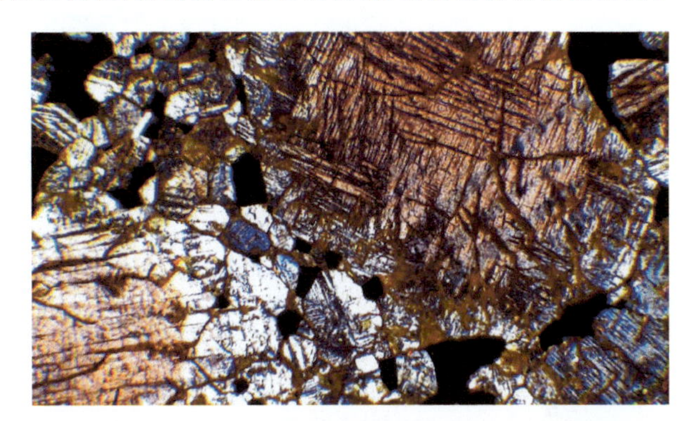

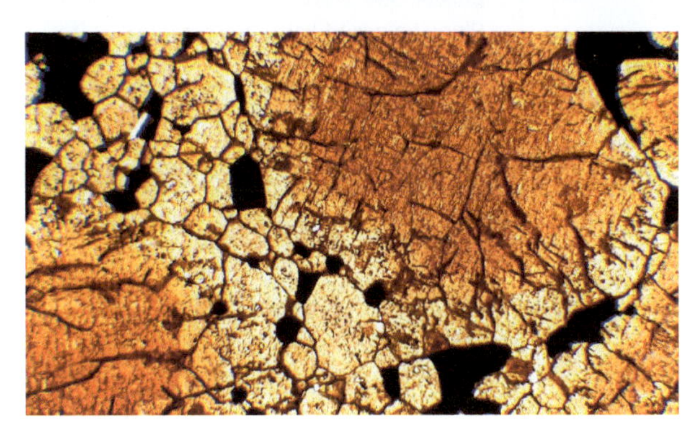

Figure 1.81 Pervoskite (BH250-208c) in xenolith from unknown location (Uganda). (*Top*) Perovskite crystals (golden to red–brown in plane-polarized light) form granular clusters. (*Bottom*) Same image as above but with crossed polarizers showing diagnostic twinning and birefringence of perovskite. 10× magnification, field of view = 2 mm.

Images courtesy of B. Haileab.

BOX 1.5 The Fe–Ti Oxides

Iron–titanium oxide minerals in the $FeO–Fe_2O_3–TiO_2$ system are present in most rocks and are easily recognized, being black (to darkly colored) in plane-polarized transmitted light and isotropic in cross-polarized transmitted light. Specific mineral species can sometimes be identified by, for example, color at the crystal edges in transmitted light (a distinctive "blood red" is associated with hematite) or by association (ilmenite alters to grayish-white leucoxene). However, much more information is obtained for these minerals from reflecting light microscopy, where variations in reflectance are diagnostic of specific minerals and is a necessary first step before pursuing more expensive analytical methods.

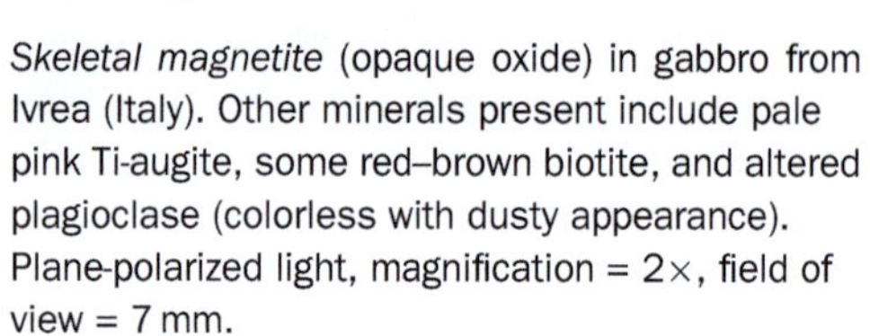

Skeletal magnetite (opaque oxide) in gabbro from Ivrea (Italy). Other minerals present include pale pink Ti-augite, some red–brown biotite, and altered plagioclase (colorless with dusty appearance). Plane-polarized light, magnification = 2×, field of view = 7 mm.

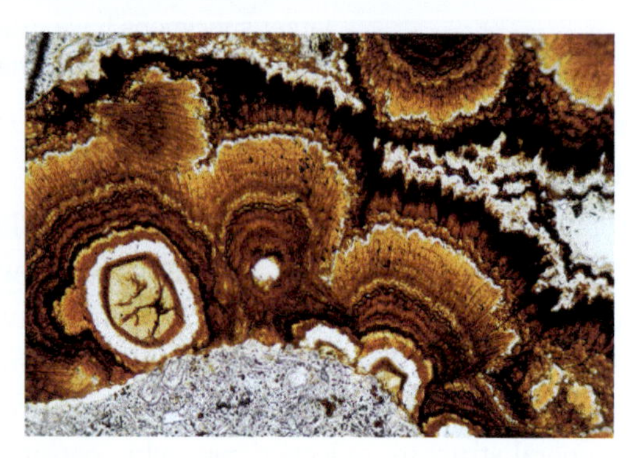

Hematite-bearing basalt from Iblei (Italy). Hematite (red and orange) has grown in a cavity following the circulation of iron-rich fluids with a concentric, laminated structure. Plane-polarized light, magnification = 10×, field of view = 2 mm.

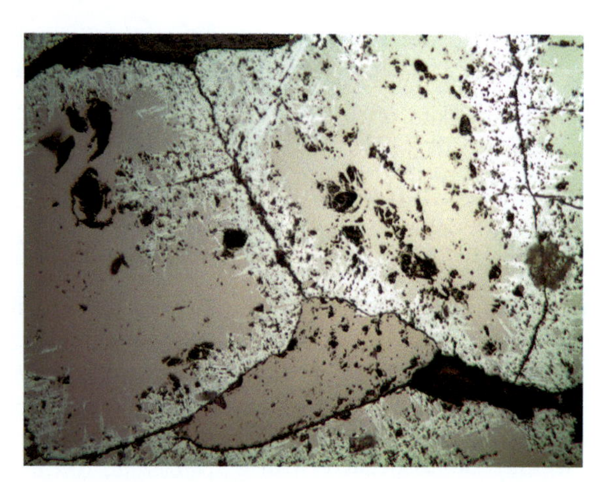

Fe–Ti oxides from Killingi (Sweden). This image shows magnetite (tan) with exsolution of hematite lamellae (bright) radiating from fractures and grain boundaries. Ilmenite (slightly darker phase, central lower part of image without hematite lamellae) is also present. Reflected light, magnification = 20×, field of view = 1.1 mm.

Image courtesy of I. Pitcairn

Rutile TiO_2

Optical Properties

Color in thin-section:	Reddish-brown or yellowish
Pleochroism:	From weak pale yellow to pale brown and brown–red
Birefringence:	0.286–0.296; interference colors usually masked by mineral color
Relief:	Exceptionally high
Cleavage:	Good {110}, moderate {100}
Crystal system/optic sign:	Tetragonal / (+)
Other features:	Its deep red–brown color and very high relief are distinctive. It forms elongate prisms.

Paragenesis

Rutile is one of three polymorphs (along with brookite and anatase) of TiO_2 and is the most common form of TiO_2. It is the high-temperature polymorph and tends to occur in high P–T assemblages. It occurs as small grains with other Fe–Ti–O phases in plutonic rocks, from SiO_2 undersaturated to acidic rocks and pegmatites. Rutile exsolves as fine needle-like crystals in some igneous minerals like quartz or Ti-biotite ("sagenite"). It forms as an alteration product after other Ti-bearing minerals and alters to leucoxene.

Rutile is a common accessory mineral in many metamorphic rocks such as shales, quartzites, mica-schists, gneiss, amphibolites, blueschist, and eclogites. It is common in some amphibolites and eclogites, and in metamorphosed limestones. Rutile can also be produced at the expense of ilmenite during hydrothermal alteration (greisenization).

Figure 1.82

A Rutile in andesite from Amiata volcano (Italy). Exsolved rutile (small black needles) and apatite (colorless) included in red biotite (hexagonal basal section). Biotite is in contact with colorless plagioclase and pale green pyroxene crystals (lower-right corner). Plane-polarized light, 2× magnification, field of view = 7 mm.

B Rutile in orthogneiss from an unknown Alpine locality (Italy). Small, dark rutile crystals (almost black) with high relief associated with biotite (medium to light brown) and quartz (colorless). Plane-polarized light, 2× magnification, field of view = 7 mm.

BOX 1.6 The Sulfides

Sulfides comprise a vast group of minerals with the general formula M_aS_b, in which sulfur bonds with metal (M). The sulfide family is large (hundreds of minerals), but only a few are truly common: pyrite, pyrrhotite, sphalerite, chalcopyrite, and galena. Given the large number of metals involved, phase relations of sulfides are particularly complex; furthermore, numerous reactions occurring at relatively low temperatures (100–300 °C) produce complex intergrowths, making sulfides suitable for low-temperature geothermometry. Sulfides also have economic significance, being the most important group of ore minerals and the main sources of a wide range of metals. Under reducing conditions, sulfide ore deposits are formed in two main ways:

1. Liquid immiscibility: separation of an immiscible sulfide-rich melt during the early stages of crystallization of ultrabasic to basic magmas.
2. Deposition from fluids: this includes hypersaline fluids (brines) to gas-rich aqueous solutions at temperatures of 150–600 °C.

Because sulfides are formed in moderately to strongly reducing conditions, they are very sensitive to weathering (oxidation). This may release metal ions into solution and form sulfuric acid. The great majority of sulfide minerals are opaque and reflected light microscopy is needed for correct identification.

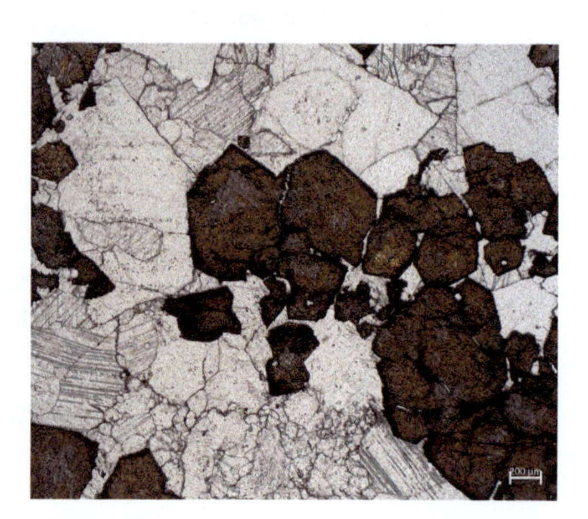

Idiomorphic sphalerite crystals from Gorno (Italy). Zoned sphalerite crystals are associated with clear, colorless fluorite, and pale brown calcite with visible cleavage traces. Sphalerite zoning is due to small variations in the Fe content. Plane-polarized light, scale bar = 200 μm.

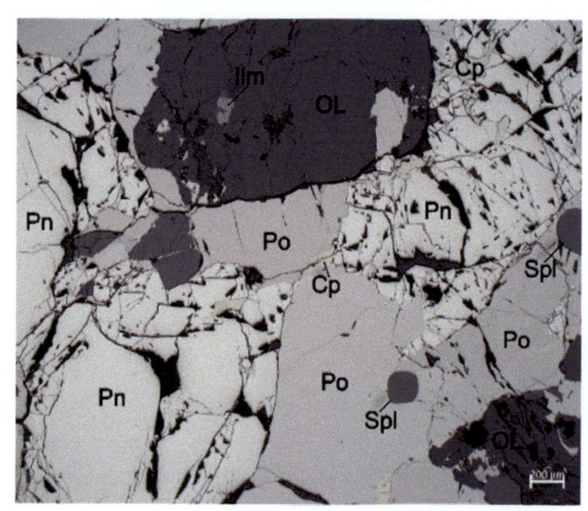

Sulfide crystals in ultramafic cumulate from Valle Strona (Italy). The image is dominated by coexisting sulfide minerals: Po = pyrrhotite; Pn = pentlandite; Cp = chalcopyrite. Other minerals present include olivine (OL), ilmenite (Ilm), and spinel (Spl). Reflected light image, scale bar = 200 μm.

Images courtesy of M. Moroni

Spinel Group $A^{2+}B^{3+}{}_2O_4$

Optical Properties

Color in thin-section:	Variable (reds, browns, blues, greens) to opaque
Pleochroism:	Absent to very weak
Birefringence:	Isotropic
Relief:	High
Cleavage:	None
Twinning:	Common on {111} (spinel law)
Crystal system/optic sign:	Cubic
Other features:	Isotropic nature, high relief, and lack of cleavage are characteristic; typical octahedral form distinguishes it from garnet.

Paragenesis

Minerals of the spinel group are common accessory minerals in both igneous and metamorphic rocks. The spinel group is divided into three series: the spinel, magnetite, and chromite series. Each series has potential solid solution with Al_2O_3 and Fe_2O_3 in the spinel series, and Fe–Mg in the magnetite and chromite series. Maghemite (γFe_2O_3) and ulvospinel ($Fe^{2+}{}_2TiO_4$) are included in the spinel group because they have spinel structure. The general descriptor "*opaque oxides*" may be used since the composition of these often darkly colored minerals is difficult to determine without a chemical analysis.

The spinel series. The most common minerals in the spinel series are spinel and hercynite. Spinel is a common high-temperature mineral in metamorphic rocks, and in contact-metamorphosed limestone and SiO_2-poor argillaceous rocks. In thin-section the spinel series varies widely in color from almost colorless to red (Cr), blue (Fe^{2+}), brown (Fe^{3+}), yellow, and pink. Hercynite is dark green to black and gahnite dark bluish-green. Galaxite is uncommon but mahogany-red to black in color.

The magnetite series. Magnetite and magnesioferrite are associated with most igneous rocks, as well as a metasomatic product in skarn deposits. Magnetite is the more common and magmatic segregations or crystal settling accumulations can be ore-forming (e.g. Bushveld complex). Magnesioferrite (brownish-black to black) is typically intergrown with hematite. Magnetite series minerals are opaque in transmitted light but are distinguishable via reflecting light.

The chromite series. Chromite and magnesiochromite have similar parageneses, with ferroan magnesiochromite being the most common of the two. They are common in ultrabasic igneous rocks and can be locally abundant as monomineralic masses, pods, or layers. Such occurrences are of economic importance. They are opaque and chemical analysis is needed for specific identification.

Spinel series (Al)	Magnetite series (Fe^{3+})	Chromite series (Cr)
Spinel, $MgAl_2O_4$	Magnesioferrite, $MgFe^{3+}{}_2O_4$	Magnesiochromite, $MgCr_2O_4$
Hercynite, $Fe^{2+}Al_2O_4$	Magnetite, $Fe^{2+}Fe^{3+}{}_2O_4$	Chromite, $FeCr_2O_4$
Gahnite, $ZnAl_2O_4$	Franklinite, $ZnFe^{3+}{}_2O_4$	
Galaxite, $MnAl_2O_4$	Jacobsite, $MnFe^{3+}{}_2O_4$	
	Trevorite, $NiFe^{3+}{}_2O_4$	

Figure 1.83

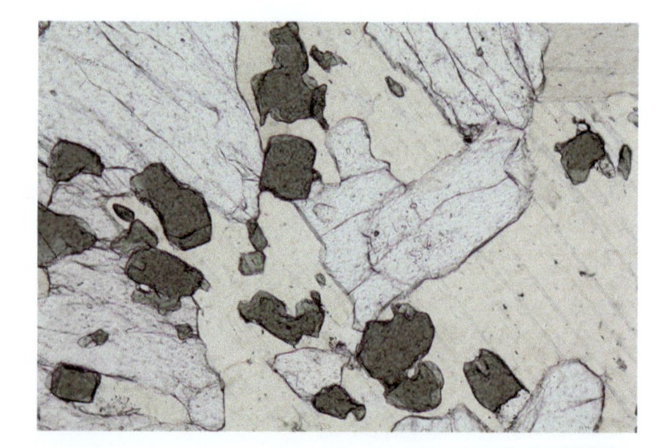

A Hercynite in a metamorphic xenolith from Vesuvius (Italy). Hercynite (green) occurs with clinopyroxene (high relief) and white mica (low relief, slightly yellowish). Plane-polarized light, 10× magnification, field of view = 2 mm.

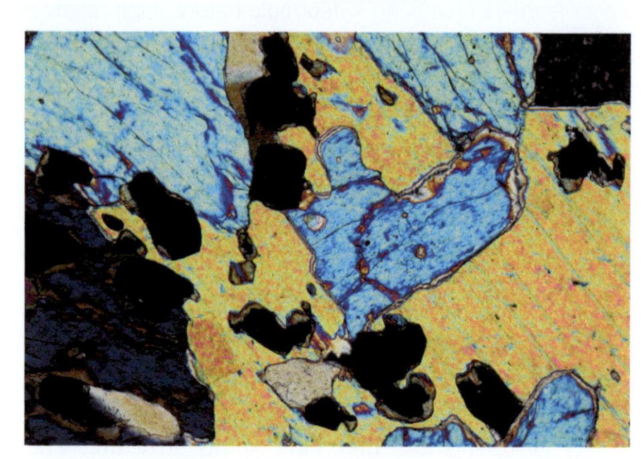

B Hercynite in a metamorphic xenolith from Vesuvius (Italy). Same image as (A) but with crossed polarizers. Note the isotropic nature of the spinel group minerals. Cross-polarized light, 10× magnification, field of view = 2 mm.

C Chromite in kimberlite dike from Lac De Gras (Canada). Red chromite (central image) in a groundmass of altered olivine and carbonate. The chromite and other crystals have oxidation rims, the result of oxygenation during rapid ascent from mantle depths. Plane-polarized light, 10× magnification, field of view = 2 mm.

Image courtesy of D. Gainer

Apatite $Ca_5(PO_4)_3(OH,F,Cl)$

Optical Properties

Color in thin-section:	Colorless; can appear cloudy due to inclusions
Pleochroism:	Very weak
Birefringence:	0.001–0.007
Relief:	Moderate
Cleavage:	Poor on {0001}
Crystal system/optic sign:	Hexagonal / (–)
Other features:	Apatite forms short columnar crystals with straight extinction (parallel to the c-axis) and in basal section shows hexagonal outline; in carbonatites can have higher birefringence; it may be confused with nepheline (which is also uniaxial) but apatite has a higher relief and generally forms euhedral, unaltered crystals. Sillimanite is biaxial, has higher birefringence, and one good cleavage.

Paragenesis

Apatite is a ubiquitous phosphate mineral found in many rock types. It is a common accessory mineral in almost all magmatic rocks, from basic to acid, but also in metamorphic and sedimentary rocks (e.g. phosphorites). Apatite is a general term for the calcium–phosphate minerals of the apatite group, which forms an isomorphous series with the end-members:

fluorapatite: $Ca_5(PO_4)_3F$

chlorapatite: $Ca_5(PO_4)_3Cl$

hydroxylapatite: $Ca_5(PO_4)_3OH$

Fluorapatite is the most common apatite group mineral, and "apatite" is used synonymously for "fluorapatite."

Igneous

Apatite can be an early or late crystallizing phase from a magma. It is an accessory mineral (0.1–1%) in almost all magmatic rocks and is very common in Na- and K-rich magmatic rocks (nepheline syenite, nepheline monzosyenite, olivine nephelinite, etc.). It is more common in carbonatites, lamprophyres, and alkaline/agpaitic rocks, where it can reach high enough concentrations to be of economic interest (e.g. the alkaline complexes of the Kola Peninsula [Russia] or the Transvaal [South Africa]). In granitic pegmatites, bluish manganese-bearing apatite is common and it also occurs in hydrothermal veins and cavities.

As an accessory mineral, apatite can be used to constrain igneous processes in magma chambers. Improvements in microanalytical techniques (e.g., SIMS, LA-ICP-MS) allow analysis of single crystals. The fact that apatite is a ubiquitous mineral in virtually all rocks, together with the fact that it can incorporate magmatic water, halogens, S, C, trace elements including Sr, U, Th, as well as rare earth elements, means it is useful for age-dating and fluid inclusion studies.

Metamorphic

Apatite is a common mineral in contact and regionally metamorphic rocks such as chlorite schists, amphibole-bearing schists, gneisses, and calc-silicate rocks. In metasomatized calc-silicate rocks and impure limestones, fluorapatite is associated with chondrodite and phlogopite. In rocks affected by chlorine metasomatism, chlorapatite is associated with scapolite.

Figure 1.84

A Apatite in tonalite, location unknown (Antarctica). Small, prismatic apatite crystals (colorless, high relief), feldspar (colorless), magnetite (black), and biotite (brown pleochroism). Plane-polarized light, 2× magnification, field of view = 7 mm.

B Apatite in nepheline syenite from Bohemia (Czech Republic). Distinctive, colorless hexagonal basal (and prismatic) sections enclosed in pale Ti-augite. Plane-polarized light, 20× magnification, field of view = 1 mm.

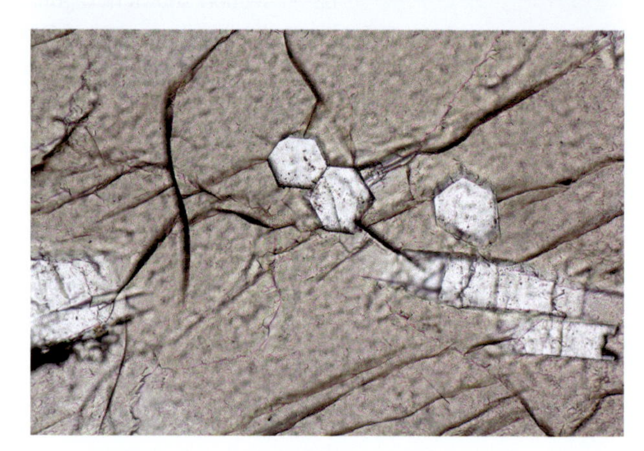

C Apatite in plutonic carbonatite (sövite) from Sao Vicente island (Cape Verde). Large and fragmented apatite crystals (blue birefringence) occur with calcite (fourth-order pale pink birefringence). Higher birefringence is associated with apatite from carbonatites. Plane-polarized light, 2× magnification, field of view = 7 mm.

D Apatite in Alpine chlorite schist, location unknown (Italy). Small, prismatic apatite crystals (colorless, with high relief) aligned with the foliation in sample composed of predominantly magnetite (black) and green chlorite. Plane-polarized light, 2× magnification, field of view = 7 mm.

Monazite (Ce,La,Th)PO_4

Optical Properties

Color in thin-section:	Colorless to slightly yellow
Pleochroism:	Very weak; light yellow to very light yellow
Birefringence:	0.045–0.075
Relief:	Very high
Cleavage:	Moderate on {100}, variable on {001}
Crystal system/optic sign:	Monoclinic / (+)
Other features:	Although it may be confused with zircon, apatite, or titanite, its tabular, equant habit and crystal symmetry are diagnostic. It often shows well-developed chemical zoning.

Paragenesis

Monazite is a common accessory mineral in a wide variety of rock types, hosting important incompatible trace elements (REE, Th, U) and in some cases associated with important economic REE deposits. It is commonly found in granitic and syenitic rocks, granitic pegmatites, in most metapelitic and metabasic rocks, and in dolomitic marbles. It is stable in the weathering environment.

Igneous

Monazite is a common accessory mineral in low-CaO peraluminous igneous rock from diorite to granite, pegmatite, and hydrothermal deposits. It is less frequently reported in peralkaline rocks, but is more common in carbonatites. Monazite in igneous rocks may have different origins: (1) inherited from source rock; (2) precipitated directly from melt; or (3) formed during late hydrothermal alteration. Consequently, monazite in igneous rocks can be useful to link igneous rocks and their source terrains, or to monitor the evolution of an igneous systems from magmatic through hydrothermal processes.

Metamorphic

Monazite is common in metapelites, metapsammites, and orthogneisses at amphibolite facies conditions and above. Monazite is less abundant in mafic and calcic bulk compositions, and in greenschist facies rocks where allanite would be stable. Indeed, one of the most common monazite-in reactions associated with metapelite involves the breakdown of allanite and monazite growth at around 450–550 °C. However, there are a number of monazite reactions in metamorphic rocks. Metamorphic monazite is commonly zoned, particularly with respect to REE, Y, Th, and U, and can have several compositional domains which represent different times of monazite growth. Such zoning reflects recrystallization and growth events related to monazite-forming reactions during the metamorphic cycle(s).

Figure 1.85

A Monazite in tonalite from Adamello (Italy). Idioblastic monazite is enclosed in central biotite. Monazite shows its characteristic high relief and pale yellowish colors. The colorless crystals are quartz. Plane-polarized light, 10× magnification, field of view = 2 mm.

B Monazite in tonalite from Adamello (Italy). Same image as (A) but with crossed polarizers. Monazite shows its characteristic high birefringence. Cross-polarized light, 10× magnification, field of view = 2 mm.

C Monazite in orthogneiss from Argentera (Italy). Hypidioblastic monazite porphyroblast, with metamict allanite inclusion (note the radial fracture in the surrounding monazite). Other minerals are biotite (green–brown) and quartz (colorless). Plane-polarized light, 20× magnification, field of view = 1 mm.

D Monazite in orthogneiss from Argentera (Italy). Same image as (C) but with crossed polarizers. Monazite shows its characteristic high birefringence, while metamict allanite is isotropic. Cross-polarized light, 20× magnification, field of view = 1 mm.

Xenotime $(REE,Y)PO_4$

Optical Properties

THE PHOSPHATES

Color in thin-section:	Colorless to brownish
Pleochroism:	Weak; pale pink to pale yellow or gray–brown to pale yellow–green
Birefringence:	0.096
Relief:	High
Cleavage:	Good {110} (only visible in larger crystals)
Crystal system/optic sign:	Tetragonal / (+)
Other features:	Xenotime forms short columns or granular crystals and often occurs as inclusions with pleochroic halos. It may be mistaken for zircon, monazite, titanite, anatase, or rutile. Rutile has higher relief, anatase is blueish and striated, titanite and monazite are biaxial, and zircon has lower birefringence.

Paragenesis

Xenotime is an accessory mineral in igneous, metamorphic, and sedimentary rocks. In igneous rocks it is common in alkali granites to granites, such as nepheline syenites, syenites, granites, aplites, and associated pegmatites. In metamorphic rocks it occurs during low-temperature diagenesis to upper amphibolite, granulite, and eclogite facies conditions. It is resistant to weathering and is thus a common detrital mineral in sandstone and placers deposits.

Xenotime is an REE-rich orthophosphate and is dimorphous with monazite, but monazite preferentially incorporates light REEs in an irregular polyhedron $[(REE)O_9]$, while xenotime favors heavy REE in a regular dodecahedron $[(Y)O_8]$. Xenotime is an important accessory mineral as it can be used for U/Pb geochronology and for dating diagenesis. In the presence of a fluid it may form by the dissolution–precipitation reaction:

$$\text{zircon}_1 \text{ (relict)} + \text{P-bearing fluid} = \text{zircon}_2 \text{ (metamorphic)} + \text{xenotime}$$

Figure 1.86

A Xenotime in granite from Brandberg, Namibia (Africa). Subhedral xenotime crystals have high relief and corroded faces/internal cavities; groundmass is quartz and feldspar. Plane-polarized light, 10× magnification, field of view = 2 mm.

B Xenotime in granite from Brandberg, Namibia (Africa). Same image as (A) but with crossed polarizers. Xenotime (high birefringence) in groundmass of quartz and feldspar (low birefringence). Cross-polarized light, 10× magnification, field of view = 2 mm.

Barite $BaSO_4$

Optical Properties

Color in thin-section:	Colorless
Pleochroism:	None
Birefringence:	0.011–0.012
Relief:	Moderate
Cleavage:	Perfect on {001}, good on {110} and {010}
Crystal system/optic sign:	Orthorhombic / (+)
Other features:	Commonly occurs as well-formed crystals but also as flaky or radiating aggregates. Twinning is common, with polysynthetic interpenetrating twins on {110}. Extinction tends to be straight in the direction of the cleavage trace. Barite is commonly replaced by carbonate minerals or by chalcedony.

Paragenesis

Barite is a metamorphic sulfate mineral and is the most common barium mineral in the Earth's crust. It occurs as a gangue mineral in ore-bearing hydrothermal veins, in association with fluorite, calcite, and quartz. Barite often forms concretions in different sedimentary rocks (e.g. limestones, sandstones, shales, and clays), surface deposits derived from limestone weathering, and in deposits associated with hot springs. Ba^{2+} and $(SO_4)^{2-}$ are required to form barite and can be derived from different sources: Ba^{2+} may derive from barium-rich fluids, from the alteration of micas and feldspars in granites and other igneous rocks, and from dissolution of dolomites and other sedimentary rocks. $(SO_4)^{2-}$ may be derived from sulfide deposits, from sulfur-bearing strata, or from seawater.

Figure 1.87

A Barite in basalt from Sardinia (Italy). Well-formed barite crystals showing good cleavage occur in a volcanic cavity. Cross-polarized light, 10× magnification, field of view = 2 mm.

B Barite in metalimestone from Sardinia (Italy). Barite showing typical radiating aggregates and birefringence. Cross-polarized light, 2× magnification, field of view = 7 mm.

APPLICATION 1.5

Geochronology

Geochronology is used to date rock materials in order to constrain the timing and rates of geological processes. This can be done by determining the age of a single mineral, different growth phases within a single mineral, or combining the ages of multiple minerals in a single sample. Radiometric ages are based on the scientific principles of radioactive decay, and these approximate "high," "medium," and "low" temperature processes. Radiometric ages can be used to determine the absolute age of unaltered igneous rocks, date the time of alteration in metamorphic rocks, or even both if primary minerals are present as porphyroclasts and secondary minerals have grown during metamorphism. The successful application of "*age-dating*" relies on understanding petrological relationships within the sample – that is, what process(es) is preserved, which mineral(s) record this process(es), and how did this mineral(s) form (open system/closed system, equilibrium/disequilibrium, etc.)? By linking time and temperature, we can develop a more thorough understanding of a sample's evolutionary history.

U–Pb dating. U–Pb zircon ages usually represent high-temperature crystallization/reworking ages. In this cathodoluminescence image of a ~500 μm long zircon crystal from the Amitsoq gneiss on Angisorsuaq Island (Greenland), four distinct mineral growth phases can be seen: an old core (1) that was resorbed (2), followed by two additional periods (3 and 4) of new crystal growth (see the numbered white circles in the image). The ages obtained from this crystal range from ~3.85 Ga in the older core to 3.67 Ga in the outer-most bright rim. Technological advances have made it possible to apply multiple analytical methods to a single crystal (i.e. the three dotted circles represent the locations of subsequent Hf-isotope analyses on this same grain).

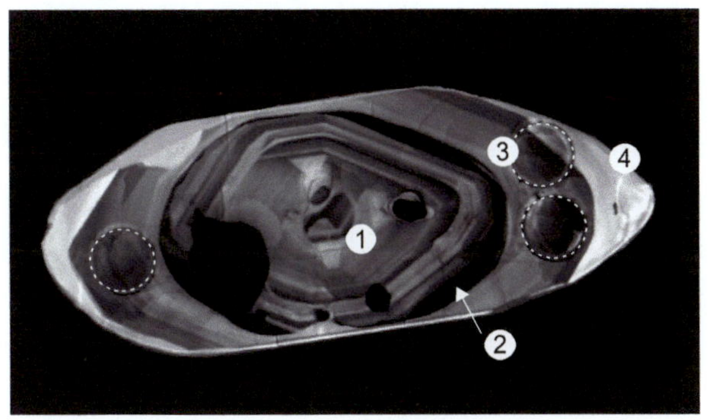

Image courtesy of A. Kemp & M. Whitehouse

Other U–Pb chronometers in common use:

- baddeleyite
- perovskite
- monazite
- rutile
- titanite
- apatite
- xenotime
- allanite
- uraninite
- thorite
- pyrochlore
- calcite.

Rb–Sr dating. The Rb–Sr internal mineral isochron method can be used to date metamorphic mineral growth, ductile deformation, and precipitation of segregations from metamorphic fluids. In the image of blueschist from Syros (Greece), coarse-grained phengitic white mica (high birefringence) + quartz occur in a finer-grained matrix of mostly blue amphibole + phengite + quartz. Different size populations of white mica may reflect different periods of growth, and these can be separated and analyzed. In this case, they yield an Rb–Sr isochron age of 53.2 ± 2.0 Ma, interpreted as the time of high-pressure deformation in which the mica grew.

Microphotograph of blueschist from Syros (Greece). Field of view = 3 mm.
Image courtesy of J. Glodny

Argon dating. The K–Ar and ^{40}Ar–^{39}Ar methods are widely used. Ages can be determined for rapidly cooled volcanic rocks using mineral separates (sanidine, biotite, hornblende) or whole rock samples. Plutonic rocks typically cool more slowly and yield "cooling" ages rather than crystallization ages. Metasedimentary rocks rich in K-bearing micas and amphiboles can be easily dated, but require careful interpretation – the apparent ages of high-grade metamorphic terranes may reflect the metamorphic crystallization history *or* the post-metamorphic cooling history, and low-grade metamorphic terranes may carry inherited argon from a previous evolutionary stage(s). In the sample of biotite schist from the Detour Lake gold deposit (Canada) below, biotite (tan), secondary amphibole (sky blue), and chlorite (blue–green) define the foliation, but all analyses yield post-deformation ages apparently documenting later, pervasive fluid-related resetting.

Image courtesy of R. Dubosq

Fission track dating. Fission of ^{238}U occurs naturally. In a crystal that contains ^{238}U, like apatite or zircon, the zone of damage due to the fission process is known as a "fission track." The number of tracks should be proportional to time if the rate of spontaneous fission is known (or can be determined) and if the daughter product of the decay process is retained. In a best-case scenario, such as an erupted volcanic rock that cools quickly, the age determined for the mineral is the same as the age of the rock.

Fission track analysis is a "low-temperature" method (i.e. the effective closure temperature (*Tc*) for the method depends on the mineral, its composition, and its cooling history, but overall is below ~300 °C). Fission track studies therefore tend to address landscape evolution, sedimentary basin analysis, and rates associated with upper crustal tectonics. Assuming reasonable geothermal gradients and surface temperatures, fission track data and the thermal histories derived from them are converted into burial and unroofing histories.

As an example, consider the mineral apatite. Apatite is a common mineral occurring in igneous and metamorphic rocks at all metamorphic grades, and is also common in most clastic sedimentary rocks. Like most minerals in rocks, apatite has a

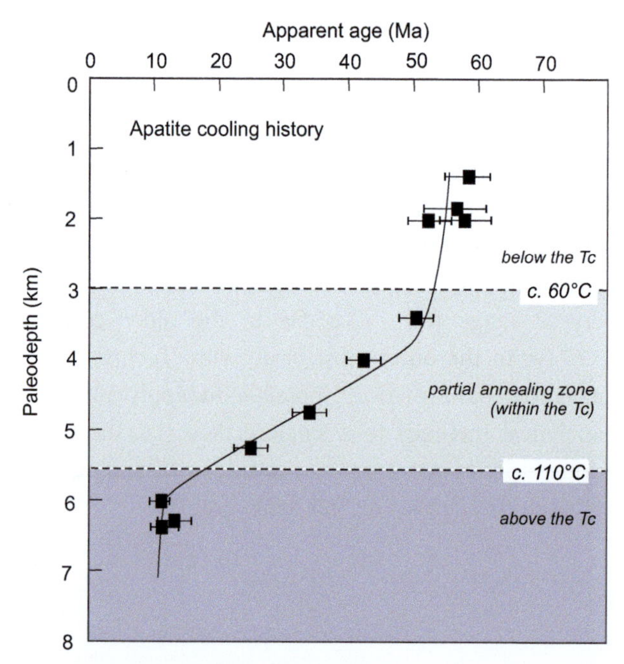

After Stockli et al. (2000), with permission.

natural variation in composition. Fission tracks in apatite are preserved at temperatures between 110 °C and 60 °C; the higher *Tc* limit is associated with chlorapatite, while the lower *Tc* limit is associated with fluorapatite (a compositional dependence).

In the diagram of apatite cooling history, depth is approximated assuming a geothermal gradient of 20 °C/km such that increasing depth = increasing *T* (°C). The darkest zone is above the *Tc*, the lighter zone is within the *Tc* (the "*partial annealing zone*"), and the white area at shallow depths is below the *Tc*. At temperatures above the *Tc* (>110 °C), the apatite fission track system is "open" and fission tracks are not retained; within the *partial annealing zone* the fission tracks are annealed at a rate dependent on composition (i.e. the fission tracks are shortened because the temperature is near the *Tc*). At temperatures below the *Tc* (<60 °C) the apatite system is "closed" and tracks are retained. The apatite fission track ages shown in this figure are from samples collected in a vertical profile along the footwall of an exhumed normal fault block in the White Mountains of eastern California (United States) (Stockli et al., 2000). The authors used these apatite fission track ages to argue that (1) the highest elevations were at low temperatures since ~55 Ma; (2) fission tracks within the partial annealing zone were shortened, yielding younger ages; and (3) at the lowest elevations, ages of ~12 Ma indicate rapid exhumation by normal faulting at that time.

Bibliography

Textbooks

Anthony, J., Bideaux, R., Bladh, K., Nichols, M. (Eds.) (n.d.). *Handbook of Mineralogy*. Mineralogical Society of America. https://handbookofmineralogy.org.

Best, M.G., 2013. *Igneous and Metamorphic Petrology*. Wiley.

Clarke, D.B., 1992. *Granitoid Rocks* (Vol. 7). Springer Science & Business Media.

Deer, W., Howie, R., Zussman, J., 2013. *An Introduction to the Rock-Forming Minerals* (3rd edn). The Mineralogical Society.

Howie, R., Zussman, J., Deer, W., et al., 2013. *Rock-Forming Minerals* (2nd edn). The Mineralogical Society.

Kline, C., Philpotts, A., 2013. *Earth Materials: Introduction to Mineralogy and Petrology* (2nd edn). Cambridge University Press.

MacKenzie, W.S., Guilford, C. 1980. *Atlas of the Rock-Forming Minerals in Thin Section*. Routledge.

McElhinny, M.W., 1979. *Palaeomagnetism and Plate Tectonics*. Cambridge University Press.

Nesse, W., 2009. *Introduction to Mineralogy*. Oxford University Press.

Pichler, H., Schmitt-Riegraf, C., 1997. *Rock-Forming Minerals in Thin Section*. Springer Science & Business Media.

Winter, J., 2014. *Principles of Igneous and Metamorphic Petrology* (2nd edn). Pearson.

Articles and Chapters

Bucher, K., Grapes, R., 2009. The eclogite-facies Allalin Gabbro of the Zermatt–Saas ophiolite, Western Alps: a record of subduction zone hydration. *Journal of Petrology* 50(8), 1405–1442.

Buddington, A.F., Lindsley, D.H., 1964. Iron–titanium oxide minerals and synthetic equivalents. *Journal of Petrology* 5(2), 310–357.

Chapman, T., Clarke, G.L., Piazolo, S., Daczko, N.R., 2017. Evaluating the importance of metamorphism in the foundering of continental crust. *Scientific Reports* 7 (1), 1–12.

Chepurov, A.I., Sonin, V.M., Zhimulev, E.I., et al. 2018. Dissolution of diamond crystals in a heterogeneous (metal–sulfide–silicate) medium at 4 GPa and 1400°C. *Journal of Mineralogical and Petrological Sciences*, 170526.

Chinner, G.A., 1962. Almandine in thermal aureoles. *Journal of Petrology*, 3(3), 316–341.

Chopin, C., 1984. Coesite and pure pyrope in high-grade blueschists of the Western Alps: a first record and some consequences. *Contributions to Mineralogy and Petrology* 86(2), 107–118.

Clarke, D.B., 1995. Cordierite in felsic igneous rocks: a synthesis. *Mineralogical Magazine* 59(395), 311-325.

Diener, J.F.A., Powell, R., White, R.W., 2008. Quantitative phase petrology of cordierite–orthoamphibole gneisses and related rocks. *Journal of Metamorphic Geology* 26 (8), 795–814.

Essene, E.J., 1989, The current status of thermobarometry in metamorphic rocks, in: J.S. Daly, R.A. Cliff, and B.W.D. Yardley (Eds.), *Evolution of Metamorphic Belts*, Geological Society, pp. 1–44.

Ferrière, L., Koeberl, C., Reimold, W.U., 2009. Characterisation of ballen quartz and cristobalite in impact breccias: new observations and constraints on ballen formation. *European Journal of Mineralogy* 21 (1), 203–217.

Filina, M.I., Sorokina, E.S., Botcharnikov, R., et al., 2019. Corundum anorthosites–kyshtymites from the South Urals, Russia: a combined mineralogical, geochemical, and U–Pb zircon geochronological study. *Minerals* 9 (4), 234.

Franz, G., Morteani, G., Rhede, D., 2015. Xenotime-(Y) formation from zircon dissolution–precipitation and HREE fractionation: an example from a metamorphosed phosphatic sandstone, Espinhaço fold belt (Brazil). *Contributions to Mineralogy and Petrology* 170, 37.

Ghiorso, M., Evans, B., 2008. Thermodynamics of rhombohedral oxide solid solutions and a revision of the Fe–Ti two-oxide geothermometer and oxygen-barometer. *American Journal of Science* 308, 957–1039.

Götze, J., 2012. Application of cathodoluminescence microscopy and spectroscopy in geosciences. *Microscopy & Microanalysis* 18, 1270–1284.

Guidotti, C.V., Cheney, J.T, Henry, D.J. 1988. Compositional variation of biotite as a function of metamorphic reactions and mineral assemblage in the

pelitic schists of western Maine. *American Journal of Science* 288A, 270–292.
Hammerstrom, J.M., Zen, E.A., 1986. Aluminum in hornblende: an empirical igneous geobarometer. *American Mineralogist* 71(11–12), 1297–1313.
Han, B., Liu, J., Zhang, L., 2008. A noncognate relationship between megacrysts and host basalts from the Tuoyun Basin, Chinese Tian Shan. *The Journal of Geology* 116(5), 499–502.
Henry, D.J., Guidotti, C.V., Thomson, J.A., 2005. The Ti-saturation surface for low-to-medium pressure metapelitic biotites: implications for geothermometry and Ti-substitution mechanisms. *American Mineralogist* 90(2–3), 316–328.
Hoiland, C.W., Miller, E.L., Pease, V., 2018. Greenschist-facies metamorphic zircon overgrowths as a constraint on exhumation of the Brooks Range metamorphic core, Alaska. *Tectonics*, 37(10), 3429–3455.
Holdaway, M.J., Mukhopadhyay, B., 1993. A reevaluation of the stability relations of andalusite: thermochemical data and phase diagram for the aluminum silicates. *American Mineralogist* 78(3–4), 298–315.
Hollister, L.S., Grissom, G.C., Peters, E.K., Stowell, H.H., Sisson, V.B., 1987. Confirmation of the empirical correlation of Al in hornblende with pressure of solidification of calc-alkaline plutons. *American Mineralogist* 72(3–4), 231–239.
Johannes, W., Puhan, D., 1971. The calcite–aragonite transition, reinvestigated. *Contributions to Mineralogy and Petrology* 31(1), 28–38.
Johnson, M.C., Rutherford, M.J., 1989. Experimental calibration of the aluminum-in-hornblende geobarometer with application to Long Valley caldera (California) volcanic rocks. *Geology* 17(9), 837–841.
Kato, T., 2001. Synthesized interference color chart with personal computer. *Journal of Geological Society of Japan*, 107, I–II.
Kohn, M.J., Spear, F.S., 1989. Empirical calibration of geobarometers for the assemblage garnet + hornblende + plagioclase + quartz. *American Mineralogist* 74(1–2), 77–84.
Kohn, M.J, Spear, F.S., 1990. Two new geobarometers for garnet amphibolites, with applications to southeastern Vermont. *American Mineralogist* 75(1–2), 89–96.
Krmíček, L., Ulrych, J., Šišková, P., et al. 2020. Geochemistry and Sr–Nd–Pb isotope characteristics of Miocene basalt–trachyte rock association in transitional zone between the Outer Western Carpathians and Bohemian Massif. *Geologica Carpathica* 71(5), 462–482.
Lindsley, D.H., Andersen, D.J., 1983. A two-pyroxene thermometer. *Journal of Geophysical Research: Solid Earth* 88(S02), A887–A906.
Messiga, B., Scambelluri, M, Piccardo, G.B., 1995. Chloritoid-bearing assemblages in mafic systems and eclogite-facies hydration of alpine Mg-Al metagabbros (Erro-Tobbio Unit, Ligurian Western Alps). *European Journal of Mineralogy* 7, 1149–1168.
Messiga, B., Tribuzio, R., Bottazzi, P., Ottolini, L., 1995. An ion microprobe study on trace element composition of clinopyroxenes from blueschist and eclogitized FeTi-gabbros, Ligurian Alps, northwestern Italy: some petrologic considerations. *Geochimica et Cosmochimica Acta*, 59(1), 59–75.
Oberti, R., Boiocchi, M., Hawthorne, F., Ball, N. Harlow, G., 2015. Eckermannite revised: the new holotype from the Jade Mine Tract, Myanmar-crystal structure, mineral data, and hints on the reasons for the rarity of eckermannite. *American Mineralogist* 100, 909–914.
Osborne, Z., Thomas, J., Nachlas, W., et al. (2019). An experimentally calibrated thermobarometric solubility model for titanium in coesite (TitaniC). *Contributions to Mineralogy and Petrology* 174, 34.
Pattison, D.R.M., 1992. Stability of andalusite and sillimanite and the Al_2SiO_5 triple point: constraints from the Ballachulish aureole, Scotland. *Journal of Geology* 100, 423–446.
Pattison, D.R.M., Tracy, R.J., 1991. Phase equilibria and thermobarometry of metapelites. *Contact Metamorphism* 26, 105–206.
Proenza, J., Gervilla, F., Melgarejo, J., Bodinier, J.L., 1999. Al- and Cr-rich chromitites from the Mayari-Baracoa ophiolitic belt (eastern Cuba); consequence of interaction between volatile-rich melts and peridotites in suprasubduction mantle. *Economic Geology* 94(4), 547–566.
Roeder, P.L., Emslie, R., 1970. Olivine-liquid equilibrium. *Contributions to Mineralogy and Petrology* 29(4), 275–289.
Schmidt, M.W., Poli, S., 1998. Experimentally based water budgets for dehydrating slabs and consequences for arc magma generation. *Earth and Planetary Science Letters* 163(1–4), 361–379.
Sen, G., MacFarlane, A., Srimal, N, 1996. Significance of rare hydrous alkaline melts in Hawaiian xenoliths. *Contribution to Mineralogy & Petrology* 122, 415–427.

Smith, D.K., 1984. Uranium mineralogy. In *Uranium Geochemistry, Mineralogy, Geology, Exploration and Resources*. Springer, pp. 43–88.

Sorokina, E.S., Botcharnikov, R.E., Kostitsyn, Y.A., et al. 2021. Sapphire-bearing magmatic rocks trace the boundary between paleo-continents: a case study of Ilmenogorsky alkaline complex, Uralian collision zone of Russia. *Gondwana Research* 92, 239–252.

Spear, F.S., Pyle, J.M., 2002. Apatite, monazite, and xenotime in metamorphic rocks. *Reviews in Mineralogy and Geochemistry* 48(1), 293–335.

Stockli, D.F., Farley, K.A., Dumitru, T.A., 2000. Calibration of the apatite (U–Th)/He thermochronometer on an exhumed fault block, White Mountains, California. *Geology* 28(11), 983–986.

Stormer, J.C., Carmichael, I.S.E., 1971. Fluorine-hydroxyl exchange in apatite and biotite: a potential igneous geothermometer. *Contributions to Mineralogy and Petrology* 31(2), 121–131.

Swamy, V., Saxena, S.K., Sundman, B., Zhang, J., 1994. A thermodynamic assessment of silica phase diagram. *Journal of Geophysical Research: Solid Earth* 99(B6), 11787–11794.

Ulrych, J., Krmíček, L., Teschner, C., et al., 2018. Chemistry and Sr–Nd isotope signature of amphiboles of the magnesio-hastingsite–pargasite–kaersutite series in Cenozoic volcanic rocks: insight into lithospheric mantle beneath the Bohemian Massif. *Lithos* 312, 308–321.

VanPeteghem, J.K., Burley, B.J., 1963. Studies on solid solution between sodalite, nosean and hauyne. *The Canadian Mineralogist* 7(5), 808–813.

Verkouteren, J., Wylie, A, 2000. The tremolite-actinolite-ferro–actinolite series: systematic relationships among cell parameters, composition, optical properties, and habit, and evidence of discontinuities. *American Mineralogist* 85, 1239–1254.

Watson, E.B., Wark, D.A., Thomas, J.B. 2006. Crystallization thermometers for zircon and rutile. *Contributions to Mineralogy and Petrology* 151, 413–433.

Whitney, D.L., Evans B.W., 2010. Abbreviations for names of rock-forming minerals. *American Mineralogist* 95, 185–187.

Williams, M.L., Jercinovic, M.J., Harlov, D.E., Budzyń, B., Hetherington, C.J., 2011. Resetting monazite ages during fluid-related alteration. *Chemical Geology* 283 (3–4), 218–225.

Winter, D.J., 2010. Thermodynamics of metamorphic reactions, in: *Geothermobarometry*, Prentice Hall, pp. 543–556.

Wu, C-M., 2019. Original calibration of a garnet geobarometer in metapelite. *Minerals* 9, 540.

2

IGNEOUS ROCKS

2.1 Introduction

Textures of Igneous Rocks

In an igneous rock it is the geometric relationship between its components (minerals and/or glass) that defines its texture, which is primarily a function of cooling rate and results in distinctive plutonic and volcanic differences. Igneous textures are described in relation to crystallinity, granularity, shape, and spatial relationships. Our images are selected to display particular textural features, but this does not preclude such textures being present in other minerals or rock types. It should also be noted that thin-sections are two-dimensional and a rock with a fabric may require thin-sections of different orientation to reveal the three-dimensional nature of its textural relationships.

Naming Igneous Rocks

Building on the knowledge of minerals gained from the previous chapter, we can begin recognizing and classifying rocks. In general, we follow the International Union of Geological Sciences (IUGS) recommendations for naming rocks. This method uses the mode – that is, the visually estimated abundance of quartz (Q) *or* feldspathoid (F), alkali feldspar (A), and plagioclase (P); these form the basis of the QAPF diagram (after Streckeisen, 1976). To use the IUGS QAPF scheme for plutonic rocks (Fig. 2.1), it is necessary to first determine whether quartz or feldspathoid (commonly shortened to *foid*) is present. If quartz is present, the rock is silica-saturated to silica-oversaturated and is named using the QAP-half of the diagram. If a feldspathoid mineral(s) is present, the rock is silica-undersaturated and named using the FAP-half of the diagram. Once the three mineral (QAP or FAP) modes are visually estimated (in the field or if the sample is not too coarse-grained using a thin-section), the relative abundance of the mineral mode is normalized to 100% and plotted on the ternary diagram to determine the rock name.

A number of mafic rocks plot in the plagioclase apex of the QAP diagram (Fig. 2.1), including anorthosite, diorite, gabbronorite, gabbro, norite, troctolite, and their hornblende-bearing equivalents. The IUGS modal classifications for mafic and ultramafic rocks (Fig. 2.2) are similar, but based on a more appropriate modal mineralogy for these rocks (e.g. olivine, pyroxene, and plagioclase).

The mode is less useful for volcanic rocks which are either fine-grained (*aphanitic*) or porphyritic (with *phenocrysts* in a fine-grained groundmass). There is an IUGS QAPF scheme for naming volcanic rocks (Fig. 2.3), but it is not always possible to accurately identify small individual minerals. Another way to name volcanic rocks is based on chemistry, primarily silica and the alkali elements Na and K – this is known as the *total alkali versus silica* (TAS) diagram (after Le Bas et al., 1986). It too has pitfalls, since the chemical composition of a rock is not always known. Our preference, therefore, is to combine several characteristics for the naming of volcanic rocks; these include the most commonly associated phenocryst phases (particularly quartz versus feldspathoid), the color index (*melanocratic*, *mesocratic*, or *leucocratic*), and the TAS diagram (Fig. 2.4). The mineralogy of a rock naturally relates to its bulk chemistry, and primary phenocryst phases are likely to reflect the degree of silica saturation and alkalinity. The TAS diagram (Fig. 2.4) identifies three broad compositional groups of volcanic rocks: the low-, medium-, and high-alkali suites. The low-alkali group includes basalt to rhyolite, the medium-alkali group includes the trachytic suite, and the high-alkali group includes the tephrite to phonolite suite.

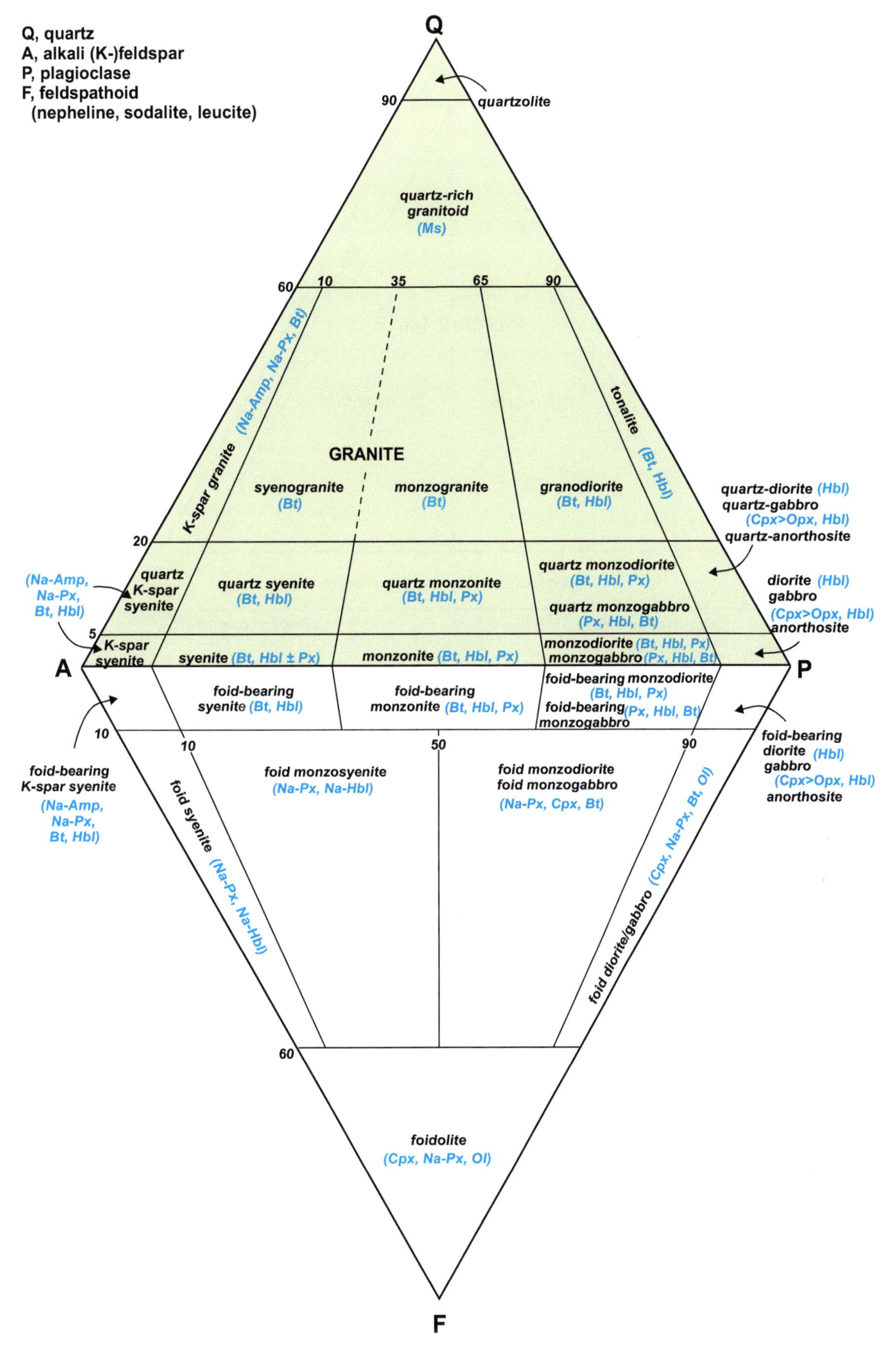

Figure 2.1 IUGS QAPF diagram for naming plutonic rocks. Silica-saturated and -oversaturated rocks are defined by the QAP portion, while silica-undersaturated rocks lie in the FAP portion. We have added common accompanying minerals (abbreviations in blue and defined in Table 1.2) which are not part of the IUGS classification scheme. Na, sodic.

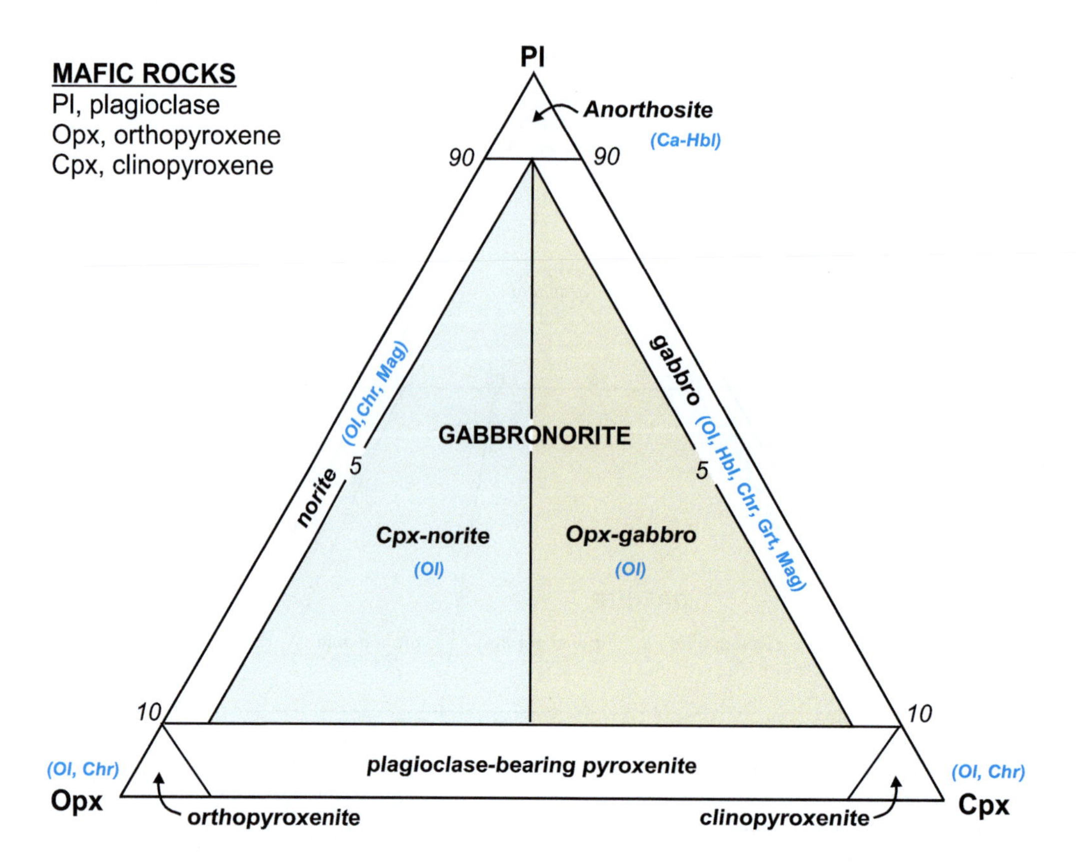

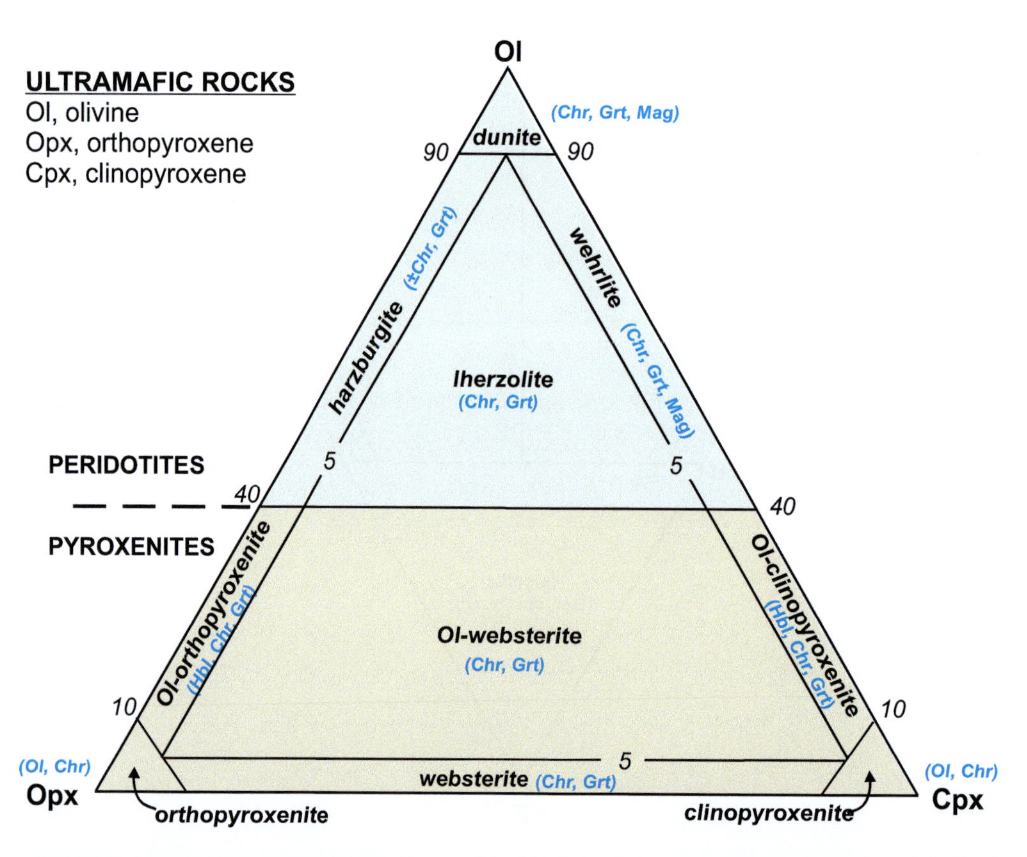

Figure 2.2 IUGS diagrams for the naming of mafic and ultramafic rocks. Essential minerals used for the mode are shown in the legends. We have added common accompanying minerals (abbreviations in blue and defined in Table 1.2) which are not part of the IUGS classification scheme.

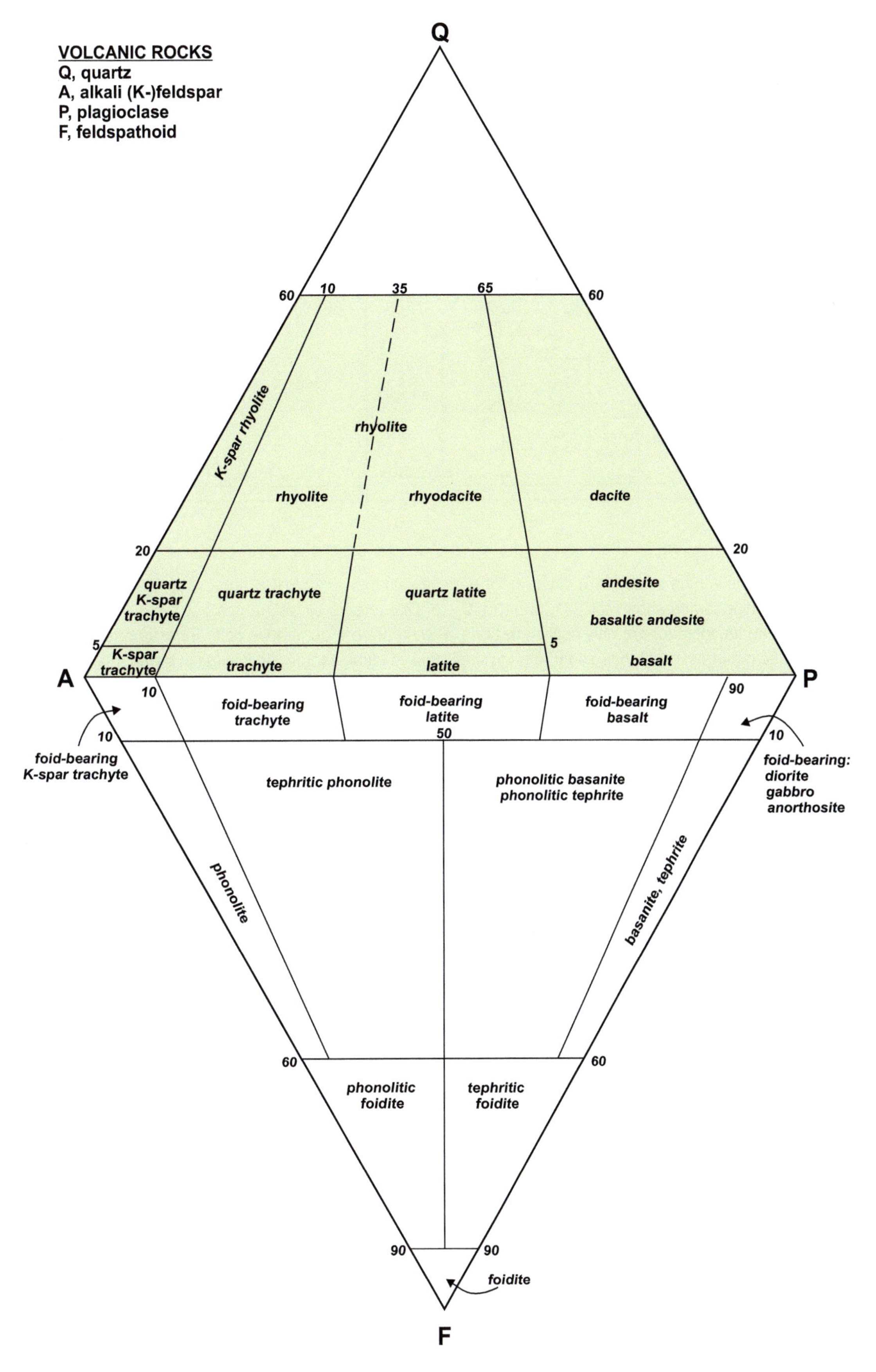

Figure 2.3 IUGS QAPF diagram for naming volcanic rocks. Silica-saturated and -oversaturated rocks are defined by the QAP portion, while silica-undersaturated rocks lie in the FAP portion. Note that it is difficult to use this scheme with aphanitic and porphyritic volcanic rocks.

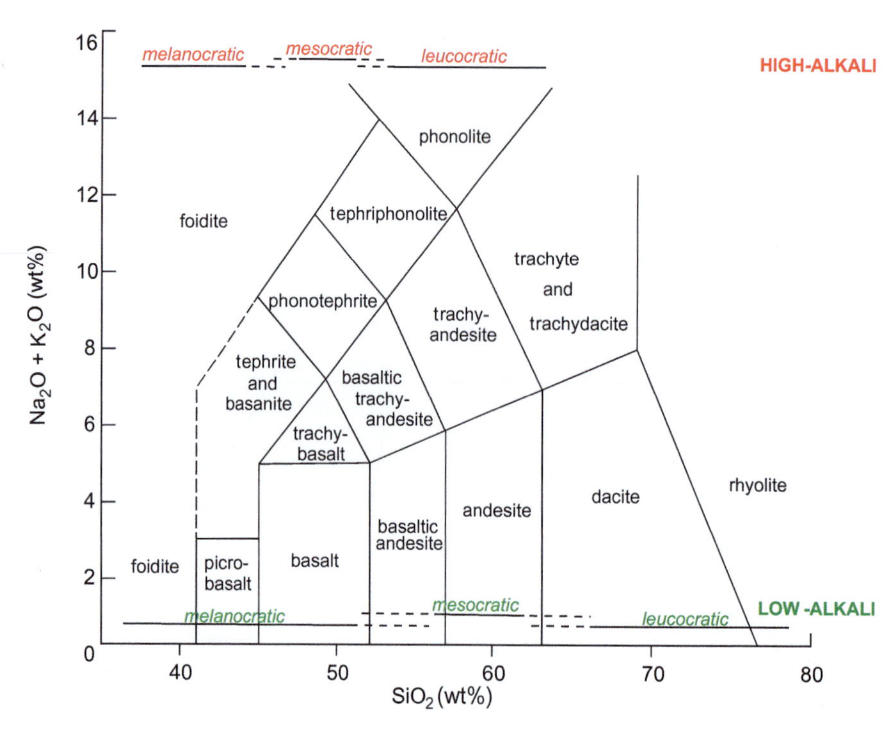

Figure 2.4 Total alkalis versus silica (TAS) diagram for naming volcanic rocks (IUGS approved; TAS boundaries after Le Bas et al., 1986). Rock names shown for SiO_2 and ($Na_2O + K_2O$) contents by weight percent (wt%). The color indices (melano-, meso-, and leucocratic) are only approximations. The somewhat qualitative color index is based on the abundance of mafic minerals, which has some relation to silica and alkali content (low alkali in green, higher alkali in red).

2.2 Crystallinity

Igneous rocks may be composed entirely of crystals, partly crystals and partly glass, or entirely glass. The degree of crystallinity is determined by the modal proportion of crystals and glass, which ranges between 0 and 100%. Based on the degree of crystallinity, igneous rocks are grouped as shown in Figure 2.5.

Figure 2.5

A *Holocrystalline*. This defines a rock composed entirely of crystals and indicates that cooling was sufficiently slow to allow complete crystallization to occur. Holocrystalline textures are typical of plutonic rocks. This holocrystalline gabbro is from the Bushveld complex (South Africa). The large pyroxene crystal (center, golden color) is surrounded by smaller plagioclase crystals with characteristic polysynthetic twinning. Plane-polarized light, 2× magnification, field of view = 7 mm.

B *Hypocrystalline/hypohaline*. A rock composed of both crystals and glass. It often indicates a period of relatively slow cooling followed by quenching of the remaining magma. This texture is mainly observed in rocks which are crystallized at shallow and intermediate (hypabyssal) depths. This hypocrystalline rhyolite from San Vincenzo (Italy) shows euhedral plagioclase crystals and small quartz phenocrysts, surrounded by a glassy (isotropic) groundmass. Cross-polarized light, 10× magnification, field of view = 2 mm.

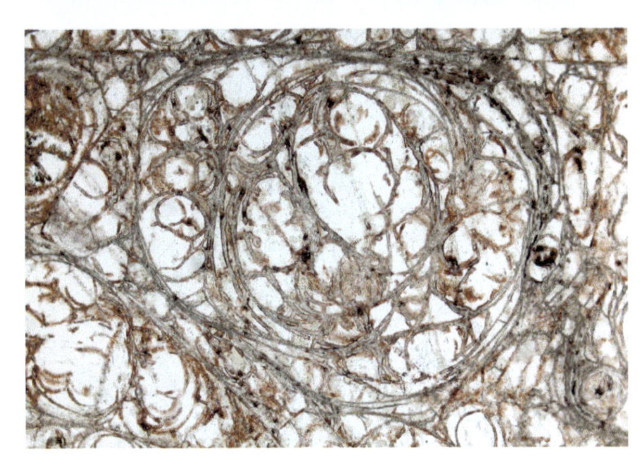

C *Holohaline*. Rocks composed entirely of glass are holohaline. No crystals are visible even with magnification, which indicates very rapid cooling. This texture is mostly seen in volcanic rocks (e.g. obsidian or pitchstone). This obsidian from Buschbad (Germany) has concentric "perlitic" cracks (fresh glass is green and devitrified glass brown). Such rocks are also known as "perlites." Plane-polarized light, 10× magnification, field of view = 2 mm.

2.3 Granularity

Granularity refers to the size of crystals in igneous rocks and describes what can be seen with the "naked eye" – that is, unaided by a magnifying lens or a microscope. When crystals are large enough to be recognized, they are *phaneritic*. Phaneritic textures are specifically defined according to crystal size, and terms such as *equigranular* or seriate are also applied. When crystals are small, or perhaps hardly visible, terms that describe the *aphanitic* (fine-grained) family of textures are useful.

Coarse-grained → **Fine-grained**

phaneritic ***aphanitic***

Phaneritic

Phaneritic texture is classified according to crystal size and is the result of slow cooling, which allows crystals to grow large. Hypabyssal intrusions like dikes and sills, for example, typically have a fine-grained phaneritic texture. Medium-grained phaneritic texture is typical of most plutonic rocks such as granites. Intrusions formed deeper in the crust may generate coarse-grained phaneritic texture, but very coarse-grained phaneritic texture is usually restricted to pegmatites.

Phaneritic texture	Grain size
Fine-grained	<1 mm
Medium-grained	1–5 mm
Coarse-grained	5–10 mm
Very coarse-grained	>10 mm

Equigranular

Equigranular means the minerals in a crystalline rock are broadly equal in size. The rock has a granular aspect, in both hand specimen and thin-section. Equigranular texture occurs in a wide range of igneous and metamorphic rocks, but is particularly common in plutonic rocks.

Figure 2.6

A Phaneritic granite, Elba (Italy). Large orthoclase megacrysts (visible to the eye) set in a coarse-grained holocrystalline matrix. Coin ~2.5 cm diameter.

B Phaneritic hornblende-rich gabbro from the Mattoni intrusion, Adamello (Italy). Large black prismatic hornblende and white to beige plagioclase groundmass. Image is ~5 cm wide.

C Medium-grained granite from unknown location (Antarctica). Microcline (with characteristic tartan twins) and interstitial quartz. Cross-polarized light, 10× magnification, field of view = 2 mm.

D Coarse-grained gabbro from Adamello (Italy). Large, euhedral hornblende (partially altered by secondary amphibole) is surrounded by finer-grained interstitial plagioclase. Cross-polarized light, 10× magnification, field of view = 2 mm.

Figure 2.7

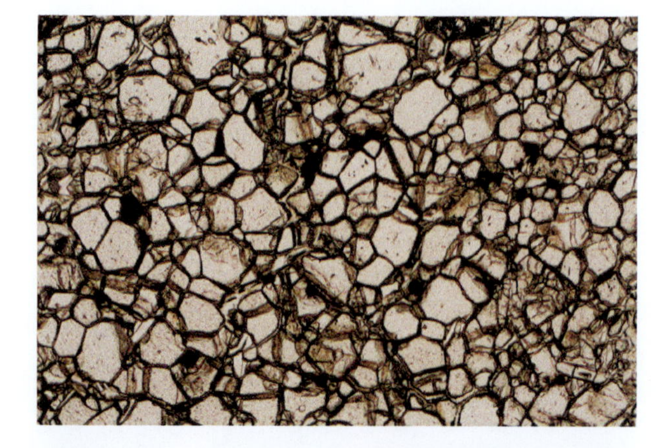

A Equigranular dunite from Perledo (Italy). Olivine crystals of uniform size and with straight grain boundaries that intersect at 120° angles. Olivine crystals are cross-cut by some serpentine veins. Plane-polarized light, 10× magnification, field of view = 2 mm.

B Equigranular dunite from Perledo (Italy). Same image as (A) but with crossed polarizers. Note typical birefringence and the mosaic (120° crystal boundaries) texture. Cross-polarized light, 10× magnification, field of view = 2 mm.

C Equigranular pyroxenite from the Bushveld complex (South Africa). Pyroxene crystals of uniform size and straight grain boundaries that intersect with 120° angles define a mosaic texture. Plane-polarized light, 2× magnification, field of view = 7 mm.

D Equigranular pyroxenite from the Bushveld complex (South Africa). Same image as (C) but with crossed polarizers. Note the typical birefringence and common twinning. Cross-polarized light, 2× magnification, field of view = 7 mm.

BOX 2.1 Nucleation and Crystal Growth

The size and number of crystals that form as a magma cools are a function of time. Crystal nucleation and growth are controlled by diffusion and the degree of undercooling; a longer time in the liquid allows ions to form nucleation sites to which additional ions bond (crystal growth). In this graph, the red curve represents crystal growth and the blue curve nucleation. Imagine a mafic magma cooling below its melting point. For a low degree of undercooling (from the melting point to the temperature T_a), crystal growth at limited nucleation sites generates a coarse-grained gabbro. Increasing the degree of undercooling (to T_b) creates more nucleation sites, allows less time for crystals to grow, and produces a texturally homogeneous basalt. For extremely rapid undercooling (supercooling; from the melting point to T_c), there is no time for nucleation sites to form or for crystals to grow, resulting in a glassy basalt.

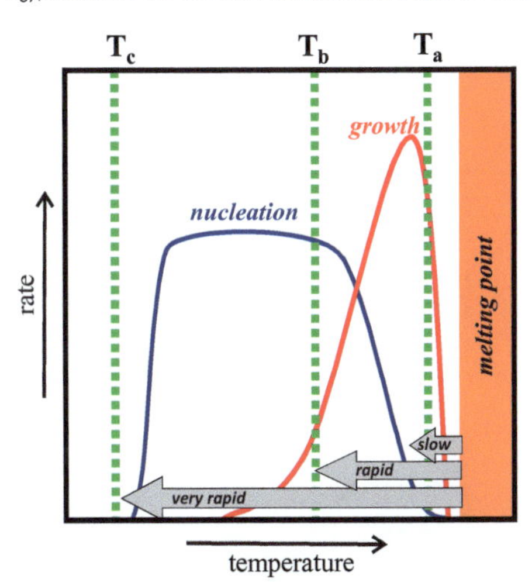

Porphyritic rocks imply a two-fold cooling history. An "early" period of slow cooling to generate large phenocrysts, followed by a later period of rapid cooling to generate the fine-grained groundmass. The former occurs deeper in the magmatic plumbing system, while the latter occurs during rapid ascent and/or eruption.

This image of olivine basalt from Etna (Italy) reflects a dual cooling history – phenocrysts of clinopyroxene (center) and feldspar (clear crystals) grew during an initial period of slower cooling, followed by rapid cooling and numerous nucleation sites, generating the microcrystalline groundmass. Plane-polarized light, 2× magnification, field of view = 7 mm.

This image shows gabbro from the Duluth Complex (Minnesota, United States). It cooled slowly, resulting in large crystals of olivine (rounded with high birefringence) and plagioclase (gray and twinned). Plane-polarized light, 2× magnification, field of view = 7 mm.

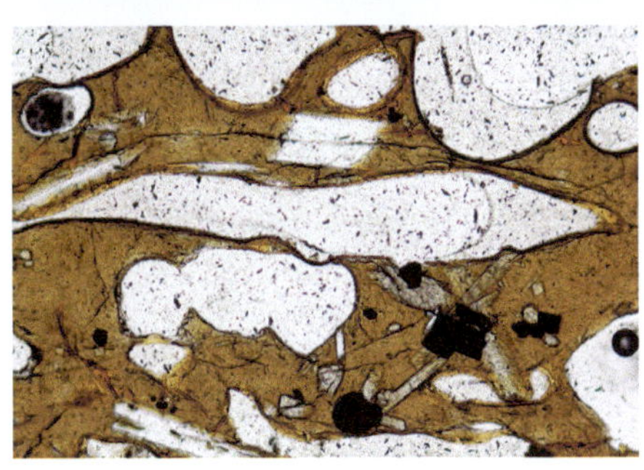

This image is mostly basaltic glass from Acicastello (Italy). The glass (brown) indicates supercooling – no time for crystal nucleation or growth. Plane-polarized light, 2× magnification, field of view = 7 mm.

Aphanitic

Aphanitic texture describes any rock in which, excluding phenocrysts, crystals are not visible without the aid of a magnifying lens.

Microcrystalline Crystals which can only be identified in a thin-section using a petrographic microscope.

Microlites Small crystals (<0.01 mm) only just large enough to show birefringence under crossed polarizers.

Figure 2.8

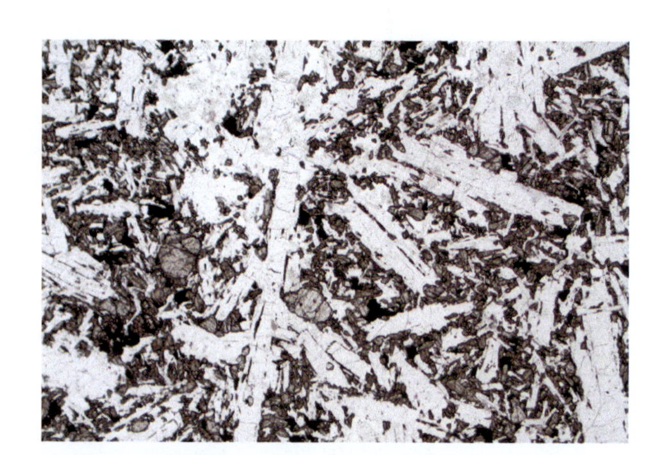

A Microcrystalline basalt from Etna (Italy). Elongate plagioclase crystals (colorless) surrounded by a fine-grained groundmass rich in augite (brown), olivine (high relief), small plagioclase crystals, and black magnetite. Plane-polarized light, 2× magnification, field of view = 7 mm.

B Microcrystalline basalt from Etna (Italy). Same image as (A) but with crossed polarizers. Plagioclase crystals show characteristic polysynthetic twinning, augite has moderate birefringence, and olivine has higher birefringence. Cross-polarized light, 2× magnification, field of view = 7 mm.

Crystallites Crystals that are too small to show polarization colors.

Cryptocrystalline The crystals are too small to be distinguished even with a microscope.

Aphyric Aphanitic rocks which lack phenocrysts.

Felsitic Siliceous rocks with an almost cryptocrystalline aggregate of quartz and alkali feldspar.

Figure 2.8 ***(cont.)***

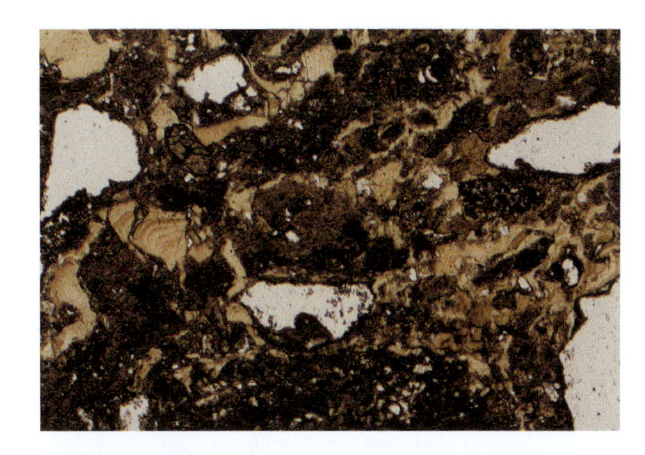

C Cryptocrystalline rhyolite from Tolfa (Italy). The rock is made of welded glass (brown material). No crystals are present except for a single pyroxene (upper-left quadrant). There are numerous cavities and vesicles (white areas). Plane-polarized light, 2× magnification, field of view = 7 mm.

D Cryptocrystalline rhyolite from Tolfa (Italy). Volcanic glass is isotropic and only small crystals are visible. The crystals with high birefringence (blue) are pyroxene and the other crystals are quartz. Cross-polarized light, 2× magnification, field of view = 7 mm.

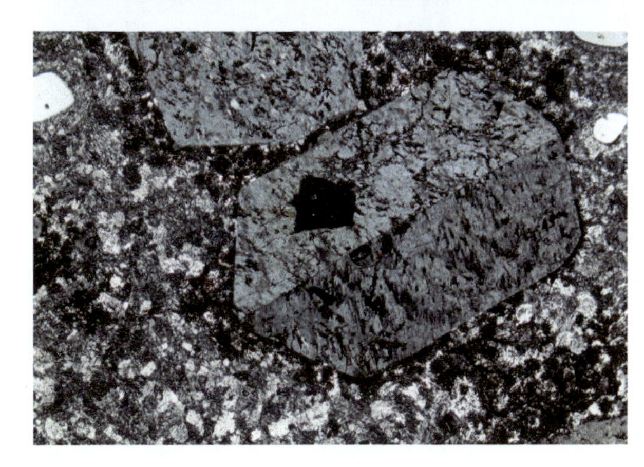

E Felsitic rhyolite from unknown location (Antarctica). Note the microcrystalline matrix of quartz and potassium feldspar hosting twinned sanidine phenocrysts. Cross-polarized light, 10× magnification, field of view = 2 mm.

BOX 2.2 Textures Related to Rapid Cooling

We have established that crystal growth is controlled by diffusion and the degree of undercooling. The growth rate of a crystal depends on the surface energy of the crystal faces and the diffusion rate of the species involved. If the cooling rate is constant, the species with the greatest diffusion (small ions with low ionic charge) will grow the largest crystals. Diffusion is faster at higher temperature and in lower viscosity melts.

Mineral textures therefore provide evidence of a crystal's relative cooling history. Textures documenting relatively rapid cooling (i.e. rapid growth and a lack of nucleation sites) include sieve, skeletal, swallow-tail, dendritic, and spinifex textures.

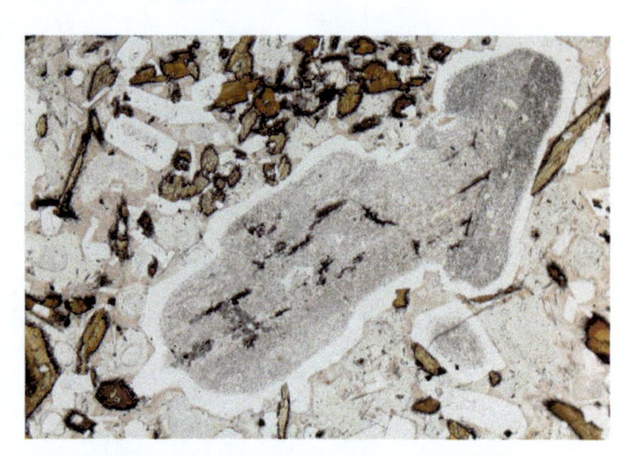

Sieve textures are the result of advanced resorption or crystal growth due to undercooling and document disequilibrium conditions. This produces crystals with internal cavities that may be crystallographically controlled. The gray region of the large, central plagioclase in this andesite from Chile is a wide zone of sieve texture. Plane-polarized light, magnification 10×, field of view = 2 mm.

Skeletal textures reflect rapid crystallization with atoms bonding more quickly to specific sites (edges and corners rather than the centers of crystal faces). This results in *branched*, tree-like forms (*dendritic*), or hollow, stepped depressions. The adjacent gray skeletal olivine in komatiite from Canada lacks a core, indicating fast growth. Plane-polarized light, magnification 10×, field of view = 2 mm.

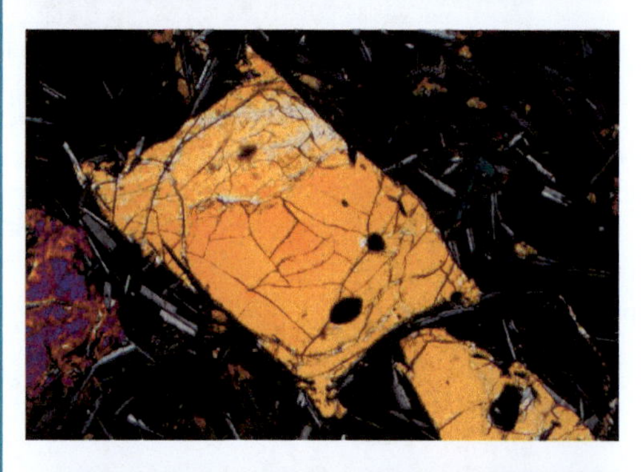

Swallow-tail textures also reflect rapid cooling and are more commonly associated with plagioclase, but this olivine in a basalt from Iceland has small tails at its corners due to preferential growth at these locations. Crossed polarized light, magnification 10×, field of view = 2 mm.

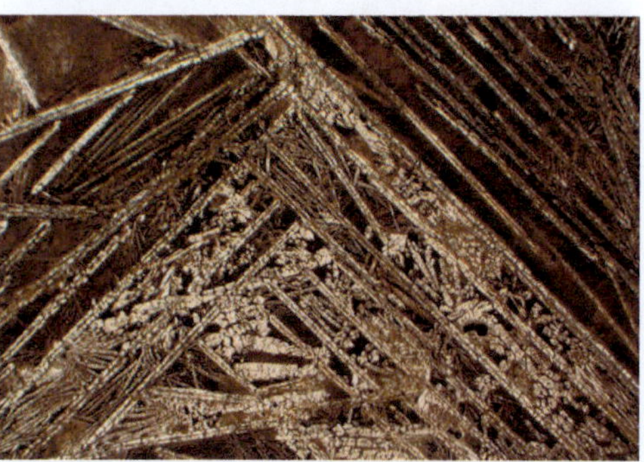

Spinifex textures represents simple skeletal, platy, or acicular arrangements. They are typical of olivine and/or pyroxene crystals in ultramafic lavas such as Precambrian komatiites, and the result of rapid cooling and rapid crystal growth in a low-viscosity melt. This image of a komatiite from Zimbabwe shows spinifex texture. Plane-polarized light, magnification 10×, field of view = 2 mm.

2.4 Crystal Form

Crystal form is typically considered with respect to the development of crystal faces and crystal shapes. The development of crystal faces may vary widely, from poorly to well-developed. Crystal shapes are considered as either equal or non-equal in dimension, and specific terms are used to describe them.

Crystal Faces

Euhedral (synonyms *idiomorphic* or *automorphic*). Minerals have well-developed crystal faces. Experiments have shown that crystals with simple structure nucleate more easily than those with more complex structure, and the formation of well-defined crystal faces decreases as Si–O polymerization increases. This explains the fact that minerals like the oxides (ilmenite and magnetite) and olivine generally nucleate more easily than minerals with more complex structures such as pyroxene or feldspar.

Subhedral (synonyms *hypidiomorphic* or *hypautomorphic*). A morphological term referring to minerals with less-developed or partly developed crystal faces, but which nevertheless show the general characteristic shape of that mineral.

Anhedral (synonyms *allotriomorphic* or *xenomorphic*). A morphological term used for a mineral lacking any of its characteristic faces.

Crystal Shapes

Equidimensional. This describes crystals that have nearly the same length, width, and height. Minerals of the cubic system, such as garnet, spinel, or leucite, have equidimensional shapes.

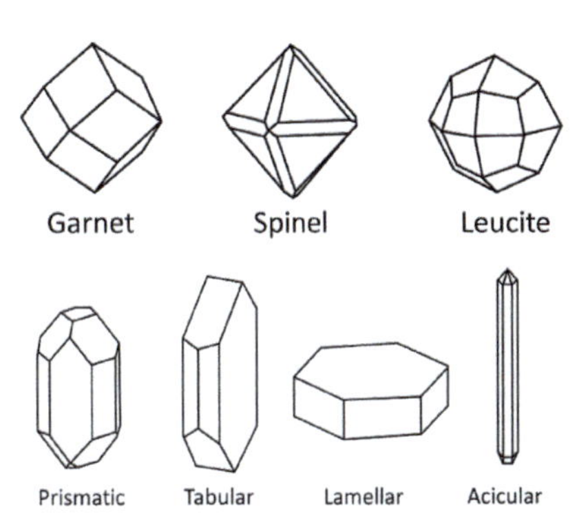

Non-equidimensional. This describes crystals that do not have uniform dimensions. These include minerals with *prismatic*, *tabular*, *lamellar*, and *acicular* shapes.

Prismatic minerals have preferred growth in one direction. They may have 3, 4, 6, 8, or 12 faces parallel to a crystallographic axis (commonly the c-axis) and produce diversely shaped cross-sections. Prismatic shapes are common in tetragonal, trigonal, and hexagonal minerals such as pyroxene, amphiboles, apatite, zircon, and corundum.

Tabular minerals are longer than they are wide and therefore appear slab-like. Tabular shapes are common in monoclinic and triclinic minerals such as orthoclase.

Lamellar (or platy) minerals are flat (longer and wider in relation to their height). Lamellar shapes are common in phyllosilicate minerals like biotite.

Acicular minerals are characterized by pronounced elongation in one direction. Acicular shapes are common in zeolites, rutile, sillimanite, tourmaline, and some amphiboles.

Figure 2.9

A Basalt from Etna (Italy). Euhedral olivine with cleavage and black magnetite inclusions, set in a glassy isotropic groundmass. Cross-polarized light, 10× magnification, field of view = 2 mm.

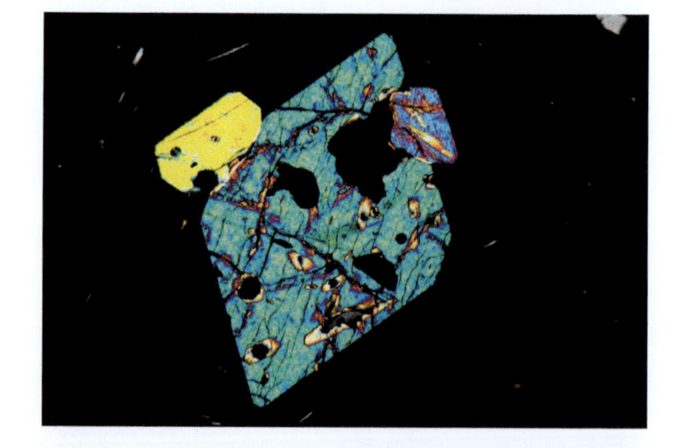

B Basalt from Vulcano (Italy). Euhedral pyroxene (basal section with eight sides) and characteristic cleavage, set in a fine-grained plagioclase-rich groundmass. Plane-polarized light, 10× magnification, field of view = 2 mm.

C Basalt from Alicudi (Italy). Fractured subhedral olivine in an intergranular groundmass. Cross-polarized light, 2× magnification, field of view = 7 mm.

D Basalt from Iblei (Italy). Large subhedral pyroxene with sieve texture and low birefringence (central grain), in an olivine-rich groundmass. Olivine has high birefringence. Cross-polarized light, 2× magnification, field of view = 7 mm.

E Prismatic aegirine in Brandberg granite (Namibia). Euhedral, highly birefringent aegirine crystal with prismatic habit in a fine-grained quartz and feldspar-rich groundmass. Cross-polarized light, 2× magnification, field of view = 7 mm.

F Tabular orthoclase in syenite from Biella (Italy). Subhedral, concentrically zoned orthoclase with a tabular shape, Carlsbad twinning, and numerous small plagioclase inclusions; surrounded by smaller orthoclase and biotite crystals. Cross-polarized light, 2× magnification, field of view = 7 mm.

G Lamellar biotite in rhyolite from val Trompia (Italy). Platy biotite crystals with rectangular habit (110 section) surrounded by devitrified glassy groundmass and irregular spherulites of quartz and feldspar. Cross-polarized light, 10× magnification, field of view = 2 mm.

H Acicular tourmaline in granite from Luxulyan (England). Radiating aggregate of acicular, fibrous tourmaline with quartz (colorless). Plane-polarized light, 2× magnification, field of view = 7 mm.

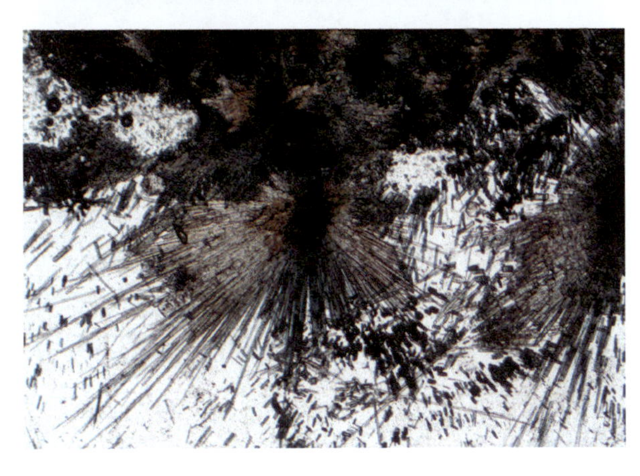

BOX 2.3 Embayment and Resorption

A change in the physical conditions of a crystal melt system can shift crystal growth from stability (equilibrium) to instability (disequilibrium). For example, mafic magma recharge or decompression during rapid magmatic ascent has the potential to increase the temperature of the melt or change its chemical composition and thus initiate magmatic corrosion (resorption, dissolution) of existing crystals. This interpretation is supported by (1) rounded (resorbed) corners of crystals, (2) truncated compositional zoning by an embayment, and (3) new minerals being deposited on the surface of the embayment due to mineral/melt reactions. Such textures are commonly associated with magma mixing/mingling.

Note that while skeletal growth may resemble "embayment," the formation of the latter typically occurs at the crystal–melt boundary while the former is linked to the crystal structure.

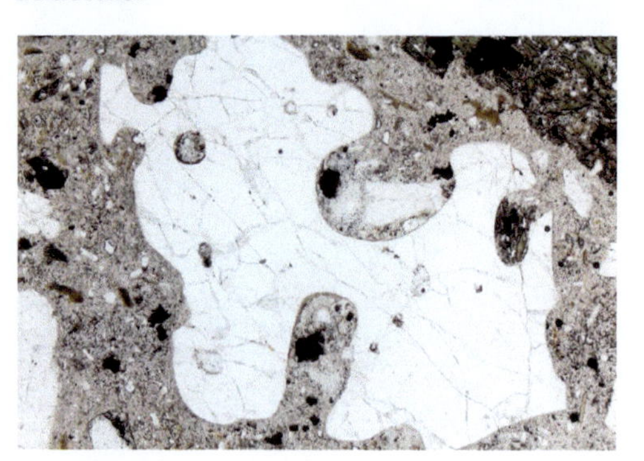

The large, central quartz crystal has a highly curvilinear and rounded outline – large "bays" into the crystal (i.e. *embayments*). This example is a volcanic rock from Elba island (Italy). Plane-polarized light, magnification = 2×, field of view = 2 mm.

This pyroxene phenocryst (in basalt from Etna, Italy) shows *resorption* in its central region. Note also the overgrowth on the edge of the crystal (thin dark rim on the left and thin lighter rim on the right). This rim has grown on an existing crystal with rounded "corners" – more evidence of resorption. Such disequilibrium reactions are often associated with an increase in temperature. Cross-polarized light, magnification = 2×, field of view = 2 mm.

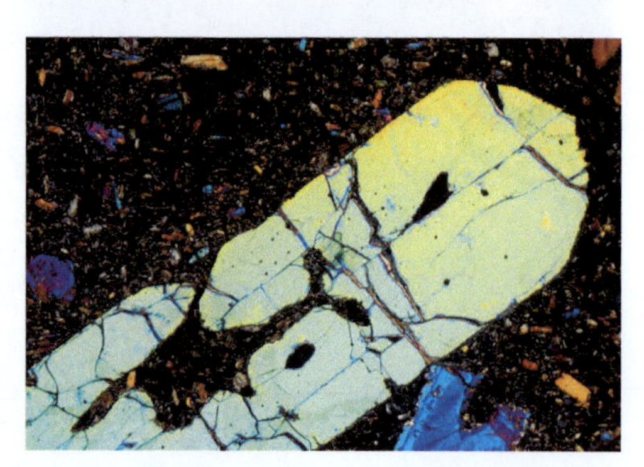

This resorbed olivine phenocryst in a basalt from Auerbach (Germany) has also lost material from its central portion, reflecting chemical or thermal disequilibrium. Cross-polarized light, magnification = 2×, field of view = 2 mm.

2.5 Mutual Relations of Crystals

Granular Textures

Granular textures refers to phaneritic rocks in which individual mineral crystals are visible and have the same general size. Three types are recognized based on the crystal arrangement and shapes:

Equigranular textures	Synonyms
Euhedral granular	Panidiomorphic
Subhedral granular	Hypidiomorphic
Anhedral granular or "granular"	Allotriomorphic or xenomorphic

Figure 2.10

A *Euhedral granular (panidiomorphic)* applies to rocks composed almost entirely of euhedral (well-formed) crystals. Rocks with this texture are extremely rare because most rocks are composed of a mixture of euhedral and subhedral minerals. This euhedral hornblendite is made mostly of euhedral hornblende crystals, surrounded by plagioclase (colorless). Plane-polarized light, 2× magnification, field of view = 7 mm.

B *Subhedral granular (hypidiomorphic)* applies to rocks composed mostly of subhedral (partly developed) crystals. Subhedral granular is the most common equigranular texture and is sometimes called *"granitic texture"* because it is common in granitic rocks. This hypidiomorphic anorthosite from the Rhum complex (Scotland) shows subhedral plagioclase with interstitial clinopyroxene (orange interference colors). Cross-polarized light, 2× magnification, field of view = 7 mm.

C *Anhedral granular (allotriomorphic)* texture defines rocks composed almost entirely of anhedral crystals (also called *"aplitic"* as it is common in aplites, the hypabyssal equivalent of granite). This anorthosite from Rogaland (Norway) is made of anhedral plagioclase (grays) and anhedral orthopyroxene (brownish) crystals. Cross-polarized light, 10× magnification, field of view = 2 mm.

Inequigranular Textures

Inequigranular is a general textural term to describe a large variation in crystal size. There are additional terms that describe specific grain size variations, such as *seriate, porphyritic, glomeroporphyritic, poikilitic, ophitic, subophitic, intersertal,* and *intergranular* textures. It is very common to find rocks that have more than one of these textures, in which case multiple terms can be used to describe the rock.

Seriate describes a rock with crystals of different sizes.

Porphyritic describes larger crystals (phenocrysts) surrounded by a finer-grained groundmass. If the phenocrysts are enclosed by glassy groundmass, the texture is called *vitrophyric.*

Glomeroporphyritic is a variety of porphyritic texture characterized by distinct clusters of individual early-formed crystals such as olivine, pyroxene, and plagioclase.

Figure 2.11

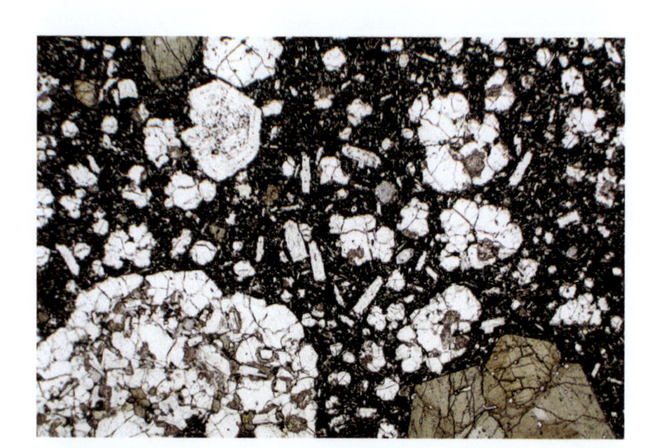

A Seriate tephrite from Bolsena Lake (Italy). Leucite (rounded and colorless), pyroxene (brown), and plagioclase (rectangular and colorless) are set in a dark, glassy groundmass. Leucite crystals show a range in size from <0.01 mm in the groundmass to >3 mm phenocrysts. Plane-polarized light, 2× magnification, field of view = 7 mm.

B Porphyritic basalt from unknown location (Iceland). Euhedral, twinned augite phenocrysts set in intergranular groundmass of olivine and plagioclase microphenocrysts. Cross-polarized light, 2× magnification, field of view = 7 mm.

C Porphyritic, seriate basalt from Ustica (Italy). Plagioclase (gray with polysynthetic twins), olivine (high birefringence), and augite (pale brown) surrounded by a finer-grained groundmass of the same mineral phases. The large olivine (center-bottom) has low-interference colors as it is cut close to the optical axis. Cross-polarized light, 2× magnification, field of view = 7 mm.

Figure 2.11 ***(cont.)***

D Seriate granite from Elba (Italy). Plagioclase and quartz both show a wide range in grain size, 0.5–3 mm (plagioclase) and 0.1–1 mm (quartz). Cross-polarized light, 2× magnification, field of view = 7 mm.

E Glomeroporphyritic basalt from Ventotene (Italy). Clusters of plagioclase crystals stick together to form clots, set in a finer-grained, intergranular groundmass. Cross-polarized light, 2× magnification, field of view = 7 mm.

Poikilitic texture is characterized by relatively large, optically continuous, but poorly formed crystals enclosing smaller, euhedral crystals of one or more type. The development of this texture is due to the simultaneous growth of crystals that have different nucleation rates. The larger "host" crystals are called *oikocrysts* and the smaller, enveloped crystals are known as *chadacrysts*. Some crystals may contain lots of small accessory mineral (apatite, magnetite, rutile, etc.) inclusions; it is possible to mistake these inclusions for the poikilitic texture.

increased crystallization with cooling

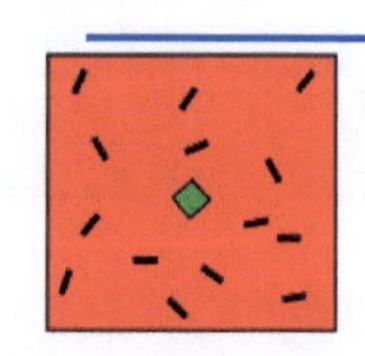

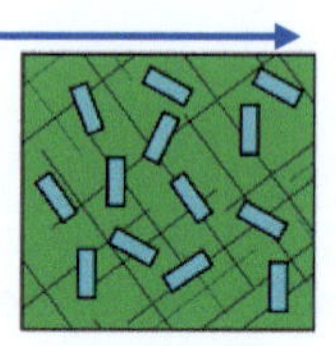

A poikilitic texture develops when crystals with different nucleation rates grow simultaneously. Pyroxene (green) has a slower nucleation rate than plagioclase (blue); as crystallization proceeds, numerous plagioclase crystals form and are incorporated while pyroxene continues to grow.

Intersertal Texture

Intersertal texture is one in which the angular interstices between plagioclase crystals are occupied wholly or partially by glass. This texture applies whether the glass is fresh or altered to chlorite, serpentine, analcime, or palagonite. If the glass surrounds the microphenocrysts, the texture is called *hyalo-ophitic*.

Intergranular Texture

Intergranular texture is one in which the angular interstices between plagioclase grains are occupied by seriate grains of pyroxene (± olivine and iron–titanium oxides) and glass is rare. Unlike ophitic texture, the interstices of intergranular texture are not in optical continuity.

Ophitic and subophitic textures are varieties of poikilitic texture. Ophitic texture describes large crystals that wholly or partly enclose smaller crystals. This is commonly lath-shaped plagioclase enclosed by large augite crystals. This texture is typical of basalts and hypabyssal rocks (like dolerite, and sometimes called "*doleritic texture*"). The smaller enclosed crystals are known as *chadacrysts* and the larger crystals are known as *oikocrysts*. Rocks with many smaller *oikocrysts* have a patchy appearance and the texture is called "*ophimottled*".

Ophitic texture has been used to establish a crystallization sequence in which pyroxene formed after plagioclase. This is generally true, but in many ultramafic complexes (such as the Skaergaard) this texture has been interpreted to reflect the simultaneous crystallization of pyroxene and plagioclase.

Subophitic implies that augite only partially encloses plagioclase laths.

Figure 2.12

A Poikilitic anorthosite, Rhum complex (Scotland). Euhedral plagioclase chadacrysts enclosed by a large, optically continuous, augite oikocryst. Cross-polarized light, 2× magnification, field of view = 7 mm.

B Poikilitic peridotite from the Eastern Layered Suite, Rhum complex (Scotland). Rounded olivine chadacrysts enclosed by plagioclase oikocryst (low birefringence). Olivine has radial cracks due to serpentinization. Cross-polarized light, 10× magnification, field of view = 2 mm.

C Poikilitic wehrlite from unknown location (Cyprus). Rounded olivine chadacryst in a large augite oikocryst (gray birefringence). Olivine has radial fractures due to serpentinization. Cross-polarized light, 10× magnification, field of view = 2 mm.

D Poikilitic gabbro from the Bushveld complex (South Africa). Rounded orthopyroxene chadacrysts in a large, twinned plagioclase oikocryst. Orthopyroxene is fractured due to alteration. Cross-polarized light, 10× magnification, field of view = 2 mm.

Figure 2.13

A Intersertal basalt from Hawaii (United States). Euhedral plagioclase and anhedral augite (brown) surrounded by amorphous, dark brown glass. Augite can be distinguished from glass by its higher relief. Plane-polarized light, 10× magnification, field of view = 2 mm.

B Intersertal basalt from Hawaii (United States). Same image as (A) but with crossed polarizers. The groundmass glass is isotopic and anhedral augite has high birefringence. Cross-polarized light, 10× magnification, field of view = 2 mm.

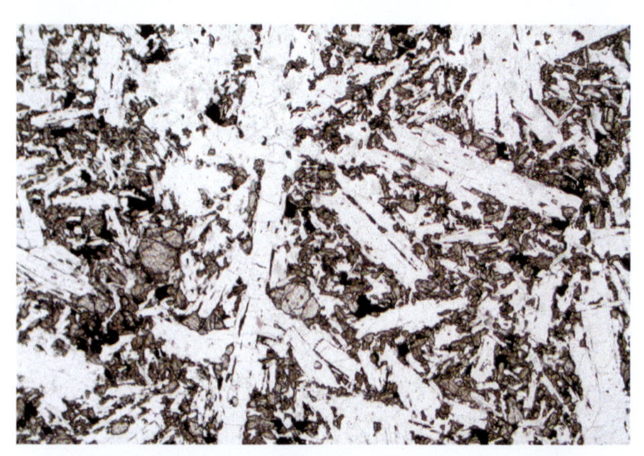

C Intergranular basalt from Alicudi (Italy). Elongate plagioclase (clear) is surrounded by a fine-grained groundmass of olivine and augite. The black crystals are magnetite. Plane-polarized light, 2× magnification, field of view = 7 mm.

D Intergranular basalt from Alicudi (Italy). Same image as (C) but with crossed polarizers. Plagioclase (gray) shows twinning and is surrounded by a seriate groundmass of olivine and augite (red and blue birefringence). Cross-polarized light, 2× magnification, field of view = 7 mm.

Figure 2.14

A Ophitic dolerite from Lessini Mount (Italy). Elongate plagioclase (gray birefringence) enclosed in a large augite (yellow–orange birefringence). Scattered within the augite crystal, small greenish olivine (lower-left quadrant) occurs; the color is due to alteration. Cross-polarized light, 10× magnification, field of view = 2 mm.

B Ophitic dolerite from Torres del Paine area (Argentina). Several large augite crystals define a texture that varies from ophitic to subophitic (some plagioclase crystals along the edges of the augite protrude into the groundmass). Olivine phenocrysts show typical third-order birefringence. Cross-polarized light, 10× magnification, field of view = 2 mm.

C Subophitic dolerite from Torres del Paine area (Argentina). Anhedral augite crystals wholly to partially enclose plagioclase laths. The brownish regions are made of altered glass (palagonite). Cross-polarized light, 10× magnification, field of view = 2 mm.

D Ophimottled dolerite from unknown location (Antarctica). The overall texture is characterized by several large anhedral crystals of augite (yellow birefringence) which give the rock its "speckled" appearance. A finer-grained intergranular groundmass of augite microphenocrysts and plagioclase laths define a magmatic foliation. Cross-polarized light, 2× magnification, field of view = 7 mm.

Oriented Textures

Comb texture (or comb layering) is defined by dendritic crystals, often curved or branching, that are oriented subparallel to each other. Dendritic crystals are commonly pyroxene and/or olivine. Comb layering occurs during crystallization associated with supercooling when crystals nucleate on a planar substrate, such as the walls of a dike. Comb-like structures are also relatively common in the uppermost parts of granitic intrusions with oriented quartz ± feldspar crystals growing downwards into the melt.

Figure 2.15

A Comb layered dolerite from Skye (Scotland). Elongate, branching augite crystals set in a fine-grained glassy groundmass. Plane-polarized light, 2× magnification, field of view = 2 mm.

B Comb layered dolerite from Skye (Scotland). Same image as (A) but with crossed polarizers. Elongate, branching augite crystals show typical high birefringence. Cross-polarized light, 2× magnification, field of view = 2 mm.

C Comb layered dolerite from Lesina (Italy). Elongate kaersutite (red–brown) set in an altered feldspar-rich groundmass. Plane-polarized light, 2× magnification, field of view = 2 mm.

Eutaxitic texture describes the parallel alignment of pumice and glass fragments. It results in the oriented fabric common in many pyroclastic rocks (ignimbrites) with sialic to intermediate composition. It forms after airfall deposits comprising hot lithic, pumice, and crystal fragments are compressed and deformed, and during consolidation of hot pyroclastic flows. Eutaxitic rocks are often rich in *fiamme* and *glass shards*. Fiamme are compressed pumice fragments that may define a deformational fabric or flow direction. Glass shards are the remains of the interstitial glass between gas bubbles that form distinctive cuspate-, spicule-, or Y-shaped structures.

Trachytic and trachytoid textures represent the alignment of elongate or tabular minerals due to flow within a melt and results in a foliated or lineated texture. Magmatic flow may produce streamlines around larger crystals and is often directed around early-formed crystals.

Trachytic texture is defined by the subparallel arrangement of lath-shaped feldspar crystals (plagioclase or sanidine) in a hypocrystalline or holocrystalline groundmass. Although the texture is typical of trachytic rocks, it can occur in other compositions. If there is no glass between the feldspar crystals, the texture is called *pilotaxitic*.

Trachytoid texture is characterized by the subparallel arrangement of crystals in any phaneritic igneous rock, regardless of the rock composition in which it occurs.

Figure 2.16

A Eutaxitic rhyolite from Pantelleria island (Italy). Feldspar (colorless) and green aegirine crystals set in a fluid, glassy groundmass. The lighter region in the center is a flattened pumice fragment. Notice how it is deformed between two feldspar crystals. Plane-polarized light, 2× magnification, field of view = 7 mm.

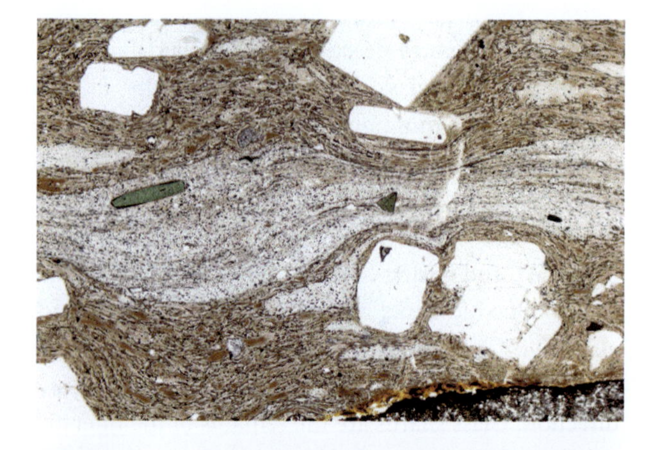

B Eutaxitic rhyolite from Psathoura (Greece). Feldspar (light brown due to sericite alteration) and quartz (clear) phenocrysts are set in a flowing, glassy groundmass with flattened pumice fragments (light-colored portions). Plane-polarized light, 10× magnification, field of view = 2 mm.

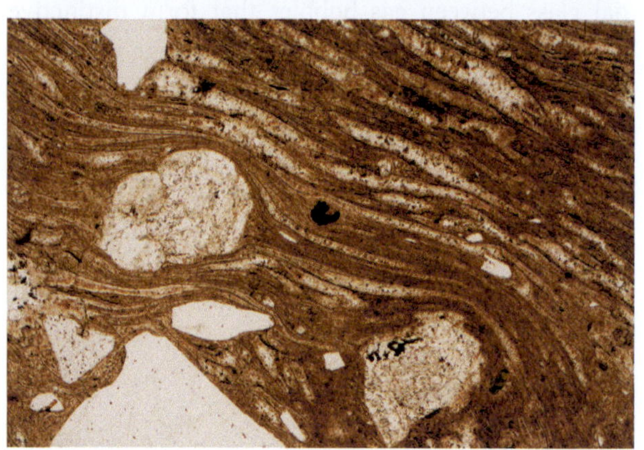

C Fiamme in rhyolite from Pantelleria (Italy). Large fiamme fragment (red–brown) containing flattened gas bubbles, surrounded by a cryptocrystalline groundmass. The colorless crystals are feldspar, and black oxides are also present. Plane-polarized light, 2× magnification, field of view = 7 mm.

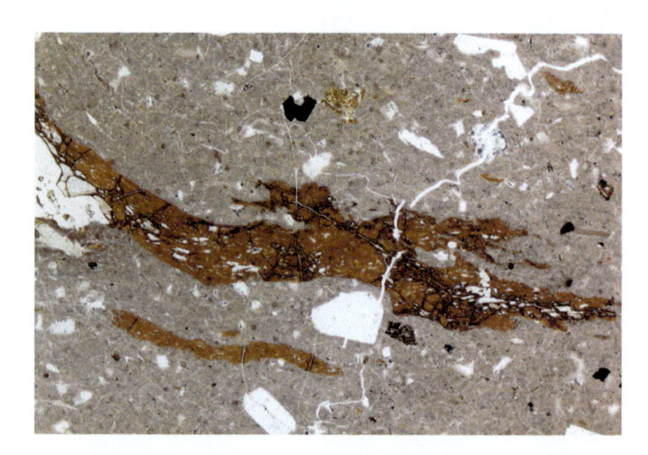

D Glass shards in rhyolite from Sardinia (Italy). Y-shaped glass fragments (colorless) in a devitrified glassy groundmass (brownish). Oxides also present. Plane-polarized light, 10× magnification, field of view = 7 mm.

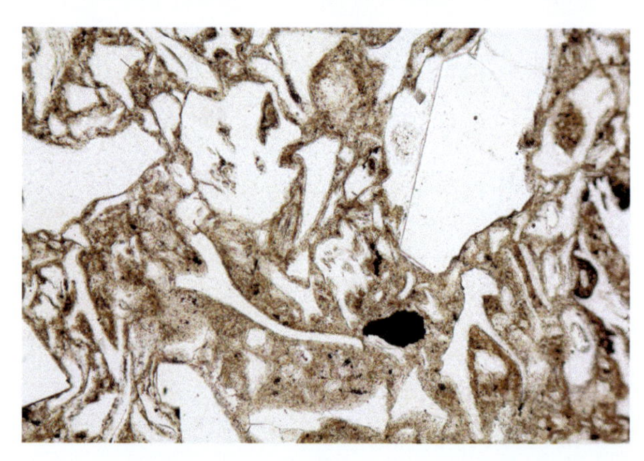

Figure 2.17

A Trachytic trachyte from Ustica (Italy). The groundmass in this picture comprises aligned alkali feldspar crystals flowing around a Carlsbad-twinned sanidine phenocryst. Cross-polarized light, 2× magnification, field of view = 7 mm.

B Trachytic trachyte from the Phlegraean Fields (Italy). Close-up view of the groundmass in which the absence of glass between alkali feldspar crystallites can be noted. Cross-polarized light, 10× magnification, field of view = 2 mm.

C Pilotaxitic trachyte from Ustica (Italy). The orientation of the crystals is irregular and more chaotic (less structured than trachytic texture). Cross-polarized light, 2× magnification, field of view = 7 mm.

D Trachytoid syenite from Biella (Italy). Trachytoid texture defined by the subparallel plagioclase crystals arranged around the large orthoclase crystal (center). Pyroxene crystals (high birefringence) are also present. Cross-polarized light, 2× magnification, field of view = 7 mm.

Hyalopilitic texture is a variety of trachytic texture in which microlites of feldspar (more or less aligned) are surrounded by glass.

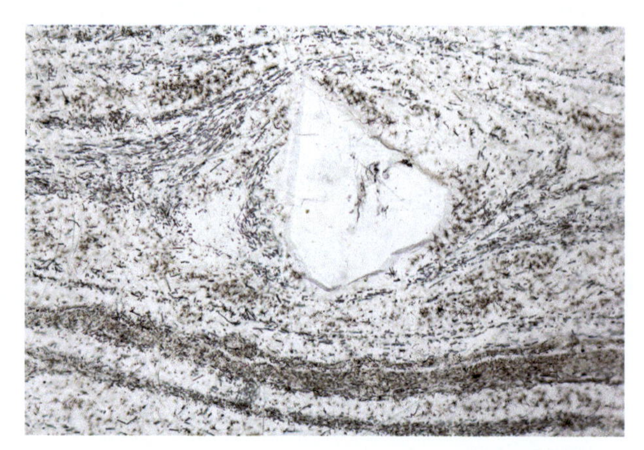

Figure 2.18

A Hyalopilitic andesite from Dubník-Červenica (Slovakia). Small plagioclase microlites weakly aligned E–W in a glass-rich groundmass. Plagioclase microlites flow around the large, central plagioclase phenocryst, defining a magmatic foliation. Cross-polarized light, 2× magnification, field of view = 7 mm.

B Hyalopilitic obsidian from Lipari (Italy). Small plagioclase crystallites aligned E–W in a glassy groundmass. Plagioclase microlites flow around the large feldspar and biotite phenocrysts. Plane-polarized light, 10× magnification, field of view = 2 mm.

Banded Textures

Layered textures are associated with mineral compositional and/or textural variation and are common in many igneous rocks. Layering ranges from centimeters to hundreds of meters and is characterized by different mineral modes and/or changing grain size, mineral orientation, and/or mineral composition. Layering is common in many mafic and ultramafic complexes due to density differences between crystals and their host melt.

Layering is most often planar, but in larger intrusions may have structures similar to those found in sedimentary rocks (cross-bedding, slump folding, etc.). Such structures are linked to a complex interaction of processes which include differential settling (or flotation) of crystals, convection within the magmatic system, and/or changes in intensive conditions (pressure, oxygen fugacity, etc.) associated with crystallization.

Figure 2.19

A Layered cumulate from the Rhum complex (Scotland). The bottom part of the image consists of a layer of euhedral and rounded olivine; the upper part comprises a layer rich in plagioclase with some interstitial pyroxene. Between the two layers is an accumulation of black chromite crystals. Plane-polarized light, 2× magnification, field of view = 7 mm.

B Layered cumulate from the Rhum complex (Scotland). Same image as (A) but with crossed polarizers. Olivine shows characteristic high-interference colors (bottom layer), while plagioclase (upper layer) has polysynthetic twins and the crystals are semi-aligned. Cross-polarized light, 2× magnification, field of view = 7 mm.

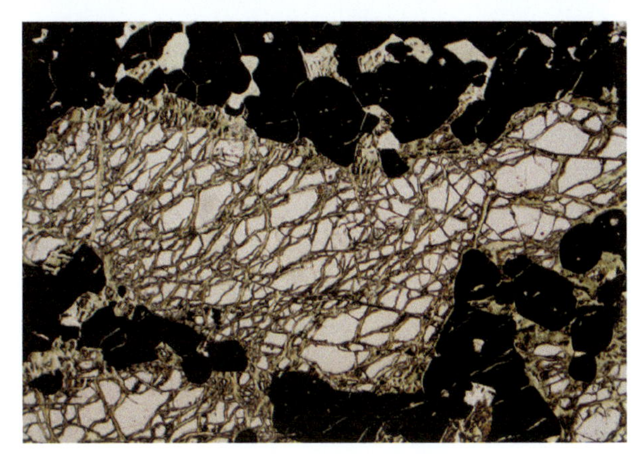

C Layered cumulate from the Bushveld complex (South Africa). Large fractured olivine crystals occur within a layer of black, euhedral chromite crystals. Plane-polarized light, 2× magnification, field of view = 7 mm.

Orbicular textures are usually restricted to coarser-grained, silicic to mafic and ultramafic rocks. The rounded or spherical "orbs" (orbicules) have a distinctive, concentrically banded structure consisting of thin, irregular, but sharply defined layers, each with a different structure and/or composition. The various layers have a fine- to coarse-grained granular structure or may form by tabular or prismatic crystals oriented radially. The crystals that form the orbicules are generally the same ones found in the rock, but not necessarily in the same proportions. The orbicules can range from a few centimeters to >30 cm.

Figure 2.20

A Orbicular quartz monzonite from Kangasala (Finland). Larger orbicules consist of about 200 alternating layers. The main minerals in the orbicule layers include plagioclase, quartz, alkali feldspar, biotite, and oxides ± muscovite. Scale bar shows 5 cm segments.
Image courtesy of A. Heinonen, Geological Survey of Finland collections

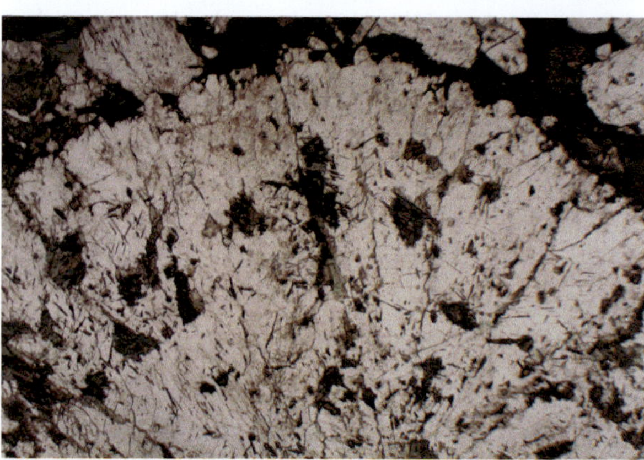

B Orbicular diorite from Sainte-Lucie-de-Tallano (France). Part of an orbicule dominates most of this image. It consists of elongate and radiating plagioclase and dark green hornblende crystals. Plane-polarized light, 1× magnification, field of view = 9 mm.

C Orbicular diorite from Sainte-Lucie-de-Tallano (France). Same image as (B) but with crossed polarizers. Radiating crystal growth of plagioclase is visible. Cross-polarized light, 1× magnification, field of view = 9 mm.

BOX 2.4 Symplectites and Myrmekites

Magmatic intergrowth textures reflect the interpenetration of distinct mineral phases during crystal growth. Intergrowth textures are diverse and numerous, including graphic, micrographic, granophyric, consertal, lamellar, and bleb-like. Here we focus on *symplectites and myrmekites*.

Symplectic texture is a general term for a fine-grained vermicular (wormy) intergrowth of any two minerals and reflects the breakdown of unstable mineral phases. Many minerals may form symplectic intergrowths (e.g. iron ore and orthopyroxene, fayalite and quartz, olivine and plagioclase [at high pressure]). Rocks with this texture are called symplectites.

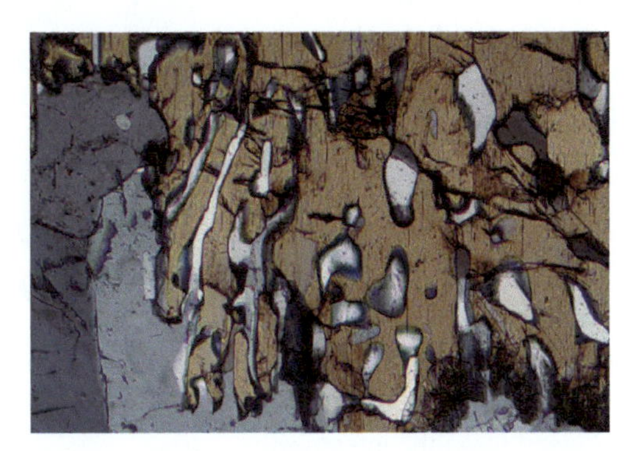

Granite from Wienebene (Austria) with symplectic intergrowth of vermicular quartz (gray) and spodumene (tan). Cross-polarized light, 10× magnification, field of view = 2 mm.

Myrmekitic texture is the irregular intergrowth of sodic plagioclase and vermicular quartz next to alkali feldspar, and it is a specific type of symplectic texture. It may represent the late stages of crystallization in the presence of a volatile phase, subsolidus exsolution, or the replacement of plagioclase during metasomatism or hydrothermal alteration. Rocks with this texture are called myrmekites.

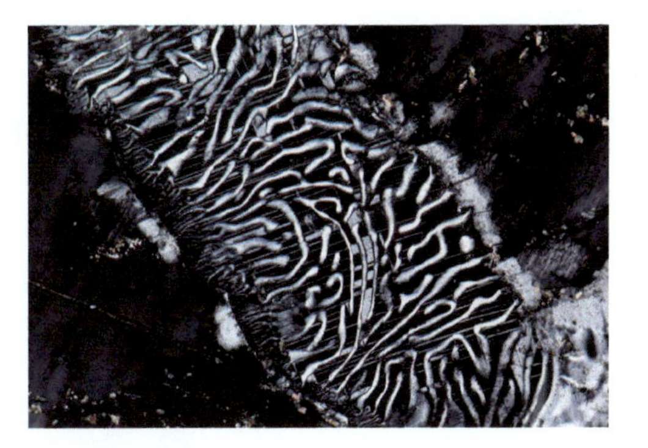

Myrmekitic granite from Braveno (Italy). Center of view shows wormy intergrowth of quartz (light) and plagioclase (dark). Upper-right and lower-left are quartz at extinction. Cross-polarized light, 10× magnification, field of view = 2 mm.

Intergrowth Textures

Graphic, micrographic, and **granophyric textures** are characterized by the intergrowth of quartz and alkali feldspar; these are especially common in shallow-level granitic intrusions. They are generated during magma ascent and decompression, as the loss of water/vapor phases at the melting point results in undercooling. The two minerals do not have time to form independent crystals, so they intergrow.

Figure 2.21

A Graphic texture reflects the intergrowth of quartz and alkali feldspar. Sometimes quartz may form rods; simultaneous extinction indicates optical continuity and that the rods are part of a single large crystal. This graphic granite pegmatite from Sondrio (Italy) shows the graphic (angular) intergrowth of quartz (light) and plagioclase (dark). Cross-polarized light, 2× magnification, field of view = 7 mm.

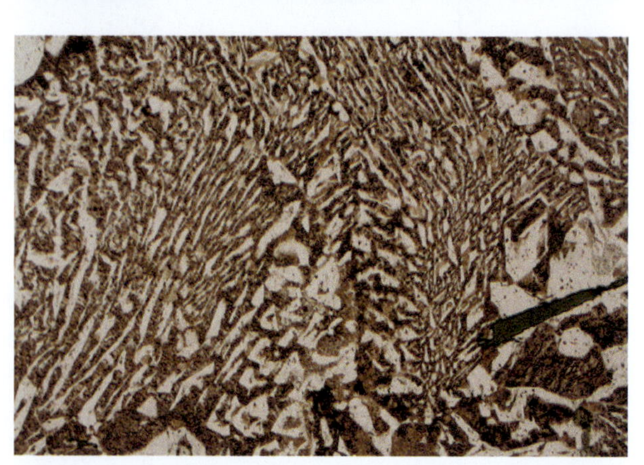

B Micrographic texture is a finer-scale variety of graphic texture which requires a microscope to be seen. In this micrographic granite from Sassari (Italy) the intergrowth of quartz and alkali feldspar looks like cuneiform writing in which orthoclase (dusty brown due to sericite alteration) and quartz (colorless) are intergrown. The dark green crystal to the right is biotite. Plane-polarized light, 10× magnification, field of view = 2 mm.

C Granophyric texture is similar to micrographic and graphic textures, but the intergrowths of quartz and alkali feldspar are typically in a radial arrangement and the result of eutectic crystallization. This granophyre from Valganna (Italy) is made of fine- to coarse-grained quartz and feldspar intergrowths, radially arranged around orthoclase. Cross-polarized light, 2× magnification, field of view = 7 mm.

Consertal, **lamellar**, and **blebby textures** reflect the intergrowth of different crystals/minerals. Consertal texture describes a serrated or notched appearance of the boundaries between two crystals. Lamellar and bleb-like textures represent parallel lamellae of one mineral with consistent optical orientation, enclosed by a host of different composition.

Figure 2.22

A Consertal texture in a granite from Elba island (Italy). Quartz crystals show serrated boundaries between them. Cross-polarized light, 10× magnification, field of view = 2 mm.

B Lamellar texture in gabbro from Mutoko (Zimbabwe). Straight orthopyroxene lamellae, with low-interference colors enclosed in a large, twinned augite crystal. Cross-polarized light, 10× magnification, field of view = 2 mm.

C Bleb-like texture in gabbro from Bushveld complex (South Africa). Blebs and straight augite lamellae with yellow interference colors enclosed in anhedral orthopyroxene crystals. Cross-polarized light, 10× magnification, field of view = 2 mm.

BOX 2.5 Exsolution in Igneous Rocks

Exsolution lamellae occur in minerals of magmatic and metamorphic rocks. It describes the texture in which a former fairly homogeneous phase separates (exsolves) into two or more solid solution phases under subsolidus conditions. Exsolution lamellae are produced when the host mineral loses chemical solubility due to decreasing temperature (T-controlled exsolution) or pressure (decompression or P-controlled exsolution), which results in the exsolution of fine crystals from the host mineral along crystallographically controlled planes. Higher- to lower-temperature lamellae are typically associated with magmatic cooling.

Exsolution lamellae associated with decreasing temperature
Opx exsolves Cpx
Cpx exsolves Opx
Kfs exsolves Pl
Pl exsolves Kfs

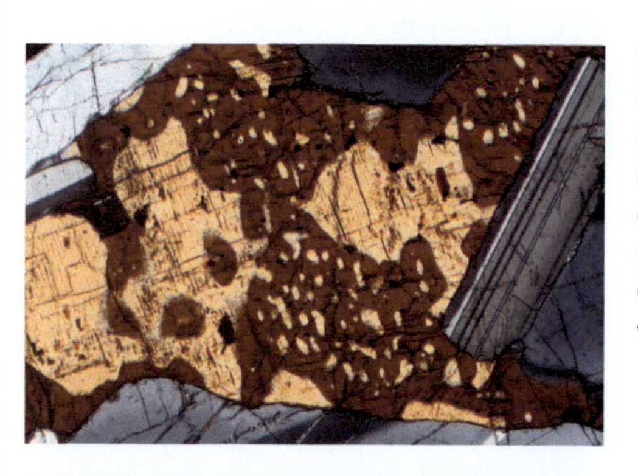

Pyroxene often has exsolution lamellae. In this image of gabbro from the Skaergaard intrusion (Greenland), plagioclase is a cumulate phase and occurs with intercumulus pyroxene. The pyroxene documents down-T exsolution of orthopyroxene (yellowish) from clinopyroxene (darker brown). Crossed polarizers, magnification = 10×, field of view = 2 mm.

Herringbone intergrowths of ilmenite in gabbro from the Bushveld complex (South Africa). Ilmenite is exsolved as small dark needles from pigeonite lamellae, which in turn are exsolved from augite. Note the orientation of ilmenite lamellae at 45° to the host lamellae is controlled by the pyroxene crystal lattice. Crossed polarizers, magnification = 20×, field of view = 1 mm.

Radiate Textures

Radiate textures are those in which crystals are arranged radially around a common nucleus. The most common are the *spherulitic* and *variolitic* aggregates.

Spherulitic texture is a radiating array of fibrous or needle-like crystals and is common in glassy volcanic rocks. A single arrangement of fibers around a common center is known as a *spherulite* and spherulitic rocks contain numerous spherulites. The formation of spherulites reflect solid state processes in which unstable volcanic glass is replaced by more stable mineral phases (*devitrification*). Commonly, glass devitrifies to form a microgranular mass of minute feldspar and silica crystals (known as *felsitic* texture), but often the devitrification process leads to the formation of spherulites. Due to their morphology and the fact that the individual crystals are parallel to the vibration directions of the polarizer and analyzer, an extinction "cross" is observed when viewed under crossed polarizers. Spherulite formation can occur even when glass is in a viscous state, as evidenced by the development of "flow" structures around many spherulites; in this case the spherulites are composed of higher-temperature polymorphs such as sanidine + tridymite or cristobalite, which later convert to quartz and low-temperature feldspar. The close growth of spherulites commonly causes interference structures such as polyhedral boundaries between them.

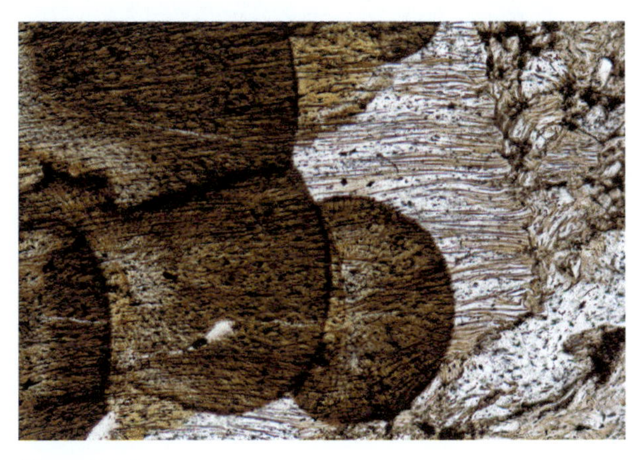

Figure 2.23

A Spherulitic ignimbrite from the Okataina volcanic center (New Zealand). Dark brown spherulites grew in competition with each other so the typical spherical shape was impeded. Note the flow of glass occurs in spherulites, denoting their late crystallization when the rock was already solidified. Plane-polarized light, 10× magnification, field of view = 2 mm.

B Spherulitic obsidian from Lipari (Italy). Note the typical extinction "cross" produced in the center spherulite is due to the alignment of crystallites under cross-polarized light, while the difference in color between the center and the edge of the spherulite is due to the different composition of the fibers. 10× magnification, field of view = 2 mm.

Variolitic texture is characterized by a radiating, fan-like arrangement of plagioclase or pyroxene. Variolitic differs from spherulitic in that there are no spherical aggregates. Variolitic texture indicates rapid cooling and is often found in submarine basalts and in the chilled margin of shallow-level, basic igneous intrusions (dikes and sills) – this is distinct from spherulites, which are formed during devitrification processes.

Figure 2.24

A Variolitic mafic dike from the Bushveld complex (South Africa). Pyroxene phenocrysts (pale brown) set in a variolitic groundmass of fan-like plagioclase and pyroxene crystals. The pyroxene in the groundmass is mixed with plagioclase but are distinguished on the basis of high relief and pale brown color. The red–brown crystals are phlogopite. Plane-polarized light, 10× magnification, field of view = 2 mm.

B Variolitic mafic dike from the Bushveld complex (South Africa). Same image as (A) but with crossed polarizers. Pyroxene phenocrysts show compositional zoning (lower birefringent core and increasing toward the rim) that indicates higher Mg in the rims. Cross-polarized light, 10× magnification, field of view = 2 mm.

BOX 2.6 Magmatic Zoning

Magmatic zoning in igneous rocks is a function of melt processes that manifest in the optical properties of solid solution minerals. Zoning textures reflect gradual or abrupt changes in crystal growth/composition as a function of disequilibrium during cooling. Zoning can be divided into two types:

1. *continuous* and *discontinuous* zoning
2. *normal, reversed, oscillatory* zoning.

Continuous and discontinuous zoning. Zoning of this type reflects gradual (continuous) or abrupt (discontinuous) changes in composition. Continuous zoning associated with, for example, plagioclase commonly reflects gradual cooling from a higher-temperature Ca-rich core, through intermediate compositions, to a lower-temperature Na-rich rim. Discontinuous zoning, on the other hand, is seen as a sharp transition from one composition to another.

Normal, reverse, and oscillatory zoning. Normal compositional zoning has a high(er)-temperature core and a lower-temperature rim. Reverse zoning has a low(er)-temperature composition in the mineral core, while the rim has a higher-temperature composition. Oscillatory zoning is a repeating compositional variation between higher- and lower-temperature compositions from core to rim and is often concentric. The crystal of plagioclase in the below image is from the Adamello tonalite and shows normal, oscillatory zoning which is also discontinuous (dark, irregular region indicated by the white arrow). Cross-polarized light, 10× magnification, field of view = 1.75 mm.

Patchy and sector zoning. Patchy (or irregular) zoning is associated with infilling after skeletal growth or resorption, whereas sector zoning refers to compositional differences between coeval growth sectors, often the result of fluid–crystal element partitioning between nonequivalent crystal faces.

(*Left*) Patchy zoning in plagioclase (cross-polarized light). (*Right*) Sector zoning in augite (plane-polarized light). Both with 2× magnification, field of view = 3 mm.

Overgrowth Textures

Corona texture is characterized by one mineral overgrowing another. It is also known as a "*reaction rim*" or "*reaction corona*" since its origin is generally due to reaction processes between early-formed crystals in disequilibrium with a melt (e.g. olivine mantled by orthopyroxene or pyroxene mantled by hornblende).

Figure 2.25

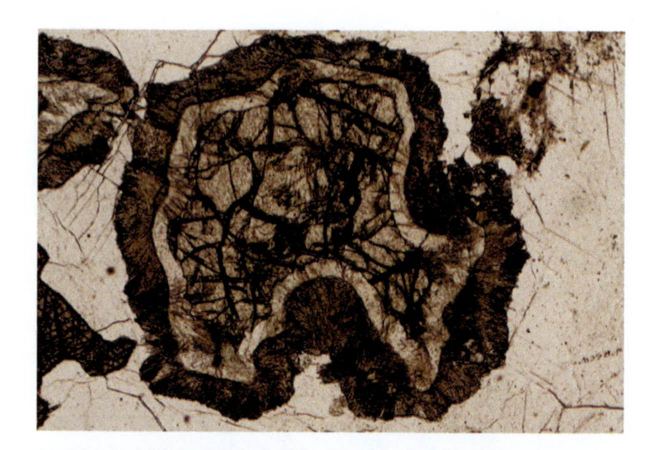

A Corona gabbro from Sondalo (Italy). Serpentinized olivine (fractured, central part) surrounded by (1) an inner corona of orthopyroxene (light color), followed by (2) a thick, fibrous hornblende-rich outer corona (brown). Plane-polarized light, 10× magnification, field of view = 2 mm.

B Corona gabbro from Sondalo (Italy). Same image as (A) but with crossed polarizers. Orthopyroxene outer rim has low-interference colors. Cross-polarized light, 10× magnification, field of view = 2 mm.

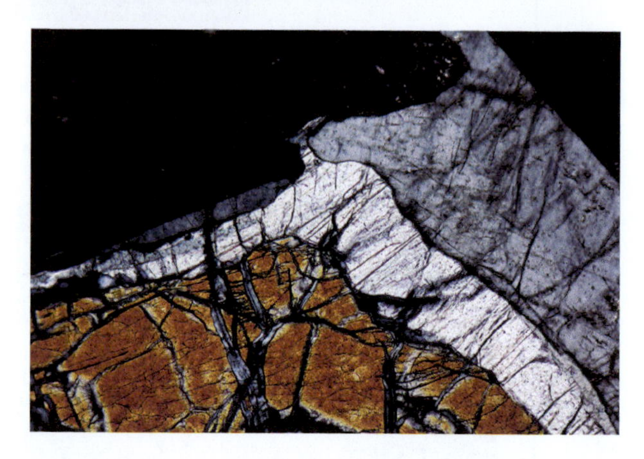

C Corona troctolite from the Bushveld complex (South Africa). Serpentinized olivine (high birefringence and cross-cut by thin serpentine veins) with an orthopyroxene corona (bright rim), surrounded by plagioclase crystals (dark gray or at extinction). Cross-polarized light, 10× magnification, field of view = 2 mm.

Rapakivi texture is a variant of corona texture. It refers to the growth of Na-plagioclase on K-feldspar in granitic rocks. The origin of the rapakivi texture is still debated and may result from: (1) magmatic disequilibrium between K-feldspar crystals and melt during magma ascent; (2) magma mixing and instability of K-feldspar crystals in the hybrid magma; and (3) subsolidus deuteric alteration of feldspar (dissolution–reprecipitation replacement processes).

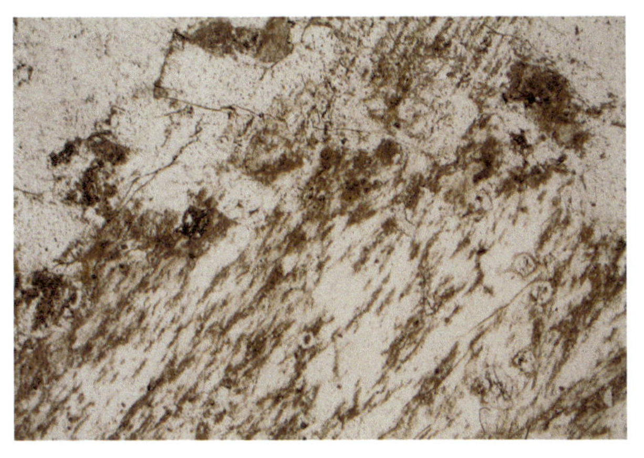

Figure 2.26

A Rapakivi granite from unknown locality (Finland). A large, rounded, and partially sericitized (brownish) orthoclase occupies most of the image. It is surrounded by less altered plagioclase. Plane-polarized light, 2× magnification, field of view = 7 mm.

B Rapakivi granite from unknown locality (Finland). Same as (A), but with crossed polarizers. Orthoclase is perthitic with small plagioclase inclusions and surrounded by twinned plagioclase. Note the irregular contact between orthoclase and plagioclase. Cross-polarized light, 2× magnification, field of view = 7 mm.

Kelyphitic texture is a variant of corona texture. It refers to the fine-grained, fibrous intergrowth typically developed around garnet or olivine crystals. Kelyphitic rims may include numerous mineral phases (e.g. pyroxene, spinel, amphibole). This texture is common in garnet–peridotite, as a retrograde metamorphic process recording the transition from high-pressure garnet–peridotite to low-pressure spinel peridotite via the reaction, at ~20 kbar:

$$\text{Gt} + \text{Ol} \longrightarrow \text{Opx} + \text{Sp} + \text{Cpx} \quad \textit{dry}$$

$$\text{Gt} + \text{Ol} + H_2O \longrightarrow \text{Opx} + \text{Sp} + \text{Am} \quad \textit{wet}$$

Figure 2.27

A Kelyphitic garnet–peridotite from Alpe Arami (Switzerland). Fractured garnet (central crystal) surrounded by a kelyphitic rim of fibrous brown–green hornblende, in turn mantled by a corona of orthopyroxene (light green). Plane-polarized light, 2× magnification, field of view = 7 mm.

B Kelyphitic garnet–peridotite from Alpe Arami (Switzerland). Same image as (A) but with crossed polarizers. Fibrous kelyphitic hornblende with low-interference colors mantled by an irregular corona of orthopyroxene (first-order gray birefringence). Cross-polarized light, 2× magnification, field of view = 7 mm.

BOX 2.7 Opacite Rims

Opacite rims represent a common reaction of hydrated minerals such as biotite and amphibole associated with the emplacement of volcanic rocks. It is therefore related to igneous and magmatic processes and is included here in the chapter on igneous rocks. During magma ascent, the reduction in pressure can result in the release of structurally bound water from biotite and amphibole; this allows the nucleation of a new phase(s) at the mineral–melt interface.

Opacite rims on amphibole often form a complex assemblage of clinopyroxene, plagioclase, magnetite, and ilmenite. Opacite rims on biotite may be associated with other interesting features, such as exsolved acicular rutile.

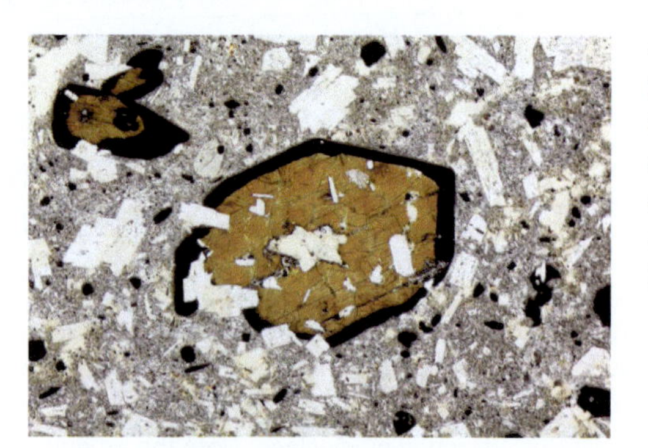

Andesite from Yanacocha (Peru). Subhedral prismatic hornblende crystal with dark opacite rim and plagioclase inclusions. The scattered black crystals in the groundmass are former hornblende crystals completely replaced by opacite. The colorless crystals are plagioclase. Plane-polarized light, 2× magnification, field of view = 7 mm.

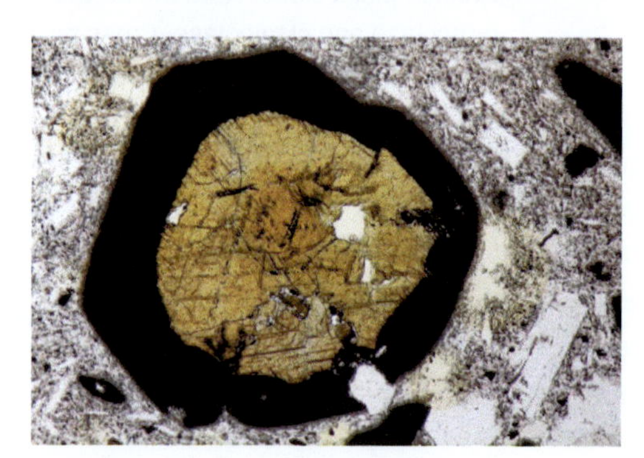

Andesite from Yanacocha (Peru). Close-up view of a resorbed hornblende crystal with a thick dark opacite rim set in a plagioclase-rich groundmass. Plane-polarized light, 10× magnification, field of view = 2 mm.

Andesite from Halkidiki (Greece). Euhedral biotite (basal section) with a thin opacite rim, and needle-like inclusions of rutile (intersecting at 60°) set in a glassy groundmass. Pyroxene (upper-right and lower-left) and plagioclase (colorless) phenocrysts also present. Small colorless, acicular crystals are apatite. Plane-polarized light, 10× magnification, field of view = 2 mm.

Cavity Textures

There are general terms used to describe the different types of cavities present in rocks.

Vesicular A vesicle is a rounded, often ovoid or elongate, hole formed by the exsolution and expansion of dissolved gas(es) from a melt. Decompression during magma ascent releases gas, which forms bubbles that are trapped during rapid solidification. It is a common texture in many basaltic rocks (vesicular basalts).

Amygdaloidal Amygdales are former vesicles now partially or completely filled with late-stage or post-magmatic secondary minerals (e.g. zeolite, calcite, quartz, epidote, analcime, chlorite, chalcedony). It is a common texture in many basaltic rocks (amygdaloidal basalts).

Ocellar Ocelli, unlike amygdales, are not secondary. They are spherical or ellipsoidal leucocratic minerals genetically associated with their more mafic host rocks. They may represent small pockets of solidified immiscible fluid (liquid or gas), the products of incongruent melting associated with magma mixing, or the segregation of late-stage liquids.

Miarolitic These irregular cavities, mostly associated with granites and granitic pegmatites, are lined with crystals of rock-forming minerals (usually quartz and feldspar) and often visible without the aid of magnification. They reflect mineral growth from fluids segregated by the vesiculation of granitic magma during its final stages of crystallization.

Miarolitic granite, Seward Peninsula (Alaska, United States). Coarse crystals (top) of quartz, plagioclase, and alkali feldspar growing into a cavity. Ruler in cm.
Image courtesy of E. Miller

Figure 2.28

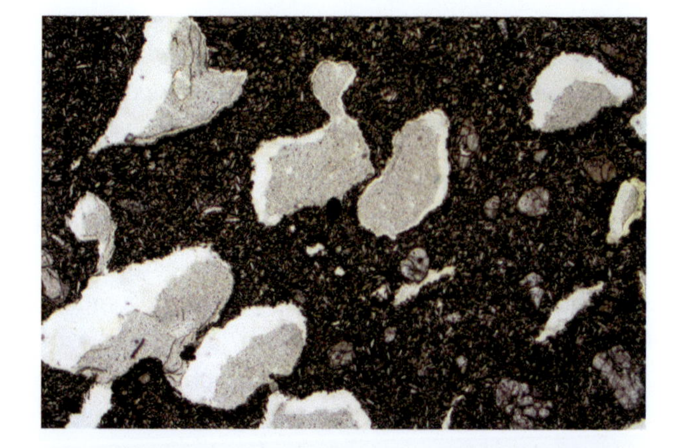

A Vesicular basalt from Radicofani (Italy). Elongate and irregular vesicles (no mineral filling; thin-section glue is seen inside them) are surrounded by a fine-grained groundmass. The three vesicles in the lower-left have coalesced. Plane-polarized light, 2× magnification, field of view = 7 mm.

B Amygdaloidal basalt from San Venanzo (Italy). Calcite amygdales with an ultra-thin layer of chalcedony (thin bright rings) occur along the vesicle edges. Plane-polarized light, 10× magnification, field of view = 2 mm.

C Amygdaloidal basanite from Limberg (Germany). Ellipsoidal amygdales of fibrous, low birefringence phillipsite (zeolite). Phenocrysts of large, twinned augite and rectangular olivine have high birefringence and are set in a glassy (isotropic) groundmass. Cross-polarized light, 2× magnification, field of view = 7 mm.

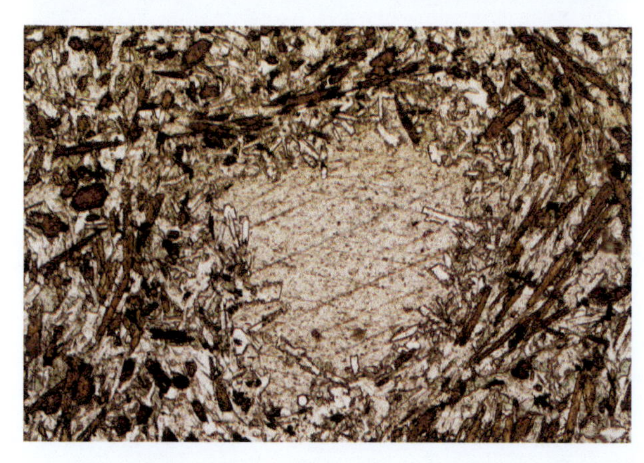

D Ocellar lamprophyre from Adamello (Italy). Spherical calcite ocelli with euhedral plagioclase crystals extending from the margins into the calcite. Kaersutite crystals in the groundmass (elongate, brown crystals) are arranged tangentially around the ocelli (right side). Plane-polarized light, 10× magnification, field of view = 2 mm.

2.6 Plutonic Rocks

Plutonic igneous rocks are formed within the Earth and are associated with relatively slow cooling that results in coarser-grained minerals – these are the *phaneritic* rocks. They include the ultramafic, gabbroic, syenitic, granitic, and feldspathoid rocks.

We use the "-ic" or "-oid" terminology (e.g. granitic/granitoid) to denote the broad compositional group (*sensu lato*) and restrict the use of "granite" (*sensu stricto*) to the specific IUGS modal composition. According to the IUGS nomenclature for rock types/names, between the syenitic and granitic rocks are the "quartz–syenitic" rocks; for convenience we treat these as more quartz-rich variants of the syenitoids and describe them within the syenitic group. The same is true for the feldspathoids – we discuss the variants of feldspathoid rocks as transitional with increasing modal abundance of feldspathoid minerals. That said, the phaneritic texture of these rocks allows them to be properly recognized and classified on the basis of their IUGS mode, and this should be the normal practice.

Our understanding of rock-forming minerals now provides a solid basis for the recognition and classification of the phaneritic plutonic rocks. We begin with the ultramafic rocks, which include the peridotites and the pyroxenites, then progress with increasing modal quartz content through the more evolved rock types, and finish with the silica-undersaturated feldspathoid rocks. Each rock type is characterized by its common mineralogy, the IUGS modal definition, and typical texture. After a brief description of the rock and its minerals, we present what is currently known about its occurrence within a plate tectonic framework. Additional reading at the end of the chapter for each rock group provides the foundation of this information, as well as a platform for further studies for the interested reader.

Ultramafic rocks
- Peridotites
  - Lherzolite
  - Wehrlite
  - Harzburgite
  - Dunite
- Pyroxenites
  - Clinopyroxenite
  - Olivine clinopyroxenite
  - Websterite
  - Olivine websterite
  - Orthopyroxenite
  - Olivine orthopyroxenite

Gabbroic rocks
- Gabbro
- Diorite
- Norite
- Troctolite
- Anorthosite

Syenitic rocks
- Monzodiorite-monzogabbro
- Monzonite
- Syenite
- Alkali feldspar syenite

Granitic rocks
- Tonalite
- Granodiorite
- Granite
- Alkali feldspar granite

Feldspathoid rocks
- Foid-bearing rocks
- Foid rocks
- Foidolite

Lherzolite

Mineralogy: *Ol, Opx, Cpx, accessory Grt, Sp, Pl, Mt, Il, Cr*
IUGS classification: *Ol (40–90%) + Opx (>5%) + Cpx (>5%)*
Texture: *When undeformed, it is typically coarse-grained, granular*

Lherzolite is an ultramafic igneous rock and a type of peridotite dominated by Mg- and Fe- rich minerals. It can have 40–90% olivine and up to 55% orthopyroxene, with generally lesser amounts of calcic, chromium-rich clinopyroxene. Accessory minerals include chromium and aluminum spinel, garnet, and/or plagioclase. Ilmenite, chromite, and magnetite may also be present. Mantle metasomatism of lherzolite can produce accessory micas and amphiboles.

Occurrence

Lherzolite is the main component of the upper mantle and usually contains Cr-diopside as the clinopyroxene and enstatite as the orthopyroxene. Lherzolite is known from the ultramafic part of ophiolite complexes, alpine-type peridotite massifs, fracture zones adjacent to mid-oceanic ridges, and occurs as xenoliths in kimberlite pipes and alkali basalts. Lherzolite can form cumulates in layered intrusions. Alpine or orogenic lherzolites represent subcontinental mantle lithosphere exhumed during continental collision. The type location of orogenic lherzolite massifs is the Lherz Massif in the Pyrenees (France).

Aluminous phases in the mantle are pressure-dependent, with plagioclase stable at low pressure, spinel at intermediate pressure, and garnet at high pressure. Plagioclase occurs in lherzolites and other peridotites at "low" pressures (relatively shallow depths of 20–30 km). At greater depth, plagioclase is unstable and is replaced by spinel through the reaction:

$$\underset{Pl}{CaAl_2Si_2O_8} + \underset{Ol}{Mg_2SiO_4} = \underset{Opx}{2MgSiO_3} + \underset{Cpx}{CaMgSi_2O_6} + \underset{Sp}{MgAl_2O_4}$$

Partial melting of spinel lherzolite is one of the primary sources of basaltic magma. At approximately 90 km depth, spinel gives way to the stable aluminous phase pyrope garnet:

$$\underset{Sp}{MgAl_2O_4} + 1.55\,\underset{Opx}{Mg_2Si_2O_6} + 0.45\,\underset{Cpx}{CaMgSi_2O_6} = 0.85\,\underset{garnet}{Mg_3Al_2Si_3O_{12}} + 0.15\,\underset{garnet}{Ca_3Al_2Si_3O_{12}} + \underset{Ol}{Mg_2SiO_4}$$

Garnet lherzolite is the major Al-bearing mineral of the Earth's upper mantle at even greater pressures (up to ~300 km depth).

This lherzolite from Finero (Italy) is dominated by olivine (pale green), clinopyroxene (grass green), and phlogopite (brown). The vertical vein is dominated by Cr-pyroxene. Coin for scale, field of view = 10 cm.

Figure 2.29

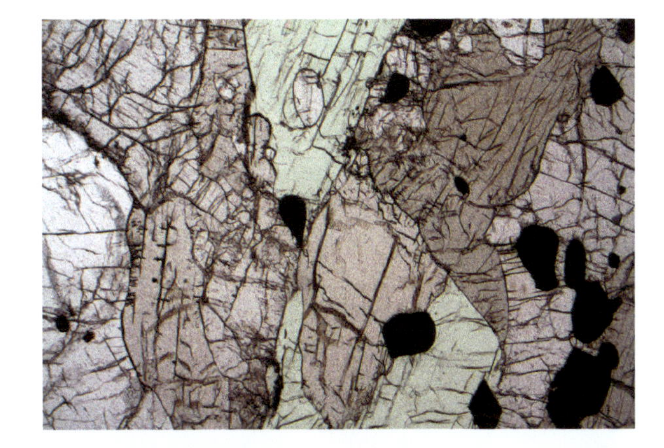

A Lherzolite from Finero (Italy). In this mantle peridotite, anhedral olivine is clear to brown (left and right of image), anhedral pyroxene is light green (center), and subhedral chromite is black. Plane-polarized light, 2× magnification, field of view = 7 mm.

B Lherzolite from Finero (Italy). Same image as (A) but with cross-polarized light. Olivine shows typical birefringence in blues (higher than pyroxene). Cumulate layering is vertical. Cross-polarized light, 2× magnification, field of view = 7 mm.

C Lherzolite from the Pyrenees (France). The image is dominated by olivine (colorless), enstatite (pale green), and anhedral spinel (brown). Both olivine and orthopyroxene are fractured and slightly deformed (a common feature in peridotites found along mountain ranges). Plane-polarized light, 2× magnification, field of view = 7 mm.

D Lherzolite from the Pyrenees (France). Same image as (C) but with cross-polarized light. Olivine and orthopyroxene both have high birefringence, but olivine is higher and spinel is isotropic. 2× magnification, field of view = 7 mm.

Wehrlite

Mineralogy: *Ol, Cpx, ±Opx, accessory Grt, Sp, Mt, Il, Cr*
IUGS classification: *Ol (40–90%) + Cpx (10–60%) ± Opx (≤5%)*
Texture: *When undeformed, it is typically protogranular*

ULTRAMAFIC ROCKS

PERIDOTITES

Wehrlite, like harzburgite, is composed of mostly forsteritic Ol (40–90%), but unlike harzburgite, wehrlite is dominated by Cpx (10–60%) rather than Opx; Opx, if present, is minor (≤5%). Accessory minerals in the peridotites are generally similar and in wehrlite also include ilmenite, chromite, and magnetite, as well as an aluminum-bearing mineral (plagioclase, spinel, or garnet).

Occurrence

Wehrlite is a minor component of the upper peridotitic mantle, where it may form by crystallization from partial melts of mantle rocks (i.e. basaltic liquids). Wehrlite can form cumulates within layered intrusions associated with gabbro and norite, or occur as mantle xenoliths within mantle-derived magmas and within the upper portions of the mantle sequence of ophiolites. Wehrlite as veins and dikes occurs as xenoliths and in ophiolites. Some meteorites are also classified as wehrlites (e.g. NWA 4797).

During the formation of oceanic crust, mantle partial melts (basalts) have the potential to produce wehrlite through early crystallization of olivine and clinopyroxene. Under wet, low-pressure conditions, crystallization of plagioclase is delayed, while crystallization of olivine + clinopyroxene are stabilized. Such wehrlite cumulates are often somewhat oxidized, consistent with high water activity (aH_2O). Under high-pressure conditions, wehrlite may crystallize from melts trapped near the crust–mantle transitional zone, forming cumulate lenses or dikes.

Wehrlites may also form as the result of metasomatism triggered by fluid/melt–peridotite interaction, a process sometimes called "wehrlitization." The metasomatic agent of wehrlitization can be (1) carbonate melt/fluid, which produce magnesian wehrlites, or (2) silicate melts which lead to the production of Fe-rich wehrlites. In the former, carbonated melts interact with mantle peridotite to form wehrlite via the reaction:

$$\underset{\text{enstatite}}{4MgSiO_3} + \underset{\text{dolomite}}{CaMg(CO_3)_2} = \underset{\text{forsterite}}{2Mg_2SiO_4} + \underset{\text{diopside}}{CaMgSi_2O_6} + \underset{\text{vapor}}{2CO_2}$$

Wehrlite hand sample from Åhem (Norway). The visible minerals include weathered olivine (orange), diopside (green clinopyroxene), and pyrope garnet (pink). The sample is 16 cm across.

Image courtesy of S. Sepp, www.sandatlas.org

Figure 2.30

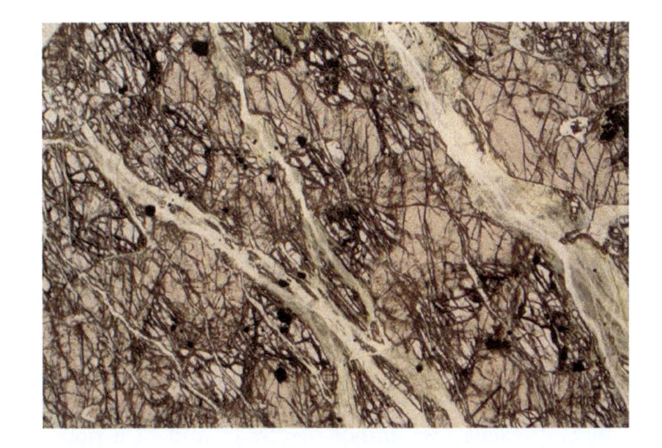

A Wehrlite xenolith from the Trodos ophiolite (Cyprus). Large poikilitic augite crystal (pale brown) with rounded and serpentinized olivine inclusions (colorless, fractured, and higher relief). The poikiolitic augite is cut by serpentine-rich veins. Plane-polarized light, 2× magnification, field of view = 7 mm.

B Wehrlite xenolith from the Trodos ophiolite (Cyprus). Same image as (A), but with crossed polarizers. Note the high third-order gray of some olivine. Cross-polarized light, 2× magnification, field of view = 7 mm.

C Wehrlite xenolith from Iblei volcanic complex (Italy). A large wehrlite xenolith (right) sits in glassy basalt (left). Clinopyroxene (pale gray) shows characteristic cleavage and olivine (colorless) is partially altered to bowlingite (pale green). The large, green olivine crystal in the groundmass (center-left) is from the xenolith. The xenolith is partially cut by later veins of magnesite (central part of image). Plane-polarized light, 2× magnification, field of view = 7 mm.

D Wehrlite xenolith from Iblei volcanic complex (Italy). Same image as (C) but with crossed polarizers. Clinopyroxene has low-interference colors (indicating a view down the optical axis), olivine has high-interference colors, and magnesite shows its characteristic very high-interference colors. Cross-polarized light, 2× magnification, field of view = 7 mm.

Harzburgite

Mineralogy: *Ol, Opx, ± Cpx, accessory Sp, Mt, Il, Cr*
IUGS classification: *Ol (40–90%) + Opx (10–60%) ± Cpx (<5%)*
Texture: *When undeformed, it is typically coarse-grained, granular*

ULTRAMAFIC ROCKS

PERIDOTITES

Harzburgite is an ultramafic igneous rock in the peridotite family. It is dominated by forsteritic Ol (40–90%) and enstatitic Opx (10–60%), and Cpx when present is minor (<5%) and Ca-poor. Accessory minerals tend to be dominated by Cr-rich spinel, ilmenite, chromite, and magnetite. Garnet is uncommon except in association with xenoliths in kimberlites. Metasomatism of harzburgite in the mantle can produce accessory micas and amphiboles.

Occurrence

Harzburgite is an igneous rock that forms in the mantle, occurs in "alpine-type" orogens, and as cumulates in the basal zones of layered mafic intrusions. In the upper mantle, harzburgite forms as the residue of partial melting of lherzolite. In ophiolites, the most common peridotite is harzburgite; it is present in the Semail ophiolite (Oman), the Troodos ophiolite (Cyprus), the Coast Range ophiolites (California, United States), and the Bay of Islands ophiolite (Newfoundland). Harzburgitic mantle xenoliths also occur within the upper portions of ultramafic mantle sections of ophiolites, whereas garnet-bearing harzburgite xenoliths occur in some kimberlite pipes and are almost exclusively associated with ancient continental cratons of Archean or Paleoproterozoic age. These mantle-derived magmas sample the thick (≥200 km) mantle lithosphere beneath the cratons and are consistent with a deeper (garnet-bearing) mantle source. Garnet harzburgite xenoliths from kimberlites in South Africa, for example, are particularly well-characterized and are less depleted in the basalt component than most ophiolite harzburgites.

Alpine peridotites represent fragments of harzburgitic mantle that have been tectonically emplaced during orogeny. Emplacement often produces deformation textures in harzburgite (e.g. undulose extinction and kink-bands in olivine or bent exsolution lamellae in orthopyroxene). Cumulate harzburgite is found in some layered mafic intrusions. At the Earth's surface, basaltic magmas typically crystallize olivine, plagioclase, and augite (a low-Ca pyroxene). Low-Ca pyroxene can only coexist with olivine at low pressure; at pressures >0.5 GPa, olivine and low-Ca pyroxene (enstatite or bronzite) crystallize and segregate from normal basaltic magmas to form cumulate harzburgite. Such conditions are common in some layered mafic intrusions.

Harzburgite from Me Maoya (New Caledonia). This outcrop is roughly equal parts olivine (pale green) and orthopyroxene (brownish). Hand for scale.

Image courtesy of A. Montanini

Figure 2.31

A Mylonitic harzburgite from Ivrea Verbano (Italy). Deformed olivine (colorless) and orthopyroxene (pale green) porphyroclasts. The fine- to very fine-grained matrix is the result of the mechanical fragmentation of the original igneous minerals. Plane-polarized light, 2× magnification, field of view = 7 mm.

B Mylonitic harzburgite from Ivrea Verbano (Italy). Same as (A) but with crossed polarizers. Note the fractures that cut the porphyroclasts are consistent with brittle failure. Cross-polarized light, 2× magnification, field of view = 7 mm.

C Harzburgite from the Bushveld complex (South Africa). Fractured olivine (pale) and orthopyroxene (light brown). The small black crystals are chromite. Plane-polarized light, 2× magnification, field of view = 7 mm.

D Harzburgite from the Bushveld complex (South Africa). Same image as (C) but with crossed polarizers. Partial re-equilibration has produced some 120° grain boundary intersections. Cross-polarized light, 2× magnification, field of view = 7 mm.

Dunite

Mineralogy: *Ol, accessory Opx, Cpx, Sp, Mt, Il, Cr*
IUGS classification: *Ol (≥90%) ± Cpx and/or Opx (≤10%)*
Texture: *When undeformed, it is typically coarse-grained, granular*

ULTRAMAFIC ROCKS

PERIDOTITES

Dunite is an intrusive ultramafic igneous rock and a type of peridotite. It contains >90% olivine and can include minor amounts (<10%) of Opx and/or Cpx. It is often accompanied by chrome spinel, but when the spinel group mineral is magnetite this rock is known as olivinite. Other accessory minerals include chromite, magnetite, spinel, ilmenite, Mg-garnet (pyrope), pyrrhotite, and, in some cases is rich in the element platinum. Dunite is easily weathered and typically undergoes retrograde metamorphism at near-surface conditions, altering to serpentinite and soapstone.

Occurrence

Dunite is named after its type locality, Dun Mountain (New Zealand); other well-known occurrences include the Skaergaard intrusion (Greenland), the Palisades Sill in New York (United States), and the Bushveld igneous complex (South Africa). Dunite forms either as an olivine cumulate associated with layered mafic intrusions or as a residue after extraction of a basaltic melt from a pre-existing ultrabasic mantle rock. In layered mafic intrusions dunite usually forms tabular sills, lenses, or pipes, with cumulate layers ranging in size from meters to hundreds of kilometers (e.g., the Great Dyke, Zimbabwe). Dunite cumulates and mantle rocks are also found as xenoliths in a wide range of mantle-derived magmas – for example in Hawaiian lavas and kimberlite dikes.

Dunite and other peridotites are considered the major constituents of the Earth's mantle above about 400 km depth. Dunite typically occurs at the base of ophiolite sequences and, when associated with continental rocks, represents slices of mantle rock from a subduction zone that have been thrust onto the continental crust by obduction during continental or island arc collision (orogeny). Dunite is also found in alpine peridotite massifs as slivers of subcontinental mantle exhumed during collisional orogeny.

This outcrop of dunite from Sesia Lanzo (Italy) has thin, pyroxene-rich layers within it. Orange to brown surface weathering such as that seen here is typical of ultramafic rocks. The coin is ~2 cm diameter.

Image courtesy of G. Borghini

Figure 2.32

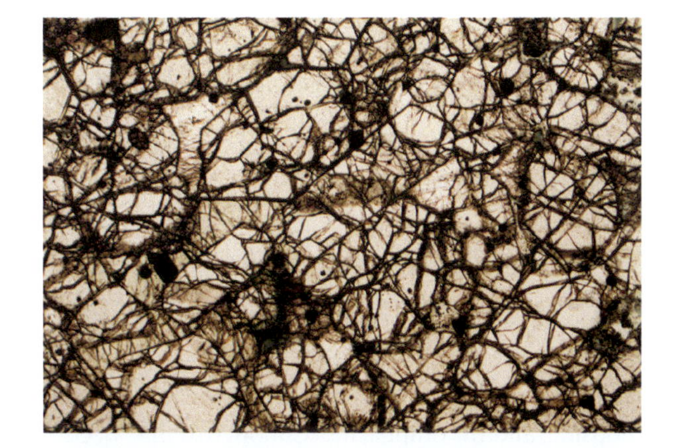

A Dunite from the Rhum complex (Scotland). In this dunite, euhedral to subhedral olivine with curvilinear fractures is the cumulate phase. Between some crystals, interstitial plagioclase occurs (low relief). Small brown crystals of chromite are also present. Plane-polarized light, 10× magnification, field of view = 2 mm.

B Dunite from the Rhum complex (Scotland). Same image as (A) but with crossed polarizers. Note some serpentinization and third-order colors of olivine. Cross-polarized light, 10× magnification, field of view = 2 mm.

C Dunite from Mount Dun (New Zealand). Olivine makes up most of the image and has a re-equilibrated mosaic texture with well-developed 120° triple junctions. The dark crystals are chromite. Plane-polarized light, 2× magnification, field of view = 7 mm.

D Dunite from Mount Dun (New Zealand). Same image as (C) but with crossed polarizers. Cross-polarized light, 2× magnification, field of view = 7 mm.

Clinopyroxenite

Mineralogy: *Cpx, minor Opx and Ol, accessory Grt ± Sp*
IUGS classification: *Cpx (≥90%) ± Opx and/or Ol (≤10%)*
Texture: *When undeformed, it typically has a coarse-grained granular texture*

ULTRAMAFIC ROCKS

PYROXENITES

Clinopyroxenite is predominantly clinopyroxene (≥90%) with minor amounts (≤10%) of orthopyroxene and/or olivine. The clinopyroxene is typically diopside and/or omphacite. Clinopyroxenite occurs in layered igneous complexes, orogenic massifs as fragments of continental lithospheric mantle, ophiolitic and related oceanic rock associations, and as nodules and xenoliths in alkali basalts and diatremes. Clinopyroxenite may form as a cumulate from basaltic magma or be a product of fluid/melt–peridotite interaction. Chlorite, epidote, amphibole, and titanite may occur as retrograde minerals in clinopyroxenite.

Occurrence

In orogenic belts associated with supra-subduction zone ophiolites (e.g. the Makran ophiolite in Iran) clinopyroxenite is a subordinate rock type and often texturally re-equilibrated (polygonal recrystallization). Olivine and/or orthopyroxene may form interstitial grains, or more rarely intercumulus plagioclase may be present. Clinopyroxenite may occur as dikes injected into the host peridotite during late magmatic phases and/or during uplift of the ultramafic complex, the result of peridotite–melt/fluid interaction. In some orogens such as the Dabi-Sulu Orogenic Belt in China, garnet clinopyroxenite forms from a clinopyroxenite protolith. If clinopyroxenite occurs in close association with dunite in orogenic belts, it typically reflects tectonic interleaving.

Ophiolites, as slivers of ancient oceanic lithosphere obducted onto continental or oceanic crust, may contain subordinate clinopyroxenite reflecting the uppermost sub-arc mantle. At shallow depths, fertilized peridotites are often characterized by selective precipitation of clinopyroxene at the expense of olivine or orthopyroxene. Garnet clinopyroxenites have been interpreted as residual rocks after eclogite melting or cumulates derived from eclogite melts.

Clinopyroxenite xenoliths have been interpreted as fragments of mantle lithosphere, cumulates from alkaline magmas, and as the reaction between carbonate melt and granitic crust. In mafic kamafugite diatremes of the East African Rift Zone, clinopyroxenite xenoliths with phlogopite and biotite are abundant and thought to sample the underlying mantle lithosphere.

Phlogopite-bearing clinopyroxenite xenolith from an alkalic basalt on Mount Morning (Antarctica). Width of sample = 4 cm.

Image courtesy of A. P. Martin

Figure 2.33

A Alkali pyroxenite from Jacupiranga (Brazil). Subhedral clinopyroxene (pinkish-brown) has visible cleavage (some crystals show two cleavages intersecting at 90°). Biotite is deep red–brown and nepheline (colorless) has a cloudy appearance due to alteration. Plane-polarized light, 2× magnification, field of view = 7 mm.

B Alkali pyroxenite from Jacupiranga (Brazil). Same image as (A) but with crossed polarizers. Pyroxene and biotite show high birefringence, while nepheline shows gray interference colors. Cross-polarized light, 2× magnification, field of view = 7 mm.

C Xenolith of clinopyroxenite from Vulsini volcanic complex (Italy). Subhedral clinopyroxene (pale green to pale brown) shows granular texture; basal sections show perfect 90° cleavage intersections. The colorless regions are holes. Plane-polarized light, 2× magnification, field of view = 7 mm.

D Xenolith of clinopyroxenite from Vulsini volcanic complex (Italy). Same image as (C) but with crossed polarizers. Clinopyroxene shows characteristic birefringence which is lower in basal sections. Cross-polarized light, 2× magnification, field of view = 7 mm.

Olivine Clinopyroxenite

Mineralogy: Cpx and Ol, ± Opx, ± accessory Grt, Sp
IUGS classification: Cpx (60–90%) + Ol (10–40%) ± Opx (≤5%)
Texture: When undeformed, it typically has a coarse-grained granular or cumulate texture

ULTRAMAFIC ROCKS

PYROXENITES

Figure 2.34 Olivine clinopyroxenite from the Luotuoshan mafic–ultramafic intrusion (China). Poikilitic olivine crystals enclosed in optically continuous intercumulus clinopyroxene. Note the high relief and incipient alteration along fractures of olivine. (*Left*) Plane-polarized light; (*right*), cross-polarized light; 2× magnification, field of view = 4 mm.
Images courtesy of B. Ma

Olivine clinopyroxenite is composed of clinopyroxene (60–90%) and olivine (10–40%), and minor orthopyroxene (≤5%). The clinopyroxene is typically augitic in lower-pressure settings or diopsidic in higher-pressure environments, with forsteritic olivine. Olivine clinopyroxenite occurs as a cumulate associated with layered igneous complexes and mantle rocks of island arcs, orogenic massifs as fragments of continental lithospheric mantle, ophiolitic and related oceanic rock associations, and as nodules and xenoliths in alkali basalts and diatremes.

Occurrence

In mafic–ultramafic complexes such as that on Duke Island, British Columbia (Canada), olivine pyroxenite crystallized from water-saturated and oxidized (log $f(O_2) \geq$ FMQ) primitive arc magmas. Ultramafic cumulates of supra-subduction zone ophiolites (e.g. Troodos [Cyrus], Pozanti-Karsanti [Turkey], or Bay of Islands [Canada]) have adcumulate textures. Highly magnesian olivine clinopyroxenite in these settings can form by crystal fractionation of primary basaltic melts at medium to high pressures (up to 10 kbar) – part of the plutonic core in an intraoceanic island arc. Alternatively, lower-Mg olivine clinopyroxenite forms at shallower depths (lower-pressure settings). In orogenic belts, olivine peridotites are associated with tectonically emplaced slices of subcontinental lithospheric mantle. They may represent products of crystal accumulation of mantle-derived magmas or melt/fluid–peridotite wall-rock interaction.

Xenoliths of olivine clinopyroxene cumulates occur in alkali olivine basalt on Takashima Island, part of the Japan volcanic arc. They contain minor orthopyroxene and accessory chrome spinel, and have mosaic equigranular to weakly porphyroclastic textures. They apparently sample arc-related lithosphere; the Mg–Fe distribution between olivine and clinopyroxene suggests subsolidus equilibration at ~800–900°C.

Websterite

Mineralogy: *Opx, Cpx, minor Ol, accessory Sp*
IUGS classification: *Opx (10–90%) + Cpx (10–90%), ± Ol (<5%)*
Texture: *When undeformed, it typically has a coarse-grained granular texture*

Websterite is the ultramafic pyroxenite comprised of orthopyroxene (enstatite) and clinopyroxene (diopside), with only minor amounts of olivine (≤5%) if present. Websterite occurs in layered igneous complexes and forms a minor component in orogenic massifs, where it may also occur as dikes and veins intruding them. Websterite may form as fragments of continental lithospheric mantle and as xenoliths in alkali basalts.

Occurrence

Cumulate websterite is common in layered mafic intrusions (e.g. the Stillwater complex). At the Penikat intrusion (Finland), clinopyroxene forms the intercumulus phase and orthopyroxene the cumulus phase. Websterite cumulates occur in supra-subduction zone ophiolites, where they are attributed to the early saturation of clinopyroxene prior to plagioclase in relatively hydrous basaltic melts at temperatures and pressures typical of the lower oceanic crust. Some are also attributed to partial melting of gabbro from the lower oceanic crustal section (e.g. partial dissolution of gabbro by infiltrating melts and subsequent recrystallization like the Bay of Islands ophiolite). Alternatively, mantle-hosted websterite of the lower oceanic crust may occur as veins in the mantle section and as sheets and lenses in the lower crustal sequence; the latter are thought to represent an early stage of melt percolation and melt–rock reaction beneath an ocean spreading center, as in the Leka ophiolite complex (Norway).

Websterites, such as those occurring in the Balmuccia massif (Italy), represent fragments of the lithospheric mantle emplaced into lower continental crust during extension. Such fragments document a complex history of partial melting, magmatic intrusion, and plastic deformation. They often occur with accessory pargasite and Ti-phlogopite, signifying late metasomatic processes.

Some websterite xenoliths from kimberlite pipes contain garnet ± olivine, making them either garnet websterites or garnet olivine websterites. They generally show subsolidus recrystallization (polygonal grain boundaries, porphyroblasts, neoblasts, etc.). Sometimes granular olivine occurs as inclusions, indicating minimum temperatures consistent with komatiitic melts (~1,500 °C).

Figure 2.35 Websterite dike from the Leka ophiolite complex (Norway). This image is dominated by clinopyroxene, which has second-order birefringence. Exsolution of orthopyroxene (upper-right) and diagnostic twinning (lower-right) are visible. The interstitial pocket (lower-left) is a fine-grained serpentine intergrowth. Cross-polarized light, 4× magnification, field of view = 7 mm.
Image courtesy of B. O'Driscoll

Olivine Websterite

Mineralogy: *Opx, Cpx, Ol, accessory Sp*
IUGS classification: *Ol (5–40%), Cpx (>5%), Opx (>5%)*
Texture: *When undeformed, it typically has a coarse granular texture*

The ultramafic pyroxenite known as olivine websterite is composed of significant orthopyroxene, clinopyroxene, and olivine. Like other ultramafic rocks, Ol-websterite occurs in layered mafic intrusions, ophiolites, and ophiolitic massifs, and as xenoliths entrained in dikes.

Occurrence

In ophiolites and ophiolitic massifs, structural and textural features of olivine websterite suggest it forms during melt–rock reactions that result in pyroxene crystallization at near-peridotite solidus conditions (e.g. the Rhonda massif of southern Spain). The localized replacement of refractory peridotite by websterite occurs when partial melts are channeled through, and react with, the lithospheric mantle during melt migration. In Rhonda, interstitial clinopyroxene, spinel, and orthopyroxene surround olivine grains, suggesting precipitation of interstitial pyroxene + spinel during olivine dissolution. Garnet-bearing pyroxenite veins melt and recrystallize as garnet-free olivine–spinel websterite, while the formation of secondary lherzolite and layered olivine–spinel websterite occurs by refertilization of harzburgite or dunite.

Figure 2.36 Olivine websterite from the Shitoukengde mafic–ultramafic complex (northwest China). Intercumulus orthopyroxene occupies most of the image (single, large crystal with low, second-order blue birefringence). Olivine crystals (second-order yellow to third-order dark blue birefringence) and lesser clinopyroxene (mostly at extinction but with exsolution lamellae just visible in the upper-right corner) are also present. Cross-polarized light, 4× magnification, field of view= 3 mm.

Image courtesy of Y. Liu

Orthopyroxenite

Mineralogy: *Opx, minor Ol and/or Cpx, accessory Grt, Sp, Mt, Il, Cr*
IUGS classification: *Opx (≥90%) ± Ol and/or Cpx (≤10%)*
Texture: *When undeformed, it typically has a massive, coarse-grained phaneritic texture*

ULTRAMAFIC ROCKS

PYROXENITES

Orthopyroxenite is an ultramafic igneous rock in the pyroxenite family that contains more than 90% orthopyroxene (usually enstatitic), with minor (≤10%) olivine and/or clinopyroxene. Orthopyroxenites are found as dikes in the mantle sections of ophiolites associated with peridotite massifs (orogenic peridotites). They also occur as xenoliths or as parts of composite xenoliths entrained in diorite, basalt, and kimberlite.

Occurrence

Mantle orthopyroxenites are associated with ophiolitic massifs (e.g. Semail in Oman, Troodos in Cyprus, and Newfoundland in Canada). Field studies and laboratory experiments invoke fluid/melt–peridotite interaction to explain the formation of orthopyroxenites in ultramafic massifs and as xenoliths; this is consistent with some orthopyroxenites containing resorbed olivine and hydrous minerals such as phlogopite and amphibole. The composition of the fluid/melt, the peridotite protolith, and the specific reaction regime control orthopyroxenite formation.

In layered igneous complexes, orthopyroxenites may be generated through fractional crystallization processes. Cumulate textures are well-documented in the orthopyroxenites of the Bushveld complex (South Africa), the Lyavaraka intrusion (Russia), and the Stillwater complex (Montana, United States). Orthopyroxenites in this setting often represent relatively "more differentiated" magmas. In orthopyroxenites associated with relatively shallow levels of emplacement, orthopyroxene and amphibole may occur together, with amphibole representing a late primary phase and/or the late deuteric alteration of orthopyroxene.

Orthopyroxenite xenoliths in kimberlite pipes may contain garnet, as well as diopside, rutile, Mg–Cr-ilmenite, and titanates. Vein-like occurrences of pyrope, diopside, and Cr-spinel interstitial to enstatite reflect subsolidus exsolution from the orthopyroxene host. Pressures and temperatures associated with orthopyroxene xenoliths in cratonic settings after exsolution has occurred record T = 700–850 °C and P = 25–37 kbar – the orthopyroxenite precursor may have formed at higher temperature and pressure.

Orthopyroxenites also occur on other planets (e.g. the Martian meteorite [ALH 84001] has been classified as an orthopyroxenite although it is the only one found to date with this composition).

Orthopyroxenite xenolith fragment from a kimberlite pipe (Udachnaya-East, Siberian craton). This polished section shows green, coarse-grained orthopyroxenite (right) in contact with red garnet (pyrope)-rich domain (left). The entire fragment is 7 cm across.

Image courtesy of D. Rezvukhin

Figure 2.37

A Orthopyroxenite from the Bushveld complex (South Africa) with partially re-equilibrated grain boundaries. Orthopyroxene (pale brown) occurs with lesser interstitial plagioclase (colorless) and olivine (easier to distinguish in crossed polarizers). The small black crystals are chromite. Plane-polarized light, 2× magnification, field of view = 7 mm.

B Orthopyroxenite from the Bushveld complex (South Africa). Same image as (A) but with crossed polarizers. Orthopyroxene has low (gray) birefringence; one crystal (northwest of center) shows clinopyroxene exsolution lamellae. Olivine (high birefringence) and plagioclase (low birefringence with deformation twins) are visible. Cross-polarized light, 2× magnification, field of view = 7 mm.

C Chromite-rich orthopyroxenite from the Bushveld complex (South Africa). This is a cumulate, chromite-rich orthopyroxenite dominated by chromite (black) and subhedral orthopyroxene (colorless). Plane-polarized light, 2× magnification, field of view = 7 mm.

D Chromite-rich orthopyroxenite from the Bushveld complex (South Africa). Same image as (C) but with crossed polarizers. Orthopyroxene has low (gray) birefringence and chromite is isotropic. Some orthopyroxene shows atypical birefringence because the section does not have uniform thickness. Cross-polarized light, 2× magnification, field of view = 7 mm.

Olivine Orthopyroxenite

Mineralogy: *Opx and Ol, ± Cpx, ± accessory Grt, Sp*
IUGS classification: *Opx (60–90%) + Ol (10–40%), ± Cpx (≤5%)*
Texture: *When undeformed, it typically has a coarse-grained granular texture*

ULTRAMAFIC ROCKS

PYROXENITES

Olivine orthopyroxenite is the pyroxenite composed of orthopyroxene (60–90%) and olivine (10–40%), with only minor amounts of clinopyroxene (≤5%). Orthopyroxene is typically enstatitic and olivine is forsteritic. Olivine orthopyroxenite occurs as a cumulate associated with layered igneous complexes and mantle rocks of island arcs, orogenic massifs as fragments of continental lithospheric mantle, ophiolitic and related oceanic rock associations, and as nodules and xenoliths in alkali basalts and diatremes. Quantitative imaging of trace element zoning and textural analysis of mineral grain size and intergrowths are commonly used to investigate the conditions associated with cumulate mineral growth.

Occurrence

In the Xiarihamu mafic–ultramafic intrusion of northern Tibet, the cumulate olivine orthropyroxenite is composed of subhedral to euhedral granular olivine and subhedral–euhedral orthopyroxene as mesocumulate. Chemical zoning and crystal size distribution analysis of olivine and orthopyroxene from the olivine orthopyroxenite suggests crystallization and growth in an open magmatic system with magma recharge/replenishment.

In orogenic belts, olivine peridotites are associated with the tectonically emplaced slices of subcontinental lithospheric mantle. Mantle orthopyroxenites occur in ophiolitic massifs (e.g. Samail in Oman, Troodos in Cyprus, and Bay of Islands in Canada). Olivine orthopyroxenite in the Havana-Matanzas massif, part of the Northern Cuban Ophiolite Belt, is interpreted to reflect Mg-rich andesitic (boninitic) affinity melt that circulated through, and interacted with, the mantle. Boninitic lavas are generally related to the initial stages of intraoceanic arc formation, thus orthopyroxenites associated with ophiolites may be interpreted as a "fingerprint" of subduction initiation.

Xenoliths of olivine orthopyroxene cumulates occur in volcanically active regions such as arcs and rift zones (e.g. Kamchatka and East Africa, respectively). In both locations they represent samples of the lithospheric mantle, but in Kamchatka this is the subarc mantle wedge, whereas in the East African Rift Zone it is subcontinental lithosphere.

Figure 2.38 Olivine orthopyroxenite from the Bushveld complex (South Africa). This sample of layer B1 shows orthopyroxene (gray) and olivine (high birefringence). Orthopyroxene and olivine have re-equilibrated textures (straight grain boundaries with 120° intersections). The small black crystals are chromite. Cross-polarized light, 2× magnification, field of view = 7 mm.

APPLICATION 2.1

Melt Generation from Mantle Rocks

The largest volumes of basalt on Earth are generated via partial melting of mantle peridotite beneath mid-ocean ridges. Mantle peridotite is mainly composed of varying amounts of clinopyroxene, orthopyroxene, and olivine. Each of these minerals has a different stability field with respect to pressure, temperature, and composition, so the mantle does not melt uniformly. When minerals with different melting points are combined, melting reactions occur at *lower temperatures* than the individual-phase melting points. Thus we generate a "partial" melt.

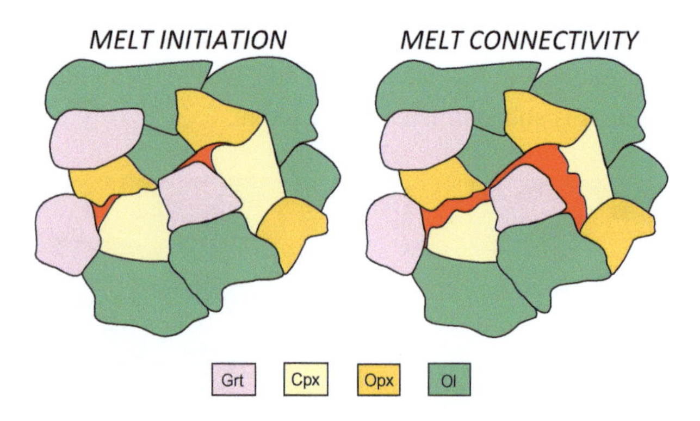

Partial melts are generally more evolved (less mafic) than the composition of their starting material. The compositional diversity of clinopyroxene (Ca, Al, Ti, etc.), relative to orthopyroxene or olivine (Fe, Mg), and the addition of an aluminous phase (Pl, Sp, Grt) lowers the melting point and facilitates partial melting of peridotite. Compositional diversity at mineral contacts, for example when Cpx is in contact with other minerals, promotes melting. Partial melting of a garnet-bearing mantle (see diagram) highlights this process. During mantle partial melting (*left*), melt pockets (red) originate at the contacts between Cpx, Opx, and Grt. As melting continues, Cpx and Grt are increasingly consumed (along with other phases) until enough melt is generated that previously isolated melt pockets are now connected (*right*). Typical melting reactions might include:

Pl peridotite: Ol + Opx + Cpx + Pl ⟶ Ol ± Opx + melt

Spl peridotite: Cpx + Opx + Spl ⟶ Ol + melt

Grt peridotite: Ol + Cpx + Grt ⟶ Opx + melt

Experiments have determined the melting relations of mantle peridotite and documented the general trends in melt composition from low- to high-pressure and variable volatile contents. Such studies have shown that partial melting of mantle peridotite will leave a residue of depleted harzburgite and produce a tholeiitic mid-ocean ridge-type basalt. The table below summarizes the effects of water and temperature on partial melting of mantle peridotite. At constant pressure, a small increase in H_2O lowers the melting point and increases the amount of melt generated from a lherzolitic mantle. With increasing temperature, water-saturated systems also produce more melt from peridotite.

Investigations of ophiolites have shown that polybaric melting of mantle peridotite involves different generations of partial melt migrating through the mantle. Multiple episodes of *depletion* (partial melt extraction and removal), as well as *refertilization*

Volatiles, temperature, and melt production

Spinel lherzoilte[1]	H_2O (wt%)	Melt (wt%)	Peridotite[2]	H_2O (wt%)	Melt (wt%)
1,300 °C/1 GPa	0.2	12	1,200 °C	0.25	3
1,300 °C/1 GPa	0.5	17	1,300 °C	0.25	12
1,300 °C/1 GPa	0.8	20			

1, Hirose & Kawamoto (1995); 2, Gaetani & Grove (1998).

(partial melt infiltrating previously depleted mantle), are typical throughout the melt column and generate distinct textures and chemistry (Koga et al., 2001; Piccardo et al., 2007; Rampone et al., 2020).

This hand sample of dunite from the Semail ophiolite (Oman) is ~5 cm across and made of Ol > Pl > Cpx + Spl. Olivine has a mosaic texture in thin-section. The sample is interpreted to have been refertilized by partial melt due to the presence of plagioclase and the relative absence of Cpx.

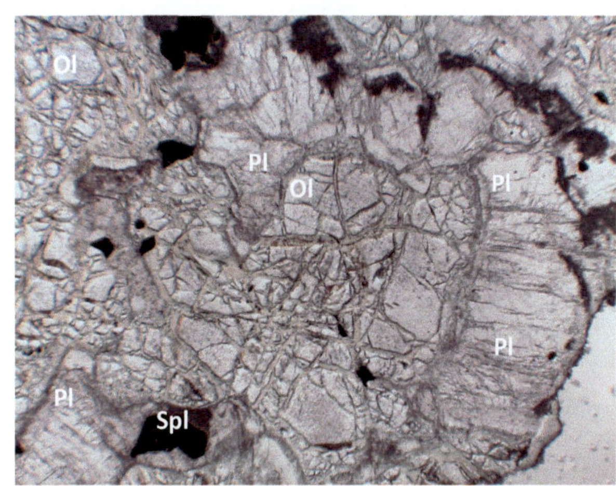

Dunite of the Semail ophiolite in thin-section. Highly fractured and partially serpentinized olivine (high relief) dominates the thin-section. Thin, irregular reaction rims occur between olivine and mantling plagioclase. Black mats are Pl altered to clay. Plane-polarized light, 4× magnification, field of view = 3 mm.

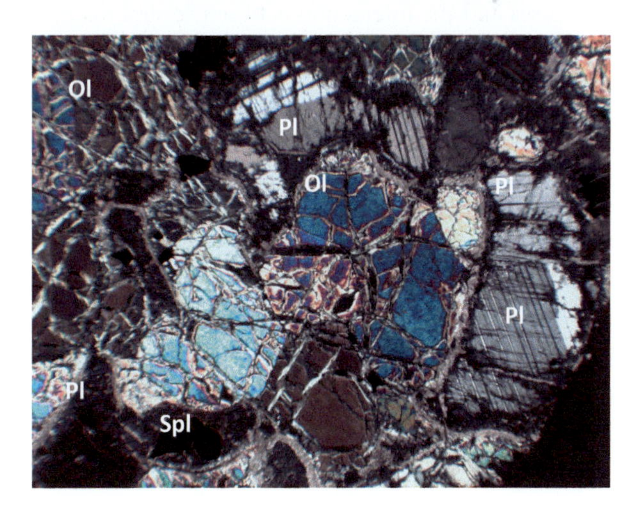

Dunite of the Semail ophiolite in thin-section. Same image as above but under crossed polarizers. Olivine has high birefringence and Pl twinning is evident. Cross-polarized light, 4× magnification, field of view = 3 mm.

Diorite

GABBROIC ROCKS

Mineralogy: *Pl, Hbl, Bt, ± Px, ± Ol*
IUGS classification: *Pl (≥90%) ± Qz (<5%) ± Ksp (<5%)*
Texture: *Phaneritic holocrystalline*

Diorite is a coarse- to medium-grained intermediate plutonic rock, the compositional equivalent of andesite. Diorite is composed of intermediate plagioclase (oligoclase to andesine) and mafic minerals such as amphibole (commonly hornblende), ± biotite, Ca-pyroxenes, and more rarely olivine. Minor amounts of alkali feldspar (<10%) and quartz (<5%) may be present. Zircon, apatite, magnetite, ilmenite, and titanite are common accessory minerals. With increasing quartz content, diorite grades into *quartz diorite* and *tonalite*; with increasing alkali feldspar content, it grades into *monzodiorite*; if feldspathoids are present (up to 10%), the rock is classified as *foid-bearing diorite*. Diorite is similar to gabbro, but contains more Na-plagioclase, whereas gabbro contains Ca-rich plagioclase (labradorite or bytownite).

Occurrence

Diorite is associated with magmatic arcs and cordilleran-like orogens. It often occurs in the marginal zones of granitic batholiths associated with these settings. The genesis of diorite involves numerous processes and the interaction between several distinct magma sources. Once the fluids derived from the subducting slab metasomatize the asthenospheric mantle wedge, primary mafic magmas are generated; these interact and re-equilibrate with the surrounding mantle peridotite. In addition to ongoing fractional crystallization processes, these mantle-derived melts may interact and assimilate crustal material en-route to shallower levels. This scenario has been hypothesized for the Corno Alto magmatic complex on the eastern border of the Adamello batholith, using major, minor, and rare earth element (REE) chemistry, isotopic analyses, and geochemical and thermodynamic modeling. In some circumstances, diorite can form an economically important resource if it is associated with Au–Cu deposits. For example, in some Archean greenstone belts, granodiorite–diorite–gabbro and tonalite–trondhjemite–diorite intrusive complexes concentrate fluids and metals in the residual melt via fractional crystallization of Cpx and Pl, leading to the formation of diorite-hosted Au–Cu disseminated sulfide deposits. In granitic complexes, diorite microgranular enclaves are common and may indicate mixing–mingling processes between magmas of different composition. Moreover, the close association between magmas of different composition is observed in numerous igneous complexes (e.g. southern Chile, western United States, etc.).

Diorite from Massachusetts (United States). The sample comprises pyroxene (dark) and plagioclase (white).
Image courtesy of J. St. John

Figure 2.39

A Diorite from Sondrio (Italy). Plagioclase crystals (colorless), hornblende (green–brown), and magnetite (black) are visible. The large hornblende crystal (central image) poikilitically encloses plagioclase. Plane-polarized light, 2× magnification, field of view = 7 mm.

B Diorite from Sondrio (Italy). Same image as (A) but with crossed polarizers. Plagioclase crystals show their characteristic gray interference colors, while hornblende shows high first-order colors (orange to brown). Cross-polarized light, 2× magnification, field of view = 7 mm.

C Diorite from Sondrio (Italy). Plagioclase crystals (colorless), hornblende (tan to green), biotite (reddish-brown), magnetite (black), and quartz (center, below the large plagioclase) are present. Note the high relief of plagioclase relative to quartz. Plane-polarized light, 2× magnification, field of view = 7 mm.

D Diorite from Sondrio (Italy). Same image as (C) but with crossed polarizers. Plagioclase shows characteristic gray interference colors, while hornblende has a tan birefringence. Quartz is anhedral and interstitial, easily distinguished from the sub- to euhedral plagioclase. Cross-polarized light, 2× magnification, field of view = 7 mm.

Gabbro

Mineralogy: *Pl, Cpx, ± Opx, ± Hbl, ± Ol*

GABBROIC ROCKS

IUGS classification: *Pl (10–90%) + Cpx (10–90%) ± Opx (≤5%)*

Texture: *Phaneritic holocrystalline*

Gabbro is a phaneritic mafic plutonic rock with a composition equivalent to a basalt. Gabbro consists of calcic plagioclase (An_{50}–An_{100}) and clinopyroxene (augite, diopside), with minor amounts of iron oxides (ilmenite, magnetite). Olivine, hornblende, biotite, orthopyroxene, and spinel may also be present as accessory minerals. Gabbro may be silica-saturated or -undersaturated: If it contains 5–20% quartz, it is classified as *quartz-gabbro*; if feldspathoids are present (≤10%), it is classified as a *foid-bearing gabbro*. With increasing Ksp, it is classified as monzogabbro and foid-bearing monzogabbro. Gabbro is often altered – plagioclase is transformed to saussurite (a mixture of various minerals such as zoisite, albite, calcite, and muscovite) and pyroxene is altered to uralite.

Occurrence

Gabbro forms a large part of the oceanic crust and "oceanic gabbro" is distinguished from continental gabbro. In ophiolitic crust gabbro occurs below the sheeted dike complex, and in such settings is divided into an upper zone of isotropic gabbro and a lower zone of layered gabbro. The gabbroic layered sequence represents crystal mush accumulated during fractional crystallization processes of mid-ocean ridge basalts. Gabbro (together with basalt) is also associated with the magmatism of subduction zones. The most common occurrence of continental gabbro is in the lower and central parts of large, layered mafic and ultramafic complexes. Layered mafic intrusions are generally characterized by ultramafic rocks (peridotites and pyroxenite) at the base, followed upwards by alternating layers of more evolved mafic rocks (norites, gabbros, and anorthosites). In such settings, gabbro forms in the central part of the intrusion via in-situ crystallization of pyroxene and plagioclase, or by crystal settling of pyroxene and plagioclase. Pseudo-sedimentary structures – such as flow banding, slumping and folding, normal or reverse graded bedding – can be formed. Some well-studied examples include the Bushveld igneous complex (South Africa) and the Skaergaard (Greenland). Other occurrences of continental gabbro include the interior parts of thick basaltic lava piles and accumulation beneath the large volumes of flood basalts associated with large igneous provinces (e.g. the Columbia River flood basalts of Washington and Oregon [United States], or the Deccan Traps [India]). Continental gabbro is less abundant than oceanic gabbro, mainly because the high density of gabbro in a continental setting lacks the necessary buoyancy to reach the surface.

Gabbro from the Stillwater complex (United States). Grayish-white plagioclase and black pyroxene. Image ~12 cm wide.

Image courtesy of J. St. John

Figure 2.40

A Gabbro from Skaergaard (Greenland). Plagioclase (colorless) partially enclosed by intercumulate clinopyroxene (pale brown). Plagioclase is fractured and exsolution in Cpx partially visible (e.g. crystal on the right). Plane-polarized light, 2× magnification, field of view = 7 mm.

B Gabbro from Skaergaard (Greenland). Same image as (A) but with crossed polarizers. Plagioclase is twinned and clinopyroxene shows high birefringence. Cross-polarized light, 2× magnification, field of view = 7 mm.

C Gabbro from the Stillwater complex, Montana (United States). Clinopyroxene has high birefringence, lamellar to bleb-like pigeonite exsolution (white or black depending of extinction position), and is surrounded by intercumulate plagioclase. Cross-polarized light, 2× magnification, field of view = 7 mm.

D Hornblende gabbro from Tappeluft (Norway). Plagioclase (gray interference colors and twinned), hornblende (brown–blue-green birefringence), and interstitial quartz are present (yellowish). Cross-polarized light, 2× magnification, field of view = 7 mm.

Norite

Mineralogy: *Pl, Opx, ± Cpx*

GABBROIC ROCKS

IUGS classification: *Pl (10–90%) + Opx (10–90%), ± Cpx (≤5%)*

Texture: *Phaneritic holocrystalline*

Norite is a coarse-grained basic plutonic rock composed of Ca-plagioclase (bytownite, labradorite, or andesine) and orthopyroxene (enstatite or hypersthene), with minor (≤5%) clinopyroxene (augite). Fe–Ti oxides, apatite, and zircon are common accessory minerals. Gabbro and norite are visually indistinguishable, and require a thin-section for proper characterization. Orthopyroxene is typical of norite (sometimes also known as "orthopyroxene gabbro"), whereas clinopyroxene is dominant in gabbro.

Occurrence

Like gabbro, norite is found in numerous tectonic settings from oceanic crust and ophiolite complexes, to layered mafic intrusions. In layered mafic intrusions, such as the Bushveld igneous complex (South Africa) or the Skaergaard igneous complex (Greenland), norite is closely associated with other mafic to ultramafic rocks. It is commonly associated with economically important ore deposits (Au–Cu, Cr, and platinum), the result of in-situ crystallization or fractional crystallization of pyroxene and plagioclase. Norite can also be found in impact melt sheets like the Sudbury complex (Canada), one of the best exposed impact melt sheets on Earth and the world's second-largest nickel deposit. In this context, norite resulted from the fractional crystallization of a homogeneous granodioritic magma produced from the fusion of crustal rocks.

Norite from the Stillwater complex, Montana (United States). Plagioclase (white–gray) and pyroxene (dark). Sulfides (gold–brown) are scattered across the sample. Image 15 cm wide.

Image courtesy of J. St. John

Figure 2.41

A Norite from the Stillwater complex, Montana (United States). Plagioclase (colorless), subhedral orthopyroxene (moderate relief, fractured, and brown) and interstitial augite (with visible cleavage). Plane-polarized light, 2× magnification, field of view = 7 mm.

B Norite from the Stillwater complex, Montana (United States). Same image as (A) but with crossed polarizers. Plagioclase (twinned and dark), orthopyroxene (low birefringence and fractured), and augite (blue birefringence) poikilitically enclosing plagioclase. Cross-polarized light, 2× magnification, field of view = 7 mm.

Troctolite

Mineralogy: *Pl, Ol, ± Cpx*
IUGS classification: *Pl (10–90%) + Ol (10–90%)*
Texture: *Phaneritic holocrystalline*

Troctolite is a coarse-grained mafic plutonic rock composed almost entirely of plagioclase and olivine; minor (<5%) clinopyroxene may be present. The plagioclase is typically labradorite or bytownite. Olivine is often partially to totally altered to serpentine, giving the rock a speckled appearance. Troctolite is also known as olivine-rich anorthosite.

Occurrence

Troctolite is commonly associated with other gabbroic and ultramafic rocks in ultramafic layered intrusions, such as the Duluth complex (Minnesota, United States), the Rhum layered intrusion (Scotland), and the Lizard complex in Cornwall (England). Troctolite also occurs in fossil sections of oceanic lithosphere (Jurassic ophiolites of the Internal Ligurian units, Italy) and the gabbroic parts of oceanic core complexes associated with slow- and ultraslow-spreading ridges (the Atlantis Massif, Mid-Atlantic Ridge), and back-arc basins (Godzilla Megamullion, Philippine Sea).

Unlike gabbro, the composition of troctolite does not correspond to any melt derived from mantle peridotite; its origin is mostly linked to fractional crystallization processes in layered igneous complexes, or from melt–rock interactions between an olivine-rich matrix and MORB-type melts at the mantle–crust transition. During this process, reactive dissolution–precipitation between melt and mineral matrix occurs, which leads to the chemical modification of migrating melts. This contributes to the formation of the lower oceanic crust through the incorporation of harzburgite–dunite mantle slivers at the base of the oceanic crust, subsequently transformed into olivine-rich gabbroic rocks by melt percolation.

Troctolite from the Stillwater complex, Montana (United States). The sample is composed of plagioclase (white–gray) and olivine (reddish to black; the red color is due to alteration). Hand lens ~2 cm wide.

Image courtesy of J. St. John

Figure 2.42

A Troctolite from the Stillwater complex, Montana (United States). Plagioclase (colorless) and olivine (high relief, pale brown, and fractured) are present. Some olivine grains form equilibrium 120° junctions. Plane-polarized light, 2× magnification, field of view = 7 mm.

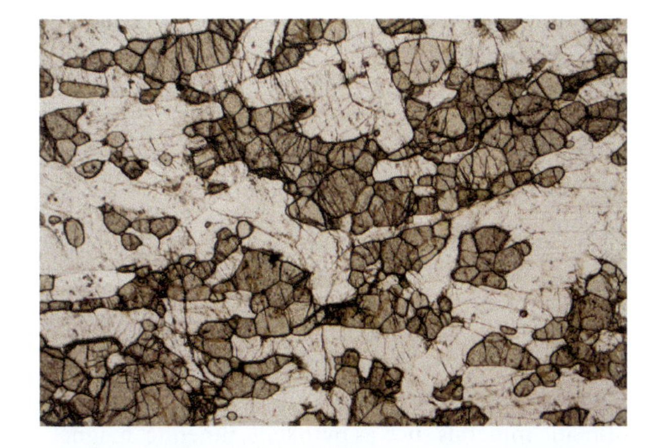

B Troctolite from the Stillwater complex, Montana (United States). Same image as (A) but with crossed polarizers. Plagioclase is twinned and olivine has high birefringence. Plagioclase crystals roughly define a magmatic foliation (horizontal). Cross-polarized light, 2× magnification, field of view = 7 mm.

C Troctolite from Rhum (Scotland). Olivine, plagioclase, and pyroxene are present. Intercumulus olivine (center) has radial fractures due to serpentinization. Note the fractures propagate into the surrounding, twinned plagioclase crystals. Augite has blue interference colors (center-left). Cross-polarized light, 2× magnification, field of view = 7 mm.

D Troctolite from Skaergaard (Greenland). Large, twinned plagioclase crystal (gray birefringence) poikilitically enclosing subhedral, high-birefringence olivine crystals. Cross-polarized light, 10× magnification, field of view = 2 mm.

Anorthosite

Mineralogy: *Pl, ± Px, ± Ilm, ± Ol*
IUGS classification: *Pl (90–100%), Px (≤10%)*
Texture: *Coarse-grained holocrystalline*

GABBROIC ROCKS

Anorthosite is a coarse-grained, leucocratic plutonic rock consisting of 90–100% plagioclase (usually labradorite or bytownite) and minor amounts (0–10%) of mafic minerals such as pyroxene, ilmenite, magnetite, and/or olivine. Labradorite often displays a characteristic iridescence or chatoyance (also known as schiller or labradorescence) that is derived from extremely slow cooling rates. In some anorthosites, plagioclase crystals reach considerable size (>1 m).

Occurrence

The distinctive composition, texture, tectonic environment, rock associations, and in some cases age restrictions, permit five types of anorthosite to be recognized.

Archean anorthosites are found in Archean greenstone belts and associated with mafic rocks; they are particularly widespread in Greenland and the Superior Province of Canada. Archean anorthosites consists of Ca-plagioclase (>An_{80}) megacrysts up to 30 cm in size, set in a mafic groundmass. They have a strong metamorphic imprint of greenschist to granulite facies. The Windimurra Complex in Australia (2,250 km^2) is also associated with mafic rocks and Ca-plagioclase megacrysts, suggesting a genetic link. The most accredited hypothesis is that anorthosites represent magma chambers from which mafic lavas have undergone fractional crystallization processes.

Massif-type Proterozoic anorthosites are the most abundant of all anorthosite types. They constitute large (pluton to batholith-sized) composite igneous complexes associated with leuco-gabbro, leuco-norite, and leuco-troctolite rocks. They are dominated by plagioclase crystals of intermediate (An_{50}) composition and mafic minerals such as pyroxene, ± olivine, Fe–Ti oxides, and apatite. The Lac-Saint-Jean complex of Quebec (20,000 km^2) is one of the largest examples. The most accepted model links the formation of large massif-type anorthosite complexes to basaltic underplating; the subsequent formation of mafic silicates presumed to sink forming cumulates, while the plagioclase-rich cumulate "float" forms an anorthositic mush. This also explains the scarcity of mafic and ultramafic rocks associated with this type of complex.

Anorthosites associated with layered mafic complexes are quite common, where they alternate with other cumulate rocks and form thick layers up to 100 m or more. These anorthosites consist of small (1 cm) plagioclase crystals with cumulate textures, and their composition commonly varies from layer to layer, reflecting fractional crystallization processes.

Anorthosites of oceanic settings are found in mid-oceanic ridge and fracture-zone settings. They represent the remains of basaltic magmatic differentiation in magmatic systems below mid-oceanic ridges.

Anorthosite xenoliths occur in many igneous rocks, from kimberlite to basalt to granite. In some cases, anorthosite xenoliths have been entrained in the magma, but in other cases they represent cognate xenoliths derived from their host magmas.

Figure 2.43

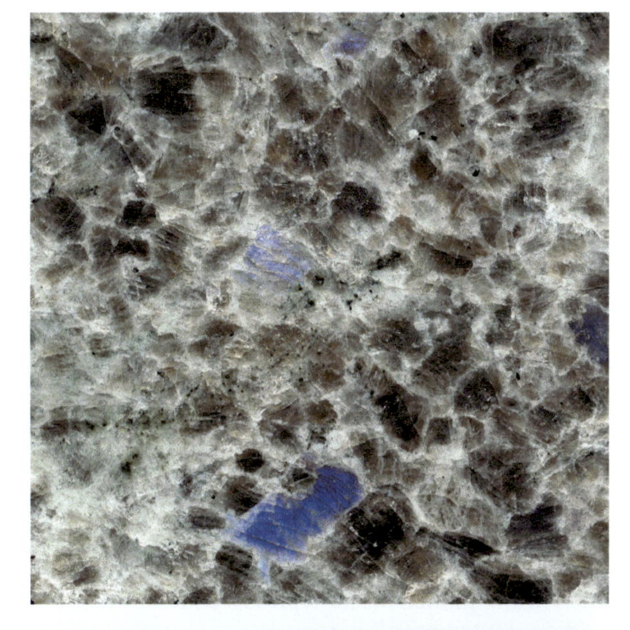

A Anorthosite from Labrador (Canada). Anorthosite is composed almost entirely of plagioclase (white, gray, and brownish) with minor oxides. Note the two crystals showing typical blue iridescence/chatoyance. Polished hand-sample, image ~4 cm wide.
Image courtesy of J. St. John

B Anorthosite (massif-type) from Rogaland (Norway). The whole rock is almost completely made up of plagioclase crystals in close contact with each other. The low-interference crystals (center) are orthopyroxene. Cross-polarized light, 2× magnification, field of view = 7 mm.

C Anorthosite (layered mafic intrusion-type) from the Stillwater complex, Montana (United States). Large, partially interpenetrating plagioclase crystals. Cross-polarized light, 2× magnification, field of view = 7 mm.

Monzogabbro–Monzodiorite

Mineralogy: *Kfs, Pl, Hbl, Bt, Qz, ± Cpx,*

IUGS classification: *Kfs (10–35%) ± Pl (65–90%), Qz (≤5%)*

Texture: *Phaneritic holocrystalline*

Monzogabbro and monzodiorite are intermediate coarse-grained plutonic rocks consisting of plagioclase, alkali feldspar, hornblende, biotite, and with or without pyroxene or quartz. Plagioclase is the dominant feldspar (65–90%) and its composition is used to discriminate between monzogabbro (>An_{50}) and monzodiorite (<An_{50}). Quartz, if present, is minor (≤5%). Monzogabbro contains 25–60% mafic minerals, while monzodiorite typically contains 15–50% mafic minerals. If quartz increases to 5–20%, monzogabbro and monzodiorite grade into *quartz–monzogabbro* and *quartz–monzodiorite.* In the absence of quartz and with up to 10% feldspathoids, monzogabbro and monzodiorite grade into *foid-bearing monzogabbro* and *foid-bearing monzodiorite.* Common accessory minerals include ilmenite, apatite, zircon monazite and sulfides, e.g. pyrrhotite, pentlandite, chalcopyrite, sphalerite, and galena.

Occurrence

Monzogabbro and monzodiorite are generally calc-alkaline to shoshonitic. They are associated with granitic to syenitic rocks and rarely form independent intrusions. Monzogabbro and monzodiorite are relatively common in collisional, post-collisional, and subduction-related tectonic settings. Their formation can be linked to different petrological processes associated with these settings: (1) the inception of continent–continent collision (e.g. the Pontide Orogenic Belt of Turkey [partial melting of a metasomatized lithospheric mantle source contaminated by upper-crustal material]); (2) post-orogenic igneous activity (e.g. the Arabian–Nubian Shield [extensive olivine and pyroxene fractional crystallization of a subduction-modified, metasomatized mantle source]); and (3) back-arc extension (e.g. monzonitic cordilleran plutonism of Patagonia [deep, high-temperature partial melting of an enriched garnet-bearing mantle source distal to the trench, or remelting of a deep underplated amphibole- + garnet-bearing, plagioclase-poor, high-K mafic source]).

Monzodiorite (orbicular-type) from North Carolina (United States). The mafic orbicules are composed of pyroxene and hornblende, while the light-colored matrix is composed of a mixture of plagioclase (andesine) and microcline. Image ~9 cm wide.

Image courtesy of J. St. John

Figure 2.44

A Monzodiorite from unknown locality (Czechia). Plagioclase dominates the thin-section and minor orthoclase is also present (both are colorless). Mafic minerals include biotite (brown) and hornblende (green). In the upper-left, a larger hornblende crystal has a relict pyroxene (pale brown) core. Plane-polarized light, 2× magnification, field of view = 7 mm.

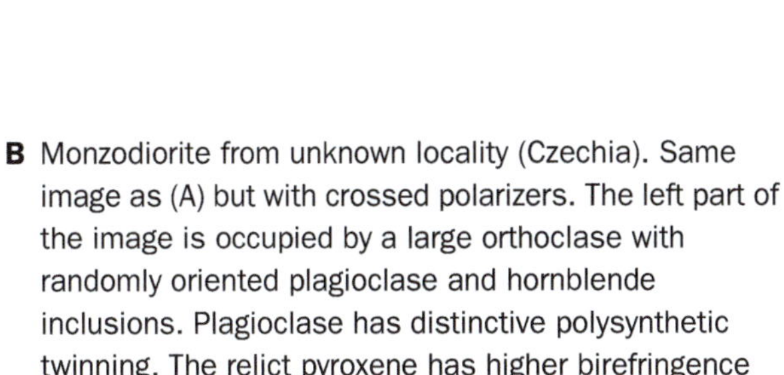

B Monzodiorite from unknown locality (Czechia). Same image as (A) but with crossed polarizers. The left part of the image is occupied by a large orthoclase with randomly oriented plagioclase and hornblende inclusions. Plagioclase has distinctive polysynthetic twinning. The relict pyroxene has higher birefringence than hornblende. Cross-polarized light, 2× magnification, field of view = 7 mm.

C Monzodiorite from Traversella (Italy). A large twinned orthoclase (center) dominates the image and is surrounded by smaller plagioclase and biotite crystals. Cross-polarized light, 2× magnification, field of view = 7 mm.

D Monzodiorite from Traversella (Italy). Polysynthetic twinned plagioclase crystals surrounded by interstitial orthoclase (without twinning). The mafic mineral is pyroxene and the central pyroxene crystal has lamellar twinning (oriented NE–SW). Cross-polarized light, 2× magnification, field of view = 7 mm.

Monzonite

Mineralogy: *Kfs, Pl, Hbl, Cpx, Bt, Qz*

IUGS classification: *Kfs (35–65%) + Pl (35–65%), Qz (≤5%)*

Texture: *Phaneritic holocrystalline*

SYENITIC ROCKS

Monzonite is an intermediate plutonic rock composed of nearly equal amounts of potassium feldspar (typically orthoclase) and sodic plagioclase (oligoclase to andesine). Potassium feldspar often forms large poikiolitic crystals enclosing euhedral plagioclase – this feature is commonly called "monzonitic texture." Quartz is rare (≤5%). Monzonite grades into *quartz monzonite* with 5–10% quartz, and if <10% feldspathoids are present into *foid-bearing monzonites.* Hornblende, biotite, and augite are the main mafic minerals, and zircon, apatite, allanite, diopside, titanite, rutile, olivine, and magnetite form accessory minerals. With increasing plagioclase content, monzonite grades into *monzodiorite*, and with increasing potassium feldspar into *syenite.* Monzonite is the intrusive compositional equivalent of latite–trachyandesite.

Occurrence

Quartz-free monzonite is uncommon and may occur as rare, independent plutons or more commonly as small heterogeneous bodies or marginal masses associated with bigger granitic intrusions. It occurs in both continental crust and ocean island environments. Monzonite and quartz monzonite occur in orogenic belts such as the Andes, where they are closely associated with granitoid and/or gabbroic rocks. The formation of granitoid and monzonitic rocks in orogenic belts is usually linked to the final stages of orogeny in the post-collisional extensional regime, or to extension in the back-arc of subduction zones; however, the generation of monzonitic rocks can also occur at the onset of orogeny. The formation of orogenic granitoids and associated monzonitic rocks is attributed to differentiation of mantle-derived magmas by fractional crystallization, partial melting of a continental crust source triggered by intrusion or underplating of mantle-derived magma, and/or mixing and mingling between coeval mantle and crustal-derived melts. Monzonite is often associated with economically important porphyry copper deposits.

Monzonite from Butte, Montana (United States). It is defined by gray to white plagioclase, pale pink alkali feldspar, and dark biotite. Image ~5 cm wide.

Image courtesy of J. St. John

Figure 2.45

A Monzonite from Monzoni (Italy). Orthoclase and plagioclase dominate the image, but both are colorless and difficult to distinguish from each other; however, orthoclase (on the left) has weak, concentric growth zoning. Hornblende (green to brown) and biotite (rectangular and tan) are also present. Plane-polarized light, 2× magnification, field of view = 7 mm.

B Monzonite from Monzoni (Italy). Same image as (A) but with crossed polarizers. Orthoclase has Carlsbad twinning. Plagioclase is zoned from core to rim and has polysynthetic twins. Hornblende shows lower birefringence and biotite orange–pink birefringence. Cross-polarized light, 2× magnification, field of view = 7 mm.

C Monzonite from Crevola (Italy). A single, large poikilitic crystal of orthoclase (colorless) encloses numerous inclusions of hornblende (green), biotite (red–brown), plagioclase (also colorless and difficult to identify), and opaque minerals (black). Plane-polarized light, 2× magnification, field of view = 7 mm.

D Monzonite from Crevola (Italy). Same image as (C) but with crossed polarizers. A single, large poikilitic orthoclase (gray birefringence) encloses small euhedral plagioclase (monzonitic texture) and hornblende (orange birefringence). Cross-polarized light, 2× magnification, field of view = 7 mm.

Syenite

Mineralogy: *Kfs, Na–Pl (>10%), Cpx, Hbl, Bt, Qz*

SYENITIC ROCKS

IUGS classification: *Kfs (65–90%) + Pl (10–35%), Qz (≤5%)*

Texture: *Phaneritic holocrystalline*

Syenite is dominated by alkali feldspar (usually orthoclase, microcline, or perthite) with minor amounts of Na-rich plagioclase (An_{20}–An_{40}) and minor (≤5%) quartz, if present. Syenite grades into *quartz-syenite* when quartz increases to 5–20%, and into *foid-bearing syenite* if feldspathoids are present instead of quartz. Syenite usually contains 10–35% mafic minerals such as hornblende, augite, or, less commonly, biotite. If Na-pyroxenes (aegirine) and/or Na-amphibole (arfvedsonite or riebeckite) are present, the term *peralkaline syenite* is used. Common accessory minerals include zircon, apatite, titanite, allanite, ilmenite, and magnetite. Syenite represents the plutonic compositional counterpart of trachyte.

Occurrence

Syenite usually occurs as relatively small single intrusions or as localized igneous bodies related to larger intrusions of different compositions (from mafic to silica-rich, including carbonatitic–kimberlitic complexes) in almost all tectonic settings. In many intrusions, syenite is comagmatic/cogenetic with granite and alkali feldspar granite, and they tend to form marginal igneous facies. In this case, syenite may evolve from the granitic magma through the assimilation of mafic or carbonate rocks and the release of volatiles with dissolved silica, which allows the removal of silica and the increase of MgO, Fe, TiO_2, CaO, and Na_2O. Syenite may form from: (1) partial melting of phlogopite-bearing metasomatized lithospheric mantle (e.g. the syenites of the Damara Orogen [Namibia]); (2) fractional crystallization of K-rich basaltic magmas, or partial melting of alkali-rich mafic rocks such as metagabbro and metamonzonite (e.g. the syenites of Sinai [Egypt]); and (3) mixing of mafic and silicic magmas followed by subsequent differentiation of the resultant hybrid melts (e.g. Transbaikalian syenites [Russia]).

Syenite from Ben Loyal (Scotland). This sample is dominated by anti-perthitic alkali feldspar (pink) and also contains pyroxene (dark green aegirine) and amphibole (arfvedsonite). Image ~6 cm wide.

Image courtesy of A. Tindle

Figure 2.46

A Syenite from Val Cervo (Italy). The image is dominated by orthoclase (colorless with minor faint brownish sericite alteration), and hornblende (brown) and magnetite (black) are also present. Plane-polarized light, 2× magnification, field of view = 7 mm.

B Syenite from Val Cervo (Italy). Same image as (A) but with crossed polarizers. Orthoclase (gray birefringence) shows concentric zoning and perthitic exsolution (oriented NW–SE). Hornblende shows low birefringence. Cross-polarized light, 2× magnification, field of view = 7 mm.

C Syenite from Biella (Italy). Orthoclase has simple Carlsbad twins, perthitic exsolution, and gray birefringence, while hornblende shows low to moderate birefringence. Plagioclase (center) has polysynthetic twinning and is surrounded by interstitial quartz. Cross-polarized light, 2× magnification, field of view = 7 mm.

D Syenite (larvikite) from Larvik (Norway). Large ternary feldspar (gray birefringence) with cleavage, surrounded by augite (green to yellow birefringence and black magnetite inclusions) and biotite (red birefringence). Cross-polarized light, 2× magnification, field of view = 7 mm.

Alkali Feldspar Syenite

SYENITIC ROCKS

Mineralogy: *Kfs, Na–Pl, Cpx, Hbl, Qz, Bt*
IUGS classification: *Ksp* ($\geq$*90%)* $\pm$ *Pl* *(*$\leq$*10%)* $\pm$ *Qz* *(*$\leq$*5%)*
Texture: *Phaneritic holocrystalline*

Alkali feldspar syenite is dominated by alkali feldspar, with little (<10%) plagioclase of Na-rich composition. Mafic minerals such as hornblende, or less commonly biotite, are also present (generally <25%). Accessory mineral phases include apatite, zircon, titanite, ilmenite, magnetite, and titanomagnetite. Olivine is rare but possible. If the rock contains alkali-bearing amphibole (arfvedsonite–riebeckite) and pyroxene (aegirine), the name *peralkaline syenite* is used. Alkali feldspar syenites may be silica-saturated or -undersaturated, and if they contain 5–20% quartz they are classified as *quartz alkali feldspar syenite*; when they contain feldspathoids ≤10% they are classified as *foid-bearing alkali feldspar syenites.*

Occurrence

Alkali feldspar syenites are not very common and are often closely associated with granites (chiefly A-type) and syenites in many igneous complexes of diverse tectonic settings including continental arcs, post-collisional arcs, intraplate volcanism, or in early- and/or late-stage oceanic arcs. In many igneous complexes, alkaline-type granite and alkali feldspar syenites are cogenetic and in some cases alkali feldspar syenites are derived from granitic magma by fractional crystallization processes.

Alkali feldspar syenite from Wausau, Wisconsin (United States). The rock consists of red alkali feldspar, black hornblende, and mafic clots (larger black spots). Image ~8 cm wide.

Image courtesy of J. St. John

Figure 2.47

A Alkali feldspar syenite from Biella (Italy). Orthoclase (colorless and slightly cloudy due to sericite alteration), light to dark green hornblende, and small black magnetite are present. Plane-polarized light, 2× magnification, field of view = 7 mm.

B Alkali feldspar syenite from Biella (Italy). Same image as (A) but with crossed polarizers. Orthoclase (gray birefringence) shows perthite exsolution. The large crystal in the center has simple Carlsbad twinning. Hornblende has lower birefringence. Cross-polarized light, 2× magnification, field of view = 7 mm

C Peralkaline syenite from Lovozero (Russia). Aegirine (high birefringence) shows typical "cusp" crystal terminations and defines a magmatic foliation. Orthoclase (gray birefringence) is also aligned with the magmatic foliation. Cross-polarized light, 2× magnification, field of view = 7 mm.

BOX 2.8 Describing Igneous Rocks

A thorough description of an igneous rock will include the following:

- **Crystallinity**
  obsidian, glass without crystals
  holohyaline, all glass
  hypohyaline, glass with crystals
  holocrystalline, all crystals
- **Crystal visibility**
  aphanitic, no crystals visible
  phaneritic, crystals visible
  pegmatitic, very coarse-grained
- **Relative grain size**
  equigranular, crystal grains are about the same size
  inequigranular, crystal grain sizes noticeably differ
  seriate, grain size of primary minerals is variable
  porphyritic, larger crystals in a finer-grained groundmass
- **Absolute grain size**
  coarse-grained, >5 mm
  medium-grained, 1–5 mm
  fine-grained, <1 mm
- **Crystal form**
  euhedral, well-formed crystal faces visible (idiomorphic)
  subhedral, some well-formed crystal faces visible (hypidiomorphic)
  anhedral, crystal faces lacking (allotriomorphic)
- **Crystal shape**
  granular, crystals are equidimensional (olivine)
  tabular or platy, crystals are flat and narrow like a table top (biotite)
  bladed, crystals are flat and thin, lath-like (feldspar laths)
  acicular, crystals are long and narrow, needle-like (fibrous sillimanite)

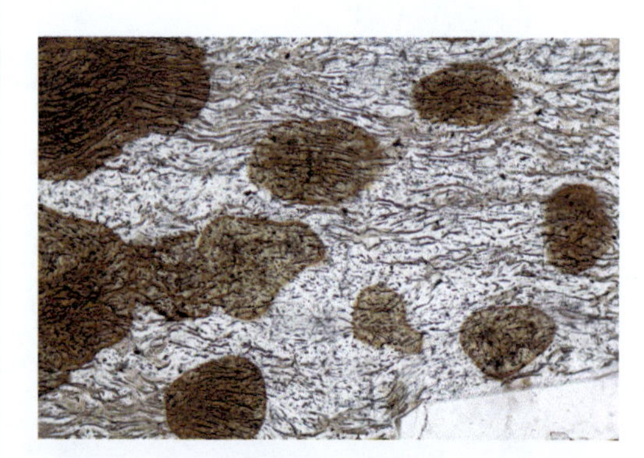

Holohyaline rhyolite (New Zealand) with rounded, brown (altered) spherulites (plane polarizers) *upper-left*. Holocrystalline alkali feldspar granite from Amparo (Brazil) (crossed polarizers), *upper-right*. Porphyritic tephrite from Bolsena volcano (Italy) (plane polarizers), *lower-left*. Granular dunite with equigranular olivine crystals and oxides (crossed polarizers, 10× magnification, field of view = 2 mm), *lower-right*. All at 2× magnification, field of view = 7 mm, unless stated otherwise.

Tonalite

Mineralogy: *Qz, Pl, Kfs, ± Amp, ± Bt, ± Px, ± other accessories (Mag, Ilm, etc.)*

GRANITIC ROCKS

IUGS classification: *Qz (20–60%) + Pl (≥90%), ± Ksp (≤10%)*

Texture: *When undeformed, it typically has a phaneritic, equigranular texture*

Tonalite is dominated by plagioclase feldspar (oligoclase or andesine) and quartz, with up to 10–40% mafic minerals (hornblende, biotite, and/or pyroxene). Potassium feldspar may also be present but is minor. Tonalite is mainly associated with: (1) the voluminous gabbro–diorite–tonalite–granodiorite–granite batholiths of continental arcs (such as Peru or California); (2) the Archean tonalite–trondhjemite–granodiorite (TTG) association (e.g. Barberton granitoid–greenstone terrain, South Africa); and (3) oceanic arcs that are dominated by gabbroic and cumulate rocks but also have lesser amounts of diorite and tonalite (Tobago, West Indies).

Occurrence

Tonalites associated with continental batholiths, and large continental arc magma systems in general, form by a variety of multi-stage processes including basaltic underplating and crustal delamination, melting of the lower crust by intrusion of basalts, crustal assimilation by basalt magmas, and magma mixing. Theoretical considerations suggest that tonalitic melt represents a liquid fractionated from a broadly quartz-dioritic magmatic source. Plagioclase (rather than amphibole) is the liquidus phase, arguing for early water undersaturation of these tonalitic melts. Textures and phase equilibria relationships for such melts indicate that hydrous mafic minerals (biotite and amphibole) precipitate late in magma evolution.

Tonalites of the Archean TTG association represent low- to medium-pressure melts of enriched mafic rocks (e.g. EMORB, or enriched mid-ocean ridge tholeiitic basalt). The compositional range of an individual pluton/intrusive suite typically varies by ~ 3 wt% SiO_2, demonstrating that differentiation processes do not play a significant role in their genesis.

Figure 2.48 Tonalite from Adamello (Italy). (*Top*) Hand sample (4.5 cm across) with plagioclase (white) and quartz (gray), plus accessory biotite and oxides. (*Bottom*) Plagioclase, quartz, and biotite dominate the thin-section. Note plagioclase has complex concentric zoning and is also twinned, with alteration in the Ca-rich core of the crystal. Cross-polarized light, 10× magnification, field of view = 2 mm.

Granodiorite

GRANITIC ROCKS

Mineralogy: Qz, Pl, Kfs, ± Amp, ± Bt, ± Px, ± other accessories (Mag, Ilm, etc.)
IUGS classification: Qz (20–60%) + Pl (65–90%) + Kfs (10–35%)
Texture: When undeformed, it typically has a phaneritic, equigranular texture

Granodiorite is dominated by plagioclase feldspar (oligoclase), quartz, and alkali feldspar. The mafic minerals hornblende and biotite are also likely to be present. Granodiorite is mainly associated with the gabbro–diorite–tonalite–granodiorite–granite suite of subduction-related continental arc batholiths (e.g. Peru or California, and the Archean TTG [tonalite–trondhjemite–granodiorite] suites).

Occurrence

Granodiorite dominates collisional continental arc settings such as the Sierra Nevada of California (United States) and is typically magnesium and calc-alkalic; lesser volumes of more siliceous compositions can also be present. Rock composition is strongly correlated with location relative to the arc, as inboard from the subduction zone the continental crust gets thicker and the resulting melts become more alkali-rich.

In arc settings, magma hybridization is commonplace and occurs to varying degrees, over numerous episodes. Re-homogenization may be so thorough that the original mafic parental magma is no longer preserved. In the Tuolumne intrusive complex (TIC), for example, where suites of amphibole and plagioclase have been analyzed, numerous episodes of hybridization involving multiple end-member melt compositions are required to explain the compositional diversity of these minerals. In addition, the mineral compositional diversity contrasts with the more uniform whole-rock geochemistry of the TIC, illustrating the preservation and sensitivity of such processes in the minerals over the homogenized whole-rock composition.

Granodiorite from the Sierra Nevada (United States). This sample of the Tuolumne intrusive complex (Kuna Crest lobe) is medium-grained with alkali feldspar (light pink), plagioclase (white), quartz (gray), and hornblende + biotite + oxides (black). Image ~25 cm wide.

Image courtesy of V. Memeti

Figure 2.49

A Granodiorite from Bohemia (Czechia). The image is dominated by orthoclase and plagioclase. Orthoclase can be distinguished by its dusty aspect due to sericite alteration. Clear quartz is also present (left edge). Biotite (brown) has small apatite and zircon inclusions. Plane-polarized light, 10× magnification, field of view = 2 mm.

B Granodiorite from Bohemia (Czechia). Same image as (A) but with crossed polarizers. Both plagioclase and orthoclase show Carlsbad twinning. The plagioclase on the left also has concentric zoning. Quartz (gray) and biotite (orange–green) are also present. Cross-polarized light, 10× magnification, field of view = 2 mm.

C Granodiorite from Bohemia (Czechia). Hornblende dominates the center. To its right there is a sericitized orthoclase, while to its left a plagioclase is also altered to sericite. To the right of the top of the hornblende crystal some clear quartz can be seen. Biotite shows reddish colors. Plane-polarized light, 10× magnification, field of view = 2 mm.

D Granodiorite from Bohemia (Czechia). Same image as (C) but with crossed polarizers. Hornblende with medium birefringence and twinning. Orthoclase is near extinction (lower-right) and plagioclase shows characteristic polysynthetic twinning. Biotite (high birefringence) is also present (pink birefringence). Cross-polarized light, 10× magnification, field of view = 2 mm.

APPLICATION 2.2

Composition, Zoning and Magmatic Processes

Mineral composition and morphology are controlled by changes in temperature, pressure, and melt composition, as well as growth, dissolution, and nucleation kinetics. Consequently, mineral compositions and textures can be used to infer the physiochemical conditions of magmatic systems. Consider plagioclase, in which the slow interdiffusion of CaAl–NaSi may prevent equilibration of adjacent compositional zones, allowing the preservation of textures over long timescales. In the adjacent image, analysis across plagioclase shows that oscillatory zoning in petrographic thin-section corresponds to fine-scale oscillations in the sodium (An) content from core to rim across the crystal (purple line). The green line documents irregular zoning with repetitive sharp increases in An content – such increases are often equated with an increase in temperature interpreted to reflect mafic recharge (input of hotter, mafic melt). The pink line documents a more complex behavior, likely the result of multiple mechanisms (e.g. mafic recharge combined with melt segregation at different levels in a complex magma chamber system). Such textures preserve an observable record of magmatic processes such as magma mixing, mush disaggregation, undercooling, and decompression, and document different morphologies resulting from different processes.

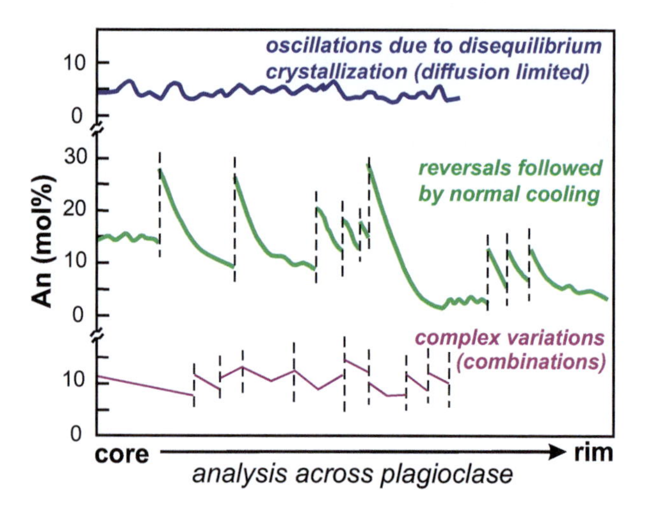

In a case study of the 1994–1995 eruption from Barren Island volcano, Renjith (2014) carefully documented up to six textural domains linked to distinct magmatic processes in plagioclase crystals:

Six textural domains in plagioclase.

Domain	Texture	Interpretation
6	Clear	Euhedral growth in quiescent equilibrium state
	- - - - - - - *Euhedral crystal growth* - - - - - - -	
5	Fine-sieve	Euhedral growth followed by interaction with hotter, Ca-rich magma
	- - - - - - - *Compositionally distinct euhedral crystal growth* - - - - - - -	
4	Fine-sieve	Euhedral growth followed by interaction with hotter, Ca-rich magma
	- - - - - - - *Major resorption phase* - - - - - - -	
3	Oscillatory zoning	Growth in a convective magma imparts fluctuating phase equilibria
	- - - - - - - *Minor resorption phase* - - - - - - -	
2	Medium-sieve	Optically distinct euhedral growth followed by decompression dissolution
	- - - - - - - *Compositionally distinct euhedral crystal growth* - - - - - - -	
1	Coarse-sieve	Euhedral growth during equilibrium conditions followed by decompression-driven dissolution

From Renjith, M.L. (2014). Micro-textures in plagioclase from 1994–1995 eruption, Barren Island Volcano: Evidence of dynamic magma plumbing system in the Andaman subduction zone. Geoscience Frontiers, 5(1), 113–126

The domains were defined by linking plagioclase microtextures and compositions. Specific growth textures associated with fluctuations at the crystal–melt interface reflect changes in temperature, vapor pressure, and/or H_2O content of the crystallizing melt; these include textures such as sieve morphology, oscillatory zoning, and resorption surfaces. Other textures (e.g. glomerocrysts, synneusis, swallow-tailed crystals, microlites, and broken crystals) reflect dynamic melt processes such as convection, turbulence, degassing, etc. and formed as a result of the dynamic behavior of the crystallizing magma.

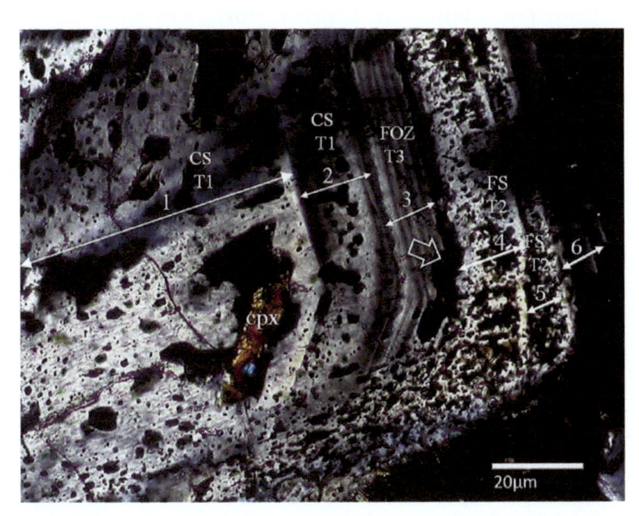

From Renjith, M.L. (2014). Micro-textures in plagioclase from 1994–1995 eruption, Barren Island Volcano: Evidence of dynamic magma plumbing system in the Andaman subduction zone. Geoscience Frontiers, 5(1), 113–126

In this image the six magmatic domains of Renjith (2014) are visible. The unfilled arrow highlights a region of resorption between domains 3 and 4; the outer rim (domain 6) has oscillatory zoning that is partially extinct.

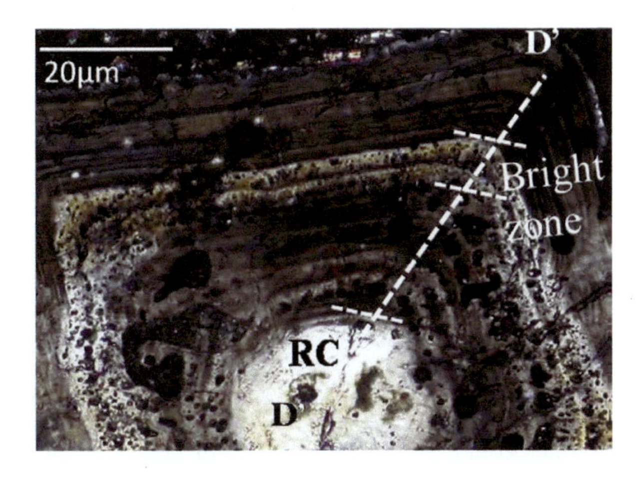

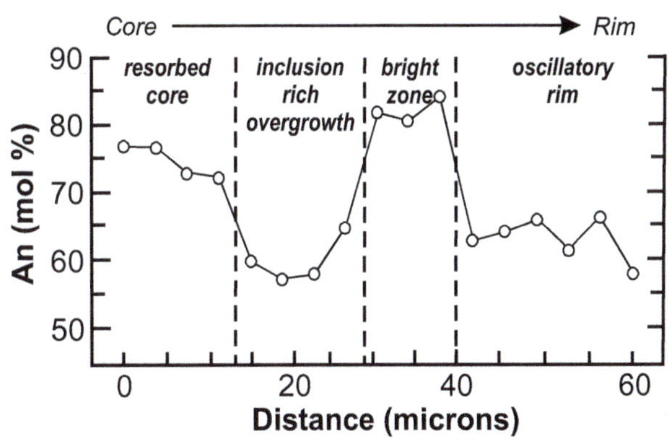

From Renjith, M.L. (2014). Micro-textures in plagioclase from 1994–1995 eruption, Barren Island Volcano: Evidence of dynamic magma plumbing system in the Andaman subduction zone. Geoscience Frontiers, 5(1), 113–126

This plagioclase (*top*) has complex textures and zoning. An electron microprobe traverse across the crystal (D–D′) shows compositional variations (*bottom*; line connecting analytical points) that can be linked to crystallization processes (Renjith, 2014). From core to rim: the resorbed core (RC) has a composition of ~An_{76-74}, the inclusion-rich oscillatory zoning ranges from ~An_{57-65}, the bright zone with ~An_{82}indicates higher temperature (influx of hotter magma?), and the darker outermost rim of oscillatory zoning has a composition of ~An_{62}.

Granite

Mineralogy: *Qz, Pl, Kfs, Bt, ± Amp, ± accessories (Mag, Ilm, Grt, Ms, etc.)*
IUGS classification: *Qz (20–60%), Pl (10–65%), Kfs (35–90%)*
Texture: *When undeformed, it is typically phaneritic and equigranular*

GRANITIC ROCKS

Granite is composed of quartz, plagioclase, and alkali feldspar in an interlocking network known as "granular" texture. Plagioclase is typically sodium-rich oligoclase and alkali feldspar is often perthitic orthoclase or microcline. Slow cooling at depth generates the coarse crystal size of most granites, so to be exposed at the surface the rock must be uplifted and have its cover eroded. Granite may contain small amounts of peritectic and/or residual minerals, and numerous accessory minerals such as zircon, monazite, and allanite are common. Enclaves and xenoliths may also be common.

Occurrence

Granites form in four main tectonic settings: magmatic arcs, continent–continent collision zones, post-collisional orogens, and intercontinental rifts. The greatest volumes of granite are generated in arc settings where large bodies (batholiths) comprise the cores of arcs. Granites play an important role in the formation and evolution of continental crust and continental shields.

The magma source of a granite is defined by its tectonic setting, but other processes determine its final composition. Granites form via fractional crystallization of mantle-derived mafic magmas, by partial melting of a wide range of crustal rock types (anatexis), and/or some combination of the two. Partial melting of the mantle generates a mafic melt with a composition equivalent to a basalt; when this basaltic melt traverses through and interacts with the mantle/crust, fractional crystallization of the basaltic magma may generate a granite. However, other processes of crystallization–differentiation, magma recharge, magma mixing, crustal assimilation/melting, and restite unmixing generally modify its composition.

Arc-related granites like the Sierra Nevada in California (United States) experience early magnetite crystallization, are relatively calcic, and are derived from potassium-poor sources. Post-collisional orogen granites such as the Caledonian granites of Great Britain are generated late in, or after, orogenesis; they tend to be the result of deeper (hotter) melting that produce limited melt volumes with higher potassium contents. The granites generated during continent–continent collision are compositionally variable, but dominated by small-volume melts high in silica and alumina. Intercontinental rift granites are iron-rich due to reducing conditions during crystal fractionation.

Granite from Lundy, Devon (England). White plagioclase, very pale pink orthoclase, gray quartz and dark brown biotite are present. Image ~7 cm wide.

Image courtesy of A. Tindle

Figure 2.50

A Granite from Baveno (Italy). The image is dominated by orthoclase. Plagioclase and quartz are also present. It is difficult to distinguish these minerals in plane-polarized light, but orthoclase is “dusty” due to sericite alteration and quartz has slightly higher relief (lower-left corner). Biotite (reddish) is also present. Plane-polarized light, 10× magnification, field of view = 2 mm.

B Granite from Baveno (Italy). Same image as (A) but with crossed polarizers. Alkali feldspar shows exsolution and Carlsbad twinning, plagioclase has concentric zoning, and quartz (lower-left corner) is a homogeneous light gray. Biotite is at extinction. Note the typical down-T (°C) progression of alkali feldspar after plagioclase. Cross-polarized light, 10× magnification, field of view = 2 mm.

C Monzogranite from Bohemia (Germany). Amphibole (tan to green) and biotite (reddish) indicate a high water content. Plane-polarized light, 2× magnification, field of view = 7 mm.

D Monzogranite from Bohemia (Germany). Same image as (C) but with crossed polarizers. Plagioclase with polysynthetic twinning dominates over alkali feldspar (yellow–gray) and interstitial quartz (homogeneous, gray). Cross-polarized light, 2× magnification, field of view = 7 mm.

Alkali Feldspar Granite

GRANITIC ROCKS

Mineralogy: *Kfs, Qz, Pl, Bt, ± Na–Amp, ± Na–Px, + accessory Mag, Ilm, Ms, Ap*
IUGS classification: *Kfs (≥90%) + Qz (20–60%) + Pl (≤10%)*
Texture: *Phaneritic equigranular*

Alkali feldspar granite is dominated by alkali feldspar and quartz, with minor plagioclase (≤10%). If sodic amphibole (richterite) and/or pyroxene (aegirine) are present, the name peralkaline granite is used. As with all granitic rocks, these minerals form an interlocking network known as "granular" texture. Alkali feldspar granites are often pink or red in color due to the abundance of alkali feldspar. Plagioclase is typically Na-rich. Alkali feldspar granites are often found with other alkali-rich granitoids, namely syenogranite and monzogranite, which vary only in the relative abundance of alkali feldspar and plagioclase.

Occurrence

Alkali granites and peralkaline granites generally represent highly fractionated melts. Some involve mantle-derived mafic precursors and most involve protracted, plagioclase-dominated fractional crystallization. In intraplate extensional settings, crystal fractionation from basaltic liquid or differentiation of a trachytic parent generated by fusion of underplated basalt may produce alkali granites. This is seen, for example, in the alkaline to peralkaline granites of Temora, Australia.

In post-collisional settings, slow cooling and protracted crystallization at depth results in pronounced differentiation. This differentiation results in an increase in viscosity, further impeding magma ascent and the production of highly fractionated melts. In the Changning-Menglian suture zone of southwest China, Triassic alkali granites and syenogranites are associated with partial melting/assimilation of mid-crustal rocks after crustal thickening due to collision. Melt genesis was followed by magma mixing, slow cooling, and extensive fractionation by Kfs + Pl + Bt + Zrn + Ti-bearing minerals.

Alkali feldspar granite, Skjeberg (Norway). Large red alkali feldspar, quartz, and oxide minerals forming granular texture. Sample is 12 cm wide.

Figure 2.51

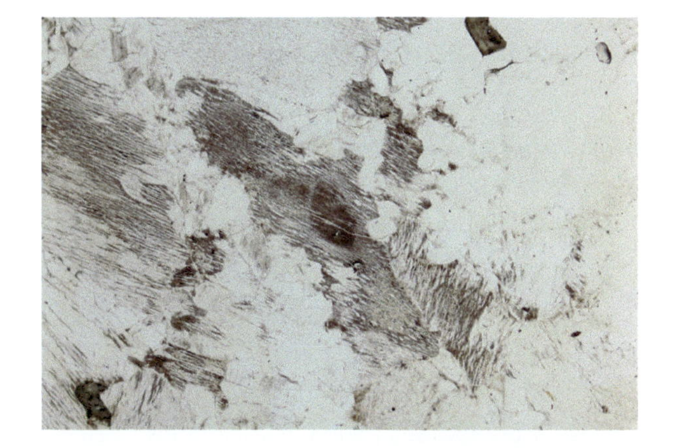

A Alkali feldspar granite from Biella (Italy). The picture is dominated by perthitic orthoclase altered by sericite (brownish) and quartz (colorless). Small biotite (green–brown) is also present. Plane-polarized light, 2× magnification, field of view = 7 mm.

B Alkali feldspar granite from Biella (Italy). Same image as (A) but with crossed polarizers. Orthoclase has gray birefringence with brownish sericite alteration, while quartz has a more uniform, darker shade of gray (upper-right). Cross-polarized light, 2× magnification, field of view = 7 mm.

C Alkali feldspar granite from Kone (Ethiopia). Sericitized orthoclase (brownish), quartz (colorless), and aegirine (green). The upper aegirine shows compositional zoning from dark to light green (from Fe-rich core to Fe-poor rim). Plane-polarized light, 2× magnification, field of view = 7 mm.

D Alkali feldspar granite from Kone (Ethiopia). Same image as (C) but with crossed polarizers. Orthoclase has perthitic exsolution and Carlsbad twins. Aegirine has high (third-order) birefringence. Cross-polarized light, 2× magnification, field of view = 7 mm.

Foid-Bearing Rocks

Mineralogy: *Opx, Cpx, Pl, Kfs, foids, ± Ol, ± Amf, ± Bt*
IUGS classification: *Pl ± Kfs + foids (<10%)*
Texture: *Phaneritic holocrystalline*

The foid-bearing rocks are alkaline, silica-undersaturated, and have ≤10% feldspathoids present. Foid-bearing gabbro and foid-bearing diorite consist of plagioclase (≥90%), feldspathoids (<10%), minor (≤10%) potassium feldspar, and mafic minerals (pyroxene, hornblende, biotite). The distinction between foid-bearing gabbro and foid-bearing diorite is determined by the abundance of mafic minerals (gabbro >35% Px ± Ol; diorite <35% Px ± Ol ± Hbl). With increasing modal potassium feldspar, foid-bearing gabbro/diorite grades into foid-bearing monzogabbro/monzodiorite (10–35% Ksp), foid-bearing monzonite (35–65% Ksp), foid-bearing syenite (65–90% Ksp), and ultimately into foid-bearing alkali feldspar syenite (90–100% Ksp). As the feldspathoid content increases, the foid-bearing rocks transition to *foid rocks* (10–60% feldspathoids): foid gabbro/foid diorite, foid monzogabbro/foid monzodiorite, foid monzosyenite, and foid syenite. The most abundant feldspathoid is used as a prefix to the rock name.

Occurrence

Alkaline magmatism is often associated with extensional and rift-related tectonic settings. Foid-bearing gabbro/foid-bearing diorite is rare and mostly found in intracontinental settings in temporal–spatial association with other silica-oversaturated and -undersaturated alkaline rocks (e.g. the Eocene–Oligocene foid-bearing plutons of northwest Iran). Their genesis, as with other alkaline rocks, requires several processes to operate simultaneously, especially in those complexes characterized by the coexistence of oversaturated and undersaturated rocks. Their origin is linked to the partial melting of a metasomatized mantle source, followed by assimilation (of predominantly crustal material) and fractional crystallization.

Figure 2.52 Analcime monzogabbro, Queensferry (Scotland). (*Top*) Plagioclase (dusty appearance due to sericite alteration) and augite pyroxene (pale brown, higher relief) dominate the thin-section. The dark, reddish crystals are biotite or phlogopite. Analcime (colorless) forms interstitial crystals between plagioclase and pyroxene crystals. Some black magnetite is also present (plane-polarized light). (*Bottom*) Augite (high birefringence), plagioclase (altered to sericite showing atypical birefringence), and analcime (almost isotropic) are seen in cross-polarized light. 2× magnification, field of view = 7 mm.

BOX 2.9 Pegmatites

Pegmatites are coarse-grained igneous rocks. They represent the final stage of magmatic crystallization, which means they are often granitic in composition (dominated by cuneiform quartz, perthitic feldspar, and micas) and rich in volatiles (H_2O, Fl, Cl, CO_2, etc.) and incompatible elements.

During differentiation of a water-bearing magma, pockets of water rich in dissolved ions can separate from the melt; these are much more mobile than the highly viscous melt and allow the ions to move and form crystals rapidly – this generates the extremely coarse grain size (>1 cm) associated with pegmatitic rocks; they are not the result of slow crystallization.

Pegmatites are most abundant in shields and mountain belts. They often occur in small pockets along the margins of batholiths or in fractures that develop on these margins (e.g. pegmatite dikes). Some pegmatites may form by melting (anatexis) during metamorphism at high temperatures and pressures. Overall, pegmatites show considerable variation, texturally, compositionally, and mineralogically. Their unusual volatile and incompatible element-rich compositions generate unusual minerals/gemstones. Common pegmatite minerals of economic importance include:

Mineral	Element
Apatite	PO_4
Beryl	Be
Lepidolite	Rb, Li
Columbite	Nb
Spodumene	Li
Tantalite	Ta
Tourmaline	Bo, K, Li,
Zircon	Zr

Below is a thin-section image of a spodumene pegmatite in Weinebene (Austria). The pegmatite is mined for lithium. Spodumene is a pyroxene rich in lithium and aluminum. It has low, first-order yellow–gray birefringence, a pronounced cleavage, is up to 5 mm long, and is associated with muscovite (blue colors). The pegmatite forms dike-like bodies in eclogitic amphibolite and kyanite-bearing micaschist.

Cross-polarized light, 2× magnification, field of view = 7 mm.

Foid Rocks

Mineralogy: *Pl, Kfs, Opx, Cpx, foids, ± Bt, ± Hbl, ± Ol*

IUGS classification: *Pl, Kfs, foids (10–60%)*

Texture: *Phaneritic holocrystalline*

FELDSPATHOID ROCKS

The "foid" rocks (foid gabbro, foid diorite, foid monzogabbro, foid monzodiorite, foid monzosyenite, and foid syenite) are alkali, silica-undersaturated plutonic rocks consisting of plagioclase, alkali feldspar, feldspathoids (10–60%), and mafic minerals (biotite, Na-hornblende, Na-pyroxene). They are similar to the foid-bearing rocks, but with greater amounts of feldspathoid being present. The composition of plagioclase is used to distinguish between foid gabbro ($>An_{50}$) and foid diorite ($<An_{50}$). As potassium feldspar content increases, foid gabbro/foid diorite gives way to foid monzogabbro or foid monzodiorite, foid monzosyenite, and ultimately to foid syenite. The most abundant feldspathoid forms a prefix to the rock name (i.e. sodalite monzogabbro or nepheline diorite).

Occurrence

As with the foid-bearing rocks, foid rocks occur mostly in intracontinental settings associated with other silica-oversaturated and -undersaturated alkaline rocks, often in temporal–spatial association, such as the Monteregian Hills and White Mountain provinces of Canada. Their genesis requires several processes to operate simultaneously, especially in those complexes characterized by the coexistence of oversaturated and undersaturated rocks. Their origin is linked to the partial melting of a metasomatized mantle source, followed by assimilation (of predominantly crustal material) and fractional crystallization.

Analcime gabbro from Lugar Sill (Scotland). Black, elongate kaersutite amphibole in a groundmass of creamy-white calcic plagioclase, abundant analcime, and accessory nepheline. Image ~5 cm wide.

Image courtesy of J. St. John

Figure 2.53

A Analcime gabbro from Lugar Sill (Scotland). Elongated kaersutite crystals (pale to strong red–brown pleochroism) and completely sericitized plagioclase (best seen under XPL) set in a fine-grained sericite + analcime-rich matrix. Small crystals of apatite (clear, lower-left) are also present. Plane-polarized light, 2× magnification, field of view = 7 mm.

B Analcime gabbro from Lugar Sill (Scotland). Same image as (A) but with crossed polarizers. Kaersutite shows high birefringence. The fine-grained matrix is composed of sericite and isotropic analcime. Cross-polarized light, 2× magnification, field of view = 7 mm.

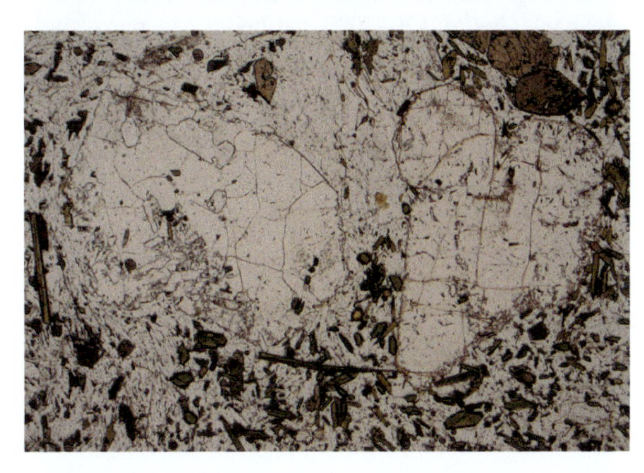

C Sodalite syenite from Lovozero (Russia). Two sodalite crystals (colorless and fractured) dominate the image. They are surrounded by aegirine (green), alkali feldspar (colorless), and titanite (brown in upper-right corner). Plane-polarized light, 2× magnification, field of view = 7 mm.

D Sodalite syenite from Lovozero (Russia). Sodalite is isotropic but has some feldspar inclusions. Aegirine has high birefringence. Alkali feldspar defines a magmatic foliation that envelops the two sodalite crystals. Cross-polarized light, 2× magnification, field of view = 7 mm.

Foidolite

Mineralogy: *Foids, ± Pl, ± Kfs, ± Cpx, ± Ol*

FELDSPATHOID ROCKS

IUGS classification: *Foids (60–100%) ± Ksp and/or Pl (≤40%)*

Texture: *Phaneritic holocrystalline*

Foidolite is a very rare silica-undersaturated plutonic rock dominated by feldspathoids (60–100%). Alkali feldspar and plagioclase are present in minor amounts, and mafic minerals can include biotite, hornblende, pyroxene, and olivine. Common accessory minerals include ilmenite, titanite, and garnet (melanite). Foidolite, like other alkaline rocks, commonly contains rare accessory minerals such as baddeleyite, perovskite, and complex Na–Ca–HFSE–REE silicates. Foidolite is the plutonic compositional equivalent of the extrusive foidite. The most abundant feldspathoid is used as a prefix in the name (e.g. nephelinolite, leucitolite). Nephelinolite is the most abundant type of foidolite and for this reason it is subdivided on the basis of mafic mineral content: *urtite* (>70% nepheline + aegirine–augite and no feldspar), *ijolite* (30–70% nepheline + pyroxene), and *melteigite* (10–30% nepheline and 70–90% mafic minerals).

Occurrence

Foidolite, like other silica-undersaturated alkaline rocks, is formed in nearly all tectonic environments and is present on most continents. Although rare, foidolite is associated with alkaline carbonatite provinces (e.g. the Kola alkaline province [Russia], the Monteregian Hills and White Mountain alkaline provinces [Canada]). In many complexes foidolite occurs on the periphery of other silica-undersaturated plutonic rocks. Foidolite is characteristically high in incompatible, large-ion lithophile, and REE. The petrogenetic association of foidolite and carbonatites is not fully understood: Does it relate to the depth and degree of mantle melting and the mineralogy of the mantle source, the fractional crystallization of CO_2-rich parent silicate melts, or the immiscible separation of a carbonate-bearing silicate melt?

Nephelinolite, made of nepheline (green to green–gray), amphibole (black), and titanite (yellowish). Image ~6 cm wide.

Image courtesy of J. St. John

Figure 2.54

A Nephelinolite from Fogo Island (Cape Verde). Contains nepheline (colorless), aegirine (green/brownish-green; lower-right crystal shows compositional zoning), and amphibole (reddish). Plane-polarized light, 2× magnification, field of view = 7 mm.

B Nephelinolite from Fogo Island (Cape Verde). Same image as (A) but with crossed polarizers. Nepheline has white–gray birefringence, aegirine has higher birefringence (orange), and amphibole shows deep red interference colors. Cross-polarized light, 2× magnification, field of view = 7 mm.

C Nephelinolite (ijolite) from Lackner Lake, Ontario (Canada). Subhedral nepheline (gray) are surrounded by aegirine (higher birefringence). Cross-polarized light, 2× magnification, field of view = 7 mm.

D Melilitolite from San Venanzo (Italy). The image is dominated by melilite with diagnostic anomalous, blue birefringence. The crystals with high-interference colors are olivine and leucite is also present (dark crystal, lower-right). Cross-polarized light, 2× magnification, field of view = 7 mm.

2.7 Volcanic Rock Types

Basalt

Mineralogy: *Pl, Cpx, Ol, accessory Mt, Ilm, foids*
IUGS/TAS classification: *Melanocratic/SiO_2 (45–52 wt%)*
Texture: *Aphanitic to porphyritic or tuffaceous to glassy*

SILICA: (OVER)SATURATED

Basalt is the most abundant igneous rock on Earth and is mainly composed of Ca-plagioclase (usually labradorite), pyroxene, ± olivine, and accessory minerals such as hornblende, nepheline, leucite, analcime, Ti-magnetite, and ilmenite. With $Na_2O + K_2O > 5$ wt%, tholeiitic (or sub-alkaline) basalt becomes trachybasalt then tephritic to foiditic. With increasing SiO_2, basalt becomes basaltic andesite.

Tholeiitic basalt is composed of plagioclase, clinopyroxene (augite), and/or low-Ca pyroxene (hypersthene and pigeonite). Olivine, if present, occurs only as phenocrysts (<5%). The groundmass commonly contains small amounts of glass (fresh or altered) and in some slowly cooled rocks the intergrowth of quartz and K-feldspar occurs. *Alkali basalt* is characterized by high Na_2O + K_2O contents relative to other basalts. Alkali basalt is composed of plagioclase, abundant olivine phenocrysts, augite, and Ti–Fe oxides. Alkali basalt does not contain low-Ca pyroxene, whereas nepheline, leucite, or analcime may be present in the groundmass. The groundmass is often *subophitic* or *intergranular*, and glass is rare. Although not very common, olivine-rich (up to 50% Ol) varieties of tholeiitic and alkaline basalts occur. Such rocks are called *picritic basalts* (or picrites), while extremely pyroxene-rich basalts are called *ankaramites*.

Occurrence

Basalt is found in three main environments: (1) oceanic divergent boundaries, (2) oceanic hotspots, and (3) hotspots beneath continents. Mid-ocean spreading ridges produce submarine volcanism known as mid-oceanic ridge basalt (MORB), which is *tholeiitic*. Experiments have shown that MORB can be produced by 10–20% partial melting of peridotitic mantle at shallow (50–85 km) depths. Seismic and geochemical studies show that the magma produced by partial melting of the mantle migrates to shallower levels while undergoing fractional crystallization; this produces tholeiitic basalt. After the extraction of tholeiitic melt, restitic olivine ± pyroxene may represent the residual harzburgitic mantle preserved in some ophiolites (e.g. Semail ophiolite [Oman]). Tholeiitic basalt is also associated with large igneous provinces (LIPs) such as the Columbia River basalts in the western United States, the Deccan Traps of India, and the Paraná Basin of South America. *Alkali basalts* are associated with oceanic islands such as Hawaii (United States) or Madeira (Portugal), and with updomed, rifted continental crust such as the Rio Grande Rift (United States) or the East African Rift. Alkali basalts are generated at greater depths and from lower (<10%) degrees of partial melting than tholeiitic basalts.

Olivine basalt, Kilauea, Hawaii (United States). Vesicles and green olivine phenocrysts are visible. Sample ~10 cm wide.

Image courtesy of J. St. John

Figure 2.55

A Tholeiitic basalt from Fagradalsfjall (Iceland). Microlites of plagioclase (colorless) and pyroxene (pale brown) microphenocrysts set in a dark, glassy groundmass. Plane-polarized light, 2× magnification, field of view = 7 mm.

B Tholeiitic basalt from Fagradalsfjall (Iceland). Same image as (A) but with crossed polarizers. Plagioclase shows characteristic birefringence and twinning, while pyroxene shows high birefringence. Groundmass glass is isotropic. Cross-polarized light, 2× magnification, field of view = 7 mm.

C Subophitic alkali basalt from Hawaii (United States). Microlites of plagioclase (colorless) partially enclosed by microphenocrysts of augite (brownish); olivine (red, altered to *iddingsite*) and ilmenite (black) are also present. Plane-polarized light, 2× magnification, field of view = 7 mm.

D Subophitic alkali basalt from Hawaii (United States). Same image as (C) but with crossed polarizers. Microlites of plagioclase (gray birefringence and twinned) partially enclosed by microphenocrysts of augite (high birefringence). Iddingsitized olivine has strong red interference colors. Cross-polarized light, 2× magnification, field of view = 7 mm.

Basaltic Andesite

Mineralogy: *Pl, Px, + accessory Ap, Mag, Ilm*

SILICA: (OVER)SATURATED

IUGS/TAS classification: *Melano- to mesocratic/SiO_2 (52–57 wt%)*

Texture: *Aphanitic to porphyritic or tuffaceous to glassy*

Basaltic andesite is melano- to mesocratic, rich in ferromagnesian minerals, has a SiO_2 content between basalt and andesite, and is mostly composed of augite and plagioclase; olivine is rare. Plagioclase often has a wider range of composition than in basalt. Magnetite, apatite, and ilmenite are common accessory minerals.

Boninite: Mg-rich basaltic andesite with phenocrysts of olivine, orthopyroxene (enstatite which inverts to clinoenstatite), and clinopyroxene; plagioclase phenocrysts are distinctively absent. Almost exclusively associated with the fore-arc of island arc settings (e.g. Izu-Bonin [Japan]) or ophiolite complexes from this setting.

Occurrence

Basaltic andesite is associated with subduction zones and forms in volcanic arc and back-arc environments (e.g. the Cascades of North America). The Columbia River basalt LIP is composed of 80% basaltic andesites on the basis of SiO_2 content. Basaltic andesite may also form in extensional settings: Basaltic andesite caps the mid-Cenozoic volcanic sequences of southeast Arizona, southwest New Mexico, and western Mexico.

Figure 2.56

A Basaltic andesite from Malinów (Poland). Phenocrysts of plagioclase (colorless), clinopyroxene (subhedral, greenish), and rare olivine (euhedral crystal, left of center) are set in a brown, glassy groundmass. Cross-polarized light, 2× magnification, field of view = 7 mm.

B Basaltic andesite from Malinów (Poland). Same image as (A) but with crossed polarizers. Plagioclase phenocrysts have typical birefringence and twinning, clinopyroxene has orange–red birefringence and is also twinned, and rare olivine (left of center, euhedral crystal with blue birefringence) is also present. Glassy groundmass is isotropic. Cross-polarized light, 2× magnification, field of view = 7 mm.

BOX 2.10 Crystal Cargo

Physical complexity within magmatic centers commonly results in the generation of diverse crystal populations. These populations are modified by processes such as magma recharge/replenishment and magma mixing/mingling that occur within the magmatic "plumbing system." The majority of magmatic systems contain four distinct crystal populations (*crystal cargo*) that may or may not be genetically related:

1. Phenocrysts – crystals that are cogenetic with their host magma.
2. Microlites – small crystals which document rapid nucleation and growth during eruption and are cogenetic with their host magma.
3. Antecrysts – crystals which are related to the magmatic system but not directly crystallized from the host magma in which they are finally entrained.
4. Xenocrysts – crystals foreign to their host magma and magmatic system.

These four components may be regarded as a tetrahedra in which a mixture of liquid (L), phenocrysts (P), antecrysts (A), and xenocrysts (X) occur (after Davidson et al., 2007).

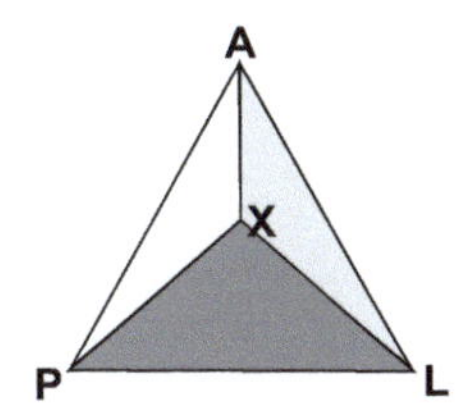

The P-A-L part of this tetrahedra illustrates how petrographic textures inform us about the role of open- and closed-system magmatic processes through, for example, direct observation of phenocryst textures (e.g. overgrowth, resorption, rims). This information in turn can be used with different analytical techniques to investigate crystal shape, size, and spatial distributions. Combining textural and micro-geochemical techniques enables us to link these textures to chemical and isotopic differences. Such data can then be used to document where crystals are sourced, how they are recycled, and to identify the involvement of multiple magmatic systems.

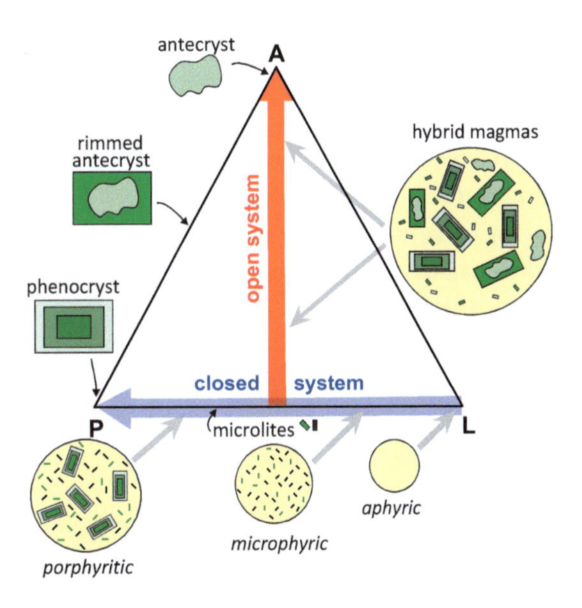

Andesite

Mineralogy: *Pl, Px, Hbl, ± Bt*

SILICA: (OVER)SATURATED

IUGS/TAS classification: *Mesocratic/SiO_2 (57–63 wt%)*

Texture: *Aphanitic to porphyritic or tuffaceous*

Andesite is mesocratic with SiO_2 content between basaltic andesite and dacite. It typically has phenocrysts of plagioclase (completely zoned and/or embayed), pyroxene(s), hornblende, and/or biotite. The phenocrysts may be *opacitic*. Andesine (An_{30-50}) is the typical plagioclase, but composition may vary from labradorite to oligoclase (An_{70-10}). Pyroxene can be augite, pigeonite, or orthopyroxene. Magnetite, apatite, ilmenite, and titanite are common accessory minerals. Andesite grades into *dacite* with increasing quartz and into *trachyandesite* with increasing alkali feldspar.

Occurrence

Andesite, because of its silica and water content, often forms explosive eruptions associated with flow breccias, mudflows, tuffs, and other fragmental rocks. To a lesser extent, andesite forms subvolcanic rocks such as dikes, sills, and plugs. The majority of andesite is produced at convergent plate boundaries in association with subduction zones, as typified by the Andes of South America or the cordillera of Central and North America. Andesite is less frequently associated with divergent plate boundaries, where it occurs with tholeiitic basalts (e.g. the Mid-Atlantic Ridge of Iceland or the intraplate setting of Hawaii).

Geophysical evidence from several volcanic arcs highlights the presence of mafic cumulates at the base of the crust. Such cumulates may form via fractional crystallization processes: As basaltic magmas crystallize, dense early-forming minerals such as olivine and pyroxene settle through the melt, forming thick cumulate layers. This can drive the residual magma toward lower iron and magnesium, and higher silica and alkali contents; if fractional crystallization continues, the melt can become even more evolved, producing the common basalt–andesite–rhyolite association of island arcs. Other hypotheses for andesite genesis include partial melting of crustal material, magma mixing between rhyolitic and basaltic magmas, and partial melting of metasomatized mantle.

Andesite from Hoggar (Algeria). Amphibole (elongate, black) and biotite (small, round, black) are difficult to distinguish. Greenish-yellow spots of sulfide oxidation present. Sample ~7.5 cm wide.

Image courtesy of Y. Cherfi

Figure 2.57

A Andesite from Citlaltépetl (Mexico). Porphyritic andesite with phenocrysts of sub- to euhedral plagioclase (colorless) and hornblende (brown), set in a fine-grained, glassy groundmass with microphenocrysts of plagioclase and hornblende. Some plagioclase crystals are sieve-textured (two crystals in the upper-right). Plane-polarized light, 2× magnification, field of view = 7 mm.

B Andesite from Citlaltépetl (Mexico). Same image as (A) but with crossed polarizers. Plagioclase phenocrysts show characteristic birefringence and twinning, and some have concentric compositional zoning (center-right). Hornblende phenocrysts have high birefringence. The glassy groundmass is isotropic. Cross-polarized light, 2× magnification, field of view = 7 mm.

C Andesite from Alaçatı (Turkey). Porphyritic andesite composed of plagioclase crystals, hornblende (the high-interference colors) and orthopyroxene (large, euhedral, central crystal, low-interference colors). Cross-polarized light, 2× magnification, field of view = 7 mm.

Trachyandesite

Mineralogy: *Pl, Afs, ± Qtz, ± Bt, ± Cpx, ± Hbl, ± Ol, and accessory Mt, Ilm, Ttn* **SILICA: (OVER)SATURATED**
IUGS/TAS classification: *Mesocratic/SiO_2 (53–63 wt%)*
Texture: *Aphanitic to porphyritic*

Trachyandesites (sometimes known as latites) belong to the medium-alkali suite of the TAS diagram. They are aphanitic to porphyritic volcanic rocks with phenocrysts in nearly equal proportions of plagioclase (often zoned) and K-feldspar; quartz, if present, is scarce. Clinopyroxene and biotite are the most common ferromagnesian minerals, but orthopyroxene, olivine, and less commonly hornblende may be present. A trachytic textured groundmass is common. Trachyandesite is the extrusive equivalent of monzonite. With increasing silica content it grades into trachyte.

Occurrence

Trachyandesite occurs as lava flows, pyroclastic deposits, and as hypabyssal intrusions (dikes, sills, laccoliths, and plugs) associated with other intermediate rocks. It is common in various tectonic settings, from regions of continental rifting to subduction-related volcanic arcs. Along active continental margins, latites are generally younger and occur stratigraphically higher and more distal to the oceanic trench than calc-alkaline arc rocks. Commonly in these tectonic settings there is a complete gradation between calc-alkaline rocks and alkaline rocks of the medium-alkali suite.

Trachyandesites may be involved in catastrophic explosive eruptions, such as the 1815 Tambora eruption which generated about 30–33 km^3 trachyandesite and tephriphonolite magma. The Tambora eruption is thought to reflect a parental trachybasaltic magma derived from ~2% partial melting of MORB-type mantle contaminated with ~3% fluids from altered oceanic crust and <1% sedimentary material. Subsequently, polybaric magmatic differentiation processes led to the formation of more evolved liquids. Trachyandesite is common in many Italian igneous centers (e.g. Monti Cimini, Monte Amiata, Roccamonfina, Vesuvius, the Phlegraean Fields, Lipari, Stromboli, and Vulcano Island). Trachyandesite magmas can be very rich in dissolved sulfur, leading to the formation of important mineral deposits, especially for porphyry Cu-(Au) type deposits.

Trachyandesite from Table Mountain, Colorado (United States). Phenocrysts of white alkali feldspar and black hornblende in a light gray glassy groundmass. Image is 11 cm wide.

Image courtesy of J. St. John.

Figure 2.58

A Trachyandesite from Vico Volcano (Italy). Subhedral to anhedral plagioclase (colorless), elongate brown biotite, and small green pyroxene (left) set in a fine-grained groundmass with trachytic texture. Plane-polarized light, 2× magnification, field of view = 7 mm.

B Trachyandesite from Vico Volcano (Italy). Same image as (A) but with crossed polarizers. Plagioclase shows gray interference colors and compositional zoning (rounded crystals in the upper-right corner), biotite and pyroxene show high-interference colors. The groundmass is composed of small, aligned sanidine crystals (trachytic texture). Cross-polarized light, 2× magnification, field of view = 7 mm.

C Trachyandesite from Amiata volcano (Italy). Euhedral plagioclase (colorless and with abundant brown glass inclusions) and brown hornblende crystals set in a fine-grained, brown glassy groundmass. Hornblende is opacitic. Plane-polarized light, 2× magnification, field of view = 7 mm.

D Trachyandesite from Amiata volcano (Italy). Same image as (C) but with crossed polarizers. Plagioclase shows gray interference colors, while hornblende is almost isotropic due to opacitic texture (some hornblende core shows high-interference colors). Cross-polarized light, 2× magnification, field of view = 7 mm.

Trachyte

Mineralogy: *Afs ± Px, ± Bt, ± Hbl, ± foids, ± Na-Pl, accessory Ap, Zrn, Mt*
IUGS/TAS classification: *Leucocratic/SiO_2 (58–70 wt%)*
Texture: *Aphanitic to porphyritic*

SILICA: (OVER)SATURATED

Trachytes are fine-grained aphanitic to porphyritic volcanic rocks, the extrusive equivalent of syenite in the IUGS classification. Trachytes are generally SiO_2-saturated rocks, but slightly oversaturated and undersaturated varieties also exist. Trachytes are part of the medium-alkali trend on the TAS diagram and are composed chiefly of alkali feldspar (sanidine) and minor mafic minerals such as biotite, hornblende, or pyroxene. Small amounts of sodic plagioclase (oligoclase) can be present. Apatite, zircon, and magnetite are often accessory minerals. Among mafic minerals, augite is the most common; brown hornblende and biotite are found in smaller quantities and are commonly surrounded by a rim of magnetite and pyroxene. The presence of quartz is rare, but if present is usually interstitial; tridymite, on the other hand, is more common and is usually found as small hexagonal plates in the groundmass. Foid-bearing trachytes commonly contain minerals such as nepheline, sodalite and leucite, aegirine–augite pyroxenes, and riebeckite-rich amphiboles. With increasing plagioclase content, trachytes grade into trachyandesite (latite, IUGS mode) and andesite; with increasing quartz they grade into rhyolites. Undersaturated trachyte grades into phonolite.

Occurrence

Trachytes occur in a wide variety of settings. They are common in regions of continental extension related to regional doming, where they can form large volumes of eruptive material as in the Ethiopian and Kenyan sections of the East African Rift System. Trachytes are common in the Rhine graben and the Eifel complex of Germany. Trachytes occur as pyroclastic deposits and lava flows in the Roman and Campania igneous provinces of Central-South Italy. In the United States, trachytes occur extensively in the Pine Canyon caldera volcano of Big Bend National Park (Texas), as well as in southern Nevada and South Dakota (Black Hills). Trachytes are known from many ocean islands (e.g. the Canary Islands, the Azores, Iceland, and Hawaii). Trachytes are also common in numerous subduction-related settings, where they commonly form in local zones of extension (e.g. Mayor Island volcano in New Zealand and the Mount Edziza Volcanic Complex in British Columbia). The formation of trachytic rocks, rather than being linked to a precise tectonic context, is linked to processes such as (1) extreme crystal fractionation of alkali basalt in a shallow reservoir; (2) partial melting of cumulate; and/or (3) partial melting of different lithospheric sources.

Trachyte from the Gallinas Mountains, New Mexico (United States). Note the large, flow-aligned sanidine crystals. Sample c. 10 cm wide.

Image courtesy of E. Owen.

Figure 2.59

A Trachyte from Ustica island (Italy). Elongate sanidine crystals (colorless) and green aegirine crystals set in a fine-grained groundmass with trachytic texture. Plagioclase crystals are also present (one with sieve texture, next to aegirine). Plane-polarized light, 2× magnification, field of view = 7 mm.

B Trachyte from Ustica island (Italy). Same image as (A) but with crossed polarizers. Sanidine shows characteristic Carlsbad twins and gray interference colors, aegirine high-interference colors, and plagioclase white–gray interference colors. Cross-polarized light, 2× magnification, field of view = 7 mm.

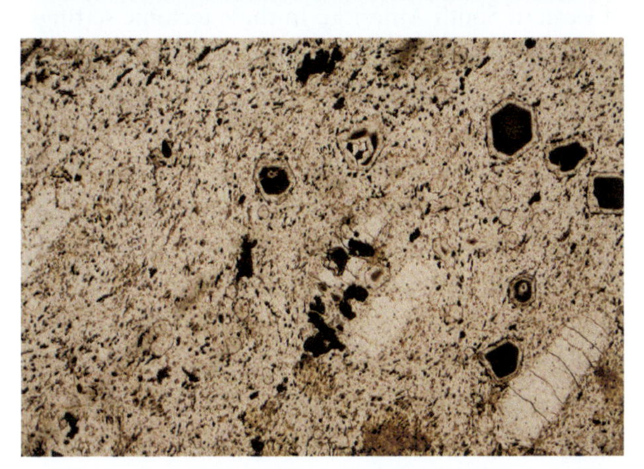

C Foid-bearing trachyte from Roccamonfina (Italy). Sanidine (colorless) and haüyne crystals (with dark core due to minute exsolutions) are set in a fine-grained trachytic-textured groundmass. Plane-polarized light, 2× magnification, field of view = 7 mm.

D Foid-bearing trachyte from Roccamonfina (Italy). Same image as (C) but with crossed polarizers. Sanidine has gray interference colors, while haüyne is isotropic. Cross-polarized light, 2× magnification, field of view = 7 mm.

Dacite

Mineralogy: *Pl, Qtz, ± Bt, ± Hbl, ± Cpx ± Afs, accessories Ttn, Ap, Mt*

SILICA: (OVER)SATURATED

IUGS/TAS classification: *Leucocratic/ SiO_2 (≥63 wt%)*

Texture: *Aphanitic to porphyritic or tuffaceous*

Dacite is the IUGS compositional extrusive equivalent of granodiorite. It consists of highly zoned Na-plagioclase (An_{70-10}) and quartz. Minor alkali feldspar, biotite, hornblende, and pyroxene (augite) may be present. Accessory minerals include titanite, apatite, and magnetite, and rarely garnet. The proportion of plagioclase and alkali feldspar is ~2:1. Quartz phenocrysts are often rounded and embayed.

Occurrence

Dacite is associated with convergent plate tectonic margins – that is, along island arcs in intraoceanic settings (the Philippines or the Aleutians) and the continental side of ocean–continent subduction zones (the Cascades of western North America or the Andes of western South America). In these tectonic settings, the close association of basalt, basaltic andesite, andesite, *dacite*, and rhyolite is often referred to as the *orogenic suite*.

Dacitic magmas have moderately high silica and volatile (mostly H_2O) contents, resulting in explosive eruptions. Dacitic to andesitic volcanoes exhibit variable eruptive styles, from explosive to domes to lava flows. Active arc volcanoes often have complicated plumbing systems. Evolved (dacitic, rhyolitic) magmas can be present at one location and later be intruded by a batch (or batches) of more primitive (basaltic) magma ascending from deeper in the system. This results in a mechanical and thermal destabilization that can trigger eruptions. For example, the 1991–1995 Unzen dacite eruption in Japan was triggered by magma mixing between an evolved phenocryst-rich, low-temperature (760–780 °C) magma and a high-temperature (1,050 °C) primitive magma. The evidence of magma mixing is deduced from chemical and textural observations including: (1) disequilibrium mineral assemblages like the coexistence of Mg-rich olivine and quartz; (2) disequilibrium crystallization textures such as reaction rims, embayments, and sieve textures; (3) disequilibrium mineral compositions with inverse, oscillatory, and normal zoning in the same sample; (4) rounded or deformed vesicular magmatic enclaves of variable size; and (5) glass inclusions trapped in minerals. The genesis of intermediate magmas such as dacite may involve one or all of the following: partial melting of hybridized mantle wedge and/or crustal material, fractional crystallization of mafic melts, and mixing between mafic and silicic magmas.

Dacite from Lassen Peak, California (United States). Phenocrysts of dark hornblende, white plagioclase, and small gray quartz in a light groundmass of quartz and plagioclase. Image 6 cm wide.

Image courtesy of J. St. John

Figure 2.60

A Dacite from Tambora (Indonesia). Porphyritic plagioclase (colorless, eu- to subhedral), hornblende (brown), and round quartz (anhedral crystal, center) set in a fine-grained glassy groundmass. Plane-polarized light, 2× magnification, field of view = 7 mm.

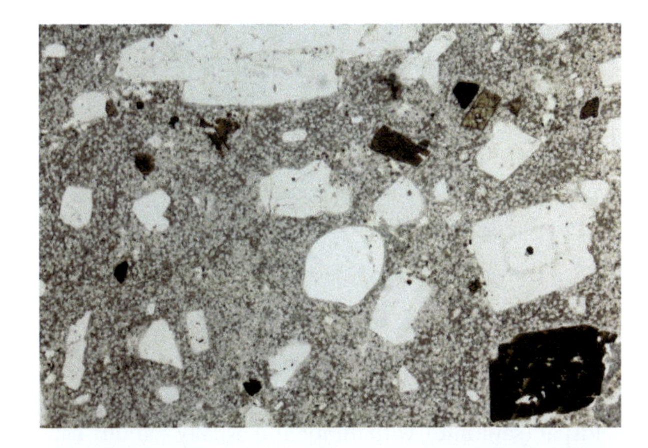

B Dacite from Tambora (Indonesia). Same image as (A) but with crossed polarizers. Plagioclase shows typical birefringence and twinning, and some concentric oscillatory growth zoning; hornblende shows high birefringence and quartz low birefringence with some fractures. The groundmass is rich in plagioclase microlites. Cross-polarized light, 2× magnification, field of view = 7 mm.

C Dacite from Capraia (Italy). Large plagioclase, rectangular biotite crystals, and twinned clinopyroxene (center) phenocrysts in an isotropic, glassy groundmass. Cross-polarized light, 2× magnification, field of view = 7 mm.

D Dacite from El Hoyazo (Spain). Large, rounded pyrope garnet (pale brown and fractured) with a narrow reaction rim, surrounded by small sieved plagioclase crystals. Rounded transparent crystals (e.g. upper-right) are quartz. Plane-polarized light, 2× magnification, field of view = 7 mm.

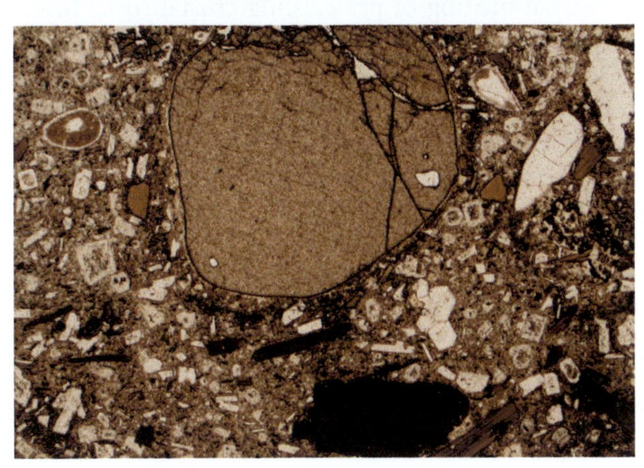

Rhyolite

Mineralogy: *Qt, + Afs, ± Na–Pl, Accessory Bt, Px, Fe–Ol, Zrn, Hbl*
IUGS/TAS classification: *Leucocratic/SiO_2 (≥69 wt%)*
Texture: *Porphyritic, aphanitic, tuffaceous, or glassy*

SILICA: (OVER)SATURATED

Rhyolite is a silica-rich volcanic rock and the extrusive compositional equivalent of granite (IUGS mode). Rhyolite consists mainly of quartz (>20%) and alkali feldspar (sanidine), with minor and usually sodium-rich plagioclase (oligoclase or andesine). Biotite, augite, fayalite, zircon, and hornblende are common accessory minerals. Some varieties with cordierite, garnet, and topaz are Al_2O_3-rich. Rhyolite is generally porphyritic, with phenocrysts set in a glassy or cryptocrystalline groundmass. High-silica rhyolitic magma is viscous and therefore generates flow textures, such as the alignment of phenocrysts and/or crystallites. When rhyolitic magma cools rapidly it generates *obsidian*; during explosive eruptions it produces *pumice*. Rhyolite grades into rhyodacite with decreasing SiO_2.

Occurrence

Rhyolite forms at convergent plate boundaries (e.g. the Andes of South America) and during continental extension (e.g. the Basin and Range Province, western United States). Rhyolite is also commonly associated with large igneous flood basalts. More rarely, rhyolite can be found in oceanic island settings (e.g. Hawaii), where it is linked to magmatic differentiation processes. In some of these settings, the close association between rhyolitic and mafic rocks (the so-called "bimodal" volcanic suite) is mostly attributed to protracted differentiation of a basaltic parental melt and/or partial melting of pre-existing crustal rocks (*anatexis*); this is supported by the fact that bimodal suites exhibit similar geochemical and isotopic signatures. On the other hand, the genesis of evolved, high-viscosity, volatile-rich, and *crystal-poor* rhyolitic melts has been explained via crystal-rich, low-melt volumes (i.e. the *crystal mush*). In this context, interstitial SiO_2-rich liquids are extracted from the crystal mush by filter-pressing processes, such as compaction or gravitational instability, following volatile exsolution. Thermomechanical models have shown that the optimal conditions for the extraction of silicate liquids from a crystal mush requires ~50–70 vol% crystals.

Welded rhyolite from Criccieth (Wales). The rock shows a eutaxitic texture characterized by flattened and elongate pumice (red) surrounded by a beige groundmass with quartz phenocrysts. Image ~7.5 cm wide.
Image courtesy of A. Tindle

Figure 2.61

A Rhyolite from Roccastrada (Italy). Euhedral biotite (brown), sanidine (parallel fractures), and several subhedral quartz phenocrysts (e.g. left of the large biotite) in a glassy (clear) groundmass. Flow banding is seen in the parallel alignment of biotite microphenocrysts (adjacent to the large biotite phenocryst). Plane-polarized light, 2× magnification, field of view = 7 mm.

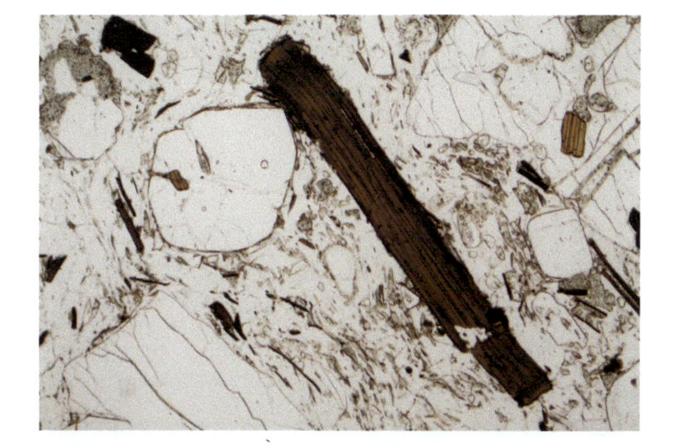

B Rhyolite from Roccastrada (Italy). Same image as (A) but with crossed polarizers. Sanidine crystals lack characteristic simple twinning but (unlike quartz) have lower birefringence and parallel fractures (lower-left). The large sanidine phenocryst is also embayed. Cross-polarized light, 2× magnification, field of view = 7 mm.

C Rhyolite from San Vincenzo (Italy). Phenocrysts include large, anhedral sanidine with Carlsbad twinning surrounded by small, subhedral quartz, rectangular biotite, and Na-rich plagioclase (center-right) in an isotropic glassy groundmass. Cross-polarized light, 2× magnification, field of view = 7 mm.

D Obsidian from Lipari (Italy). Euhedral sanidine phenocryst in a flow-banded, glassy groundmass. The brown glass is rich in microlites, while the clear glass is devoid of microlites. Plane-polarized light, 10× magnification, field of view = 2 mm.

APPLICATION 2.3

Crystal Size Distribution Analysis

During solidification and melting, crystals change in size, shape, orientation, and position; thus they convey petrological information. Crystal size distribution (CSD) theory quantitatively links this petrological information to the kinetics of crystallization. Because crystal size is a measure of growth rate, age, and nucleation density, the empirical relationship between crystal growth rate, nucleation rate, and cooling permit us to use CSD data to constrain these rates.

CSD analysis. CSD provides a method for determining the average residence time of a population of crystals by equating the slope of CSD data to $-1/$(growth rate $\times$ residence time), essentially a kinetic model in which no sorting or equilibration has occurred. Residence time is calculated from the slope, and realistic growth rates are selected. This method has been used to determine timescales of magmatic processes in a variety of settings, especially in association with volcanic systems. It allows the physical processes affecting CSD to be assessed: crystal fractionation and accumulation, mixing of populations, annealing in plutonic rocks, and nuclei destruction.

The CSD of a mineral in a rock is the number of crystals of that mineral within a series of defined size intervals per unit volume. As larger numbers of crystals are measured, the width of the size intervals can be reduced, and the CSD becomes a smooth function. CSDs are easily visualized on semi-log plots of ln (population density) versus size, where ln is the natural logarithm of the population density (number/unit volume), size is often reported as length, and the data are linear. Linear least-squares regression is then used to calculate the slope and intercept values of CSD data.

Integrated analyses. Studies combining CSD analysis with other crystal data, such as texture, composition, age, etc., provide a powerful method for understanding magmatic systems. Consider the study of Mount Lassen volcano in California, part of the Cascadia continental arc of the western United States, by Salisbury et al. (2008). The 1915 Lassen Peak eruption produced andesite and dacite. The CSD analyses of plagioclase from the eruptive products shown below indicates that each of the three rock types contain three plagioclase populations (microlites, microphenocrysts, and phenocrysts). The breaks in slope occur at the same crystal lengths, suggesting that despite being hosted in different rock types, they all share a similar cooling history.

The microlite populations have the steepest slopes and highest intercepts, implying that the highest nucleation rates in this study are associated with them. These high rates are associated with large degrees of undercooling (e.g. generated during magma ascent and/or eruption), and residence times

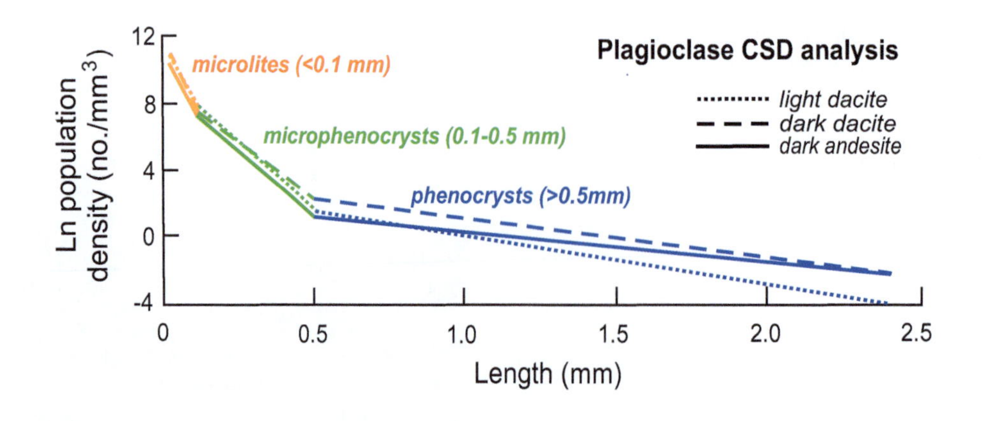

are probably on the order to hours to days. The microphenocryst populations have relatively steep slopes and also record high degrees of undercooling and nucleation, but these occur at lower rates than those associated with the microlites and are consistent with the swallow-tail and skeletal textures of the microphenocrysts. Their average residence times are on the order of months. The phenocryst populations define shallow slopes with low intercepts, implying small degrees of undercooling and low nucleation rates (e.g. modest cooling rates as in a magma chamber), with average residence times from centuries to millennia.

By analyzing the same crystals used for CSD analysis, it is possible to link crystal size, crystal growth rate, and crystal composition. The plagioclase compositions shown here are also for the 1915 Mt. Lassen eruption (Salisbury et al., 2008) and include analyses of cores (colored) and rims (gray) from all crystal size groups and rock types used in the CSD analysis. *Within* each size population, cores have similar compositions regardless of rock type. *Between* the three size groups, significant compositional differences exist, with the microlite cores having compositions intermediate (An_{40-65}) to the cores of microphenocrysts (An_{60-75}) and phenocrysts (An_{30-40}). Notably, the rims are broadly similar (An_{30-70}).

Salisbury et al. (2008) concluded that the combined CSD and plagioclase compositional data record three crystal populations with distinct chemical origins but

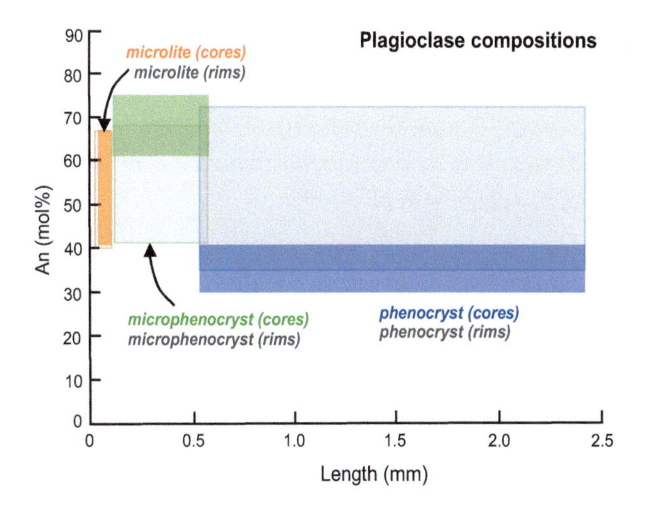

similar cooling histories that are best explained by magma recharge and mixing associated with the 1915 Mt. Lassen eruption. The phenocryst cores crystallized from slowly cooling dacitic magmas, while the microphenocrysts formed from mixing of dacite and basaltic andesite, which resulted in high degrees of undercooling and faster crystal growth rates. The microlites also reflect a mixed magma composition but with rapid nucleation and crystallization likely associated with eruption.

Today, CSD analysis is performed using semi-automated QEMSCAN instruments which combine quantitative mineral identification with shape recognition software to facilitate the identification of mineral groups by criteria such as size, shape, and composition.

Basanite/Tephrite

Mineralogy: *Pl, Cpx, Ol, foids (10–60%), ± Afs*

SILICA: UNDERSATURATED

IUGS/TAS classification: *Melano- to mesocratic/SiO_2 (41–49 wt%)*

Texture: *Aphanitic to porphyritic*

Basanite and tephrite are alkali basalts that are only distinguishable from each other on the basis of normative chemistry. Higher sodium and potassium (3–9 wt%) distinguish them from tholeiitic picrobasalts. Basanite/tephrite contains Ca-plagioclase (An_{50-70}), Ti-augite, and nepheline or leucite (10–60%) as the main feldspathoid; accessory alkali feldspar is limited to the groundmass. The feldspathoid is often seriate. The dominant feldspathoid is used in naming the rock (e.g. nepheline tephrite, leucite basanite).

Occurrence

Basanite/tephrite, as with other alkaline rocks, occur in many tectonic settings from continental rifts to subduction zones and oceanic intraplate environments. Basanite/tephrite is widespread across the Roman and Campanian volcanic provinces of Italy (e.g. the Colli-Albani volcanic complexes, or the Vesuvius and Phlegraean Fields). Basanite/tephrite also occurs in the Azores (Spain), the Tuxtla and Durango volcanic field (Mexico), and in the African Rift Valley. In many volcanic centers, the eruptive association of basanite, tephrite, phonotephrite, tephriphonolite, and phonolite is commonly interpreted as the result of magmatic differentiation. At Vesuvius (Italy), for example, phonolite, phonotephrite, and tephriphonolite can be produced from a parental basanite melt through the fractionation of clinopyroxene and olivine separation to form phonolitic tephrite. Further differentiation then produces a tephriphonolite, and finally a phonolite. The leucite-bearing lavas of the Vulsini complex (Italy) evolved in a similar way – a homogeneous basanite/tephrite parental magma, followed by fractional crystallization of clinopyroxene, leucite, and plagioclase, drove the parent magma toward phonolite compositions. Basanite also characterizes the late alkaline phase of ocean islands, such as Hawaii.

Basanite with mantle xenoliths from Peridot Mesa, Arizona (United States). Peridotite xenoliths (made of bright green olivine and pyroxene) in vesiculated basanite lava. Dark pyroxene phenocryst present above the xenoliths. Image ~5 cm wide.

Image courtesy of J. St. John

Figure 2.62

A Basanite from Vico Lake (Italy). Subhedral plagioclase (colorless), large pyroxene (left), rounded leucite (center-bottom), and two olivine phenocrysts (center-top) set in a fine-grained groundmass with scattered brown phlogopite crystals. Plane-polarized light, 2× magnification, field of view = 7 mm.

B Basanite from Vico Lake (Italy). Same image as (A) but with crossed polarizers. Plagioclase shows gray birefringence, leucite is almost isotropic, pyroxene and olivine phenocrysts show high birefringence. Olivine also shows weak compositional zoning in its changing birefringence. Cross-polarized light, 2× magnification, field of view = 7 mm.

C Leucite–tephrite from Vesuvius (Italy). Euhedral augite phenocrysts with concentric compositional zoning (light brown–green), subhedral leucite (white), and plagioclase (colorless) phenocrysts set in a dark, glassy groundmass rich in plagioclase microlites. Note the concentric alignment of inclusions in the central augite phenocryst. Plane-polarized light, 2× magnification, field of view = 7 mm.

D Leucite–tephrite from Vesuvius (Italy). Same image as (C) but with crossed polarizers. Compositional zoning in augite phenocrysts is visible as variation in birefringence. Leucite is almost isotropic and plagioclase shows typical gray birefringence. Cross-polarized light, 2× magnification, field of view = 7 mm.

Phonotephrite–Tephriphonolite

Mineralogy: *Pl, Px, foids, ± Afs, ± Amp, ± Bt*

SILICA: UNDERSATURATED

IUGS/TAS classification: *Mesocratic/SiO_2 (45–57%)*

Texture: *Aphanitic to porphyritic or tuffaceous.*

Phonotephrite is a basic alkaline volcanic rock (SiO_2 = 45–53 wt%; $Na_2O + K_2O$ = 7–12 wt%). It typically has phenocrysts of Ca-plagioclase, augite, and feldspathoid (10–60%), with minor olivine and/or sanidine. Increasing SiO_2 (48–57 wt%) and $Na_2O + K_2O$ (9–14 wt%) generates tephriphonolite which typically contains phenocrysts of alkali feldspar, Na-plagioclase, feldspathoid minerals (60–90%), and mafic minerals such as alkali amphibole/pyroxene and biotite.

Occurrence

Phonotephrite and tephriphonolite occur in many tectonic environments, but their alkaline character is most associated with intracratonic rifts such as the Asunción Rift (Paraguay) or the Rhine Graben (Germany), and intraplate settings such as the Canary Islands (Spain) or the Cape Verde Islands. Their genesis is mostly linked to extreme differentiation or partial melting of tholeiitic to alkalic mafic magmas. Phonotephrite and tephriphonolite may be spatially and temporally associated with carbonatites, which implies that both rock groups are closely related. Phonotephrites often occur as lava flows or pyroclastic deposits, while tephriphonolite are typically subvolcanic intrusions (plugs and dikes).

In numerous volcanic centers, a differentiation trend from basanite–tephrite parental liquid(s) through phonotephrite and tephriphonolite, to phonolitic compositions documents fractional crystallization processes, often repeated over time and accompanied by assimilation of crustal rocks (AFC process). For example, the fractionation of clinopyroxene, olivine, amphibole, and titanomagnetite from a parental basanitic melt can produce a residual tephritic melt. Further differentiation of the residual tephritic melt via crystal fractionation of plagioclase, amphibole, clinopyroxene, and titanomagnetite generates a tephriphonolite melt composition.

Tephriphonolite from the 1944 eruption of Vesuvius (Italy). Rounded leucite (tan) and augite (black) phenocrysts set in a dark glassy groundmass. Image ~10 cm wide.

Image courtesy of L. Melluso

Figure 2.63

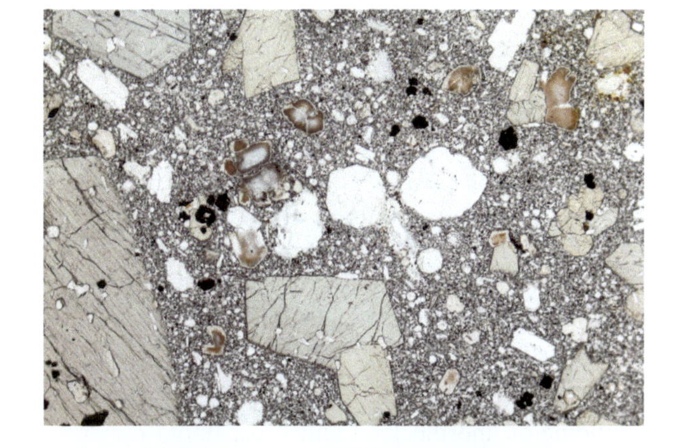

A Nosean phonotephrite from Vulture volcano (Italy). Light brown pyroxene, colorless plagioclase (with rectangular outlines), rounded seriate leucite and nosean (brown rims and pale bluish cores, just left of center), and black magnetite set in a fine-grained groundmass. Plane-polarized light, 2× magnification, field of view = 7 mm.

B Nosean phonotephrite from Vulture volcano (Italy). Same image as (A) but with crossed polarizers. Pyroxene has moderate birefringence, plagioclase typical gray with twins, while nosean is isotropic and leucite is nearly isotropic with very dark interference colors. Cross-polarized light, 2× magnification, field of view = 7 mm.

C Leucite tephriphonolite from Olbrück (Germany). The central crystal with a light brown rim is haüyne; some prismatic sanidine crystals are present in the lower part of the picture. Both are set in a fine-grained groundmass with small, rounded leucite crystals and anhedral pyroxene. Plane-polarized light, 2× magnification, field of view = 7 mm.

D Leucite tephriphonolite from Olbrück (Germany). Same image as (C) but with crossed polarizers. Haüyne is isotropic; sanidine shows characteristic gray interference colors; leucite is almost isotropic with very dark interference colors. The white birefringent crystals in the groundmass are sanidine and plagioclase. Cross-polarized light, 2× magnification, field of view = 7 mm.

Phonolite

Mineralogy: *Afs, foids, ± Px, ± Na–Pl, ± Ol, ± Amp, accessory Ap, Il, Mt, Ttn, Zrn*

SILICA: UNDERSATURATED

IUGS/TAS classification: *Leucocratic/SiO_2 (≥53 wt%)*

Texture: *Aphanitic to porphyritic*

Phonolite is the extrusive equivalent of nepheline syenite (IUGS mode). It is an aphanitic to porphyritic, alkaline volcanic rock. It is composed of alkali feldspar (sanidine or anorthoclase) and feldspathoid minerals. Biotite and alkali amphibole (riebeckite or arfvedsonite) are typical. Na-pyroxene along with Fe-rich olivine and Na-plagioclase are common minor minerals. Its groundmass may consist of elongate, flow-aligned sanidine crystals. Accessory phases include titanite, apatite, corundum, zircon, magnetite, melilite, and ilmenite, and occasionally melanite garnet. The most common feldspathoid is nepheline, but phonolite can contain any of the feldspathoid minerals. The predominant feldspathoid is added to the rock name (e.g. leucite phonolite, analcime phonolite).

Occurrence

Phonolite is found widely distributed throughout the world, from rift-related to ocean island tectonic settings, such as the East African Rift or Vesuvius (Italy). In association with the latter setting, phonolite erupts as low-volume flows together with other evolved lavas (trachytes). Phonolite genesis is much debated, but the main hypotheses for its formation are not necessarily exclusive; they include: evolution of primary basanitic magmas through fractional crystallization and/or crustal contamination/assimilation, or through low degrees of partial melting of upper mantle metasomatized peridotites. Phonolite often hosts cognate enclaves and glomerocrysts (e.g. the Teide-Pico Viejo stratovolcano [Canary Islands] or the Laacher See volcano [Germany]) that are interpreted as remnants of crystal mush. Experiments at mid- and upper-crustal conditions (850–1,150 °C, 200–500 MPa) suggest that phonolite can be produced from trachytic melt through the interplay of fractional crystallization, feldspar resorption, and melt segregation. Nevertheless, peridotitic xenoliths from the upper mantle may indicate that their genesis is linked to deeper magmatic processes. Experiments have also documented the production of 0.3–3% phonolitic liquid as K_2O concentration in the source increases from 100 to 2,000 ppm.

Phonolite, Montana (United States). Phenocrysts of subhedral leucite (white) and amphibole (black) in a fine-grained groundmass. Image ~10 cm wide.

Image courtesy of J. St. John

Figure 2.64

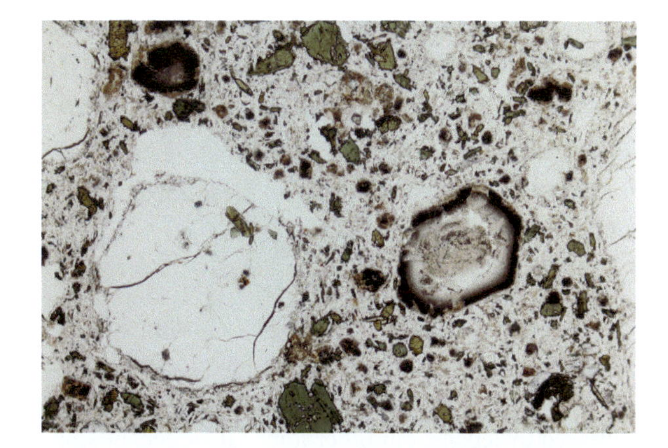

A Leucite–haüyne phonolite from Rieden, Eifel (Germany). Phenocrysts of leucite (rounded, colorless) and haüyne (with dark rims) in a fine-grained groundmass rich in aegirine (green) and sanidine microphenocrysts. Plane-polarized light, 2× magnification, field of view = 7 mm.

B Leucite–haüyne phonolite from Rieden, Eifel (Germany). Same image as (A) but with crossed polarizers. Leucite shows characteristic complex twinning and low birefringence, and haüyne is isotropic. Aegirine has high birefringence. Cross-polarized light, 2× magnification, field of view = 7 mm.

C Nosean phonolite from Eifel (Germany). A large, euhedral nosean phenocryst (left) is full of dark inclusions which give it a brownish color. Sanidine (colorless) is prismatic. Both sit in a brown, glassy groundmass. Plane-polarized light, 2× magnification, field of view = 7 mm.

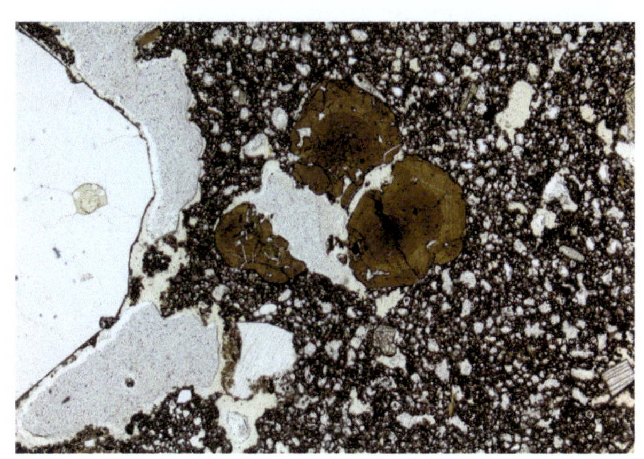

D Leucite phonolite from Vesuvius (Italy). A large leucite (left, colorless, high relief) and subhedral garnet (brown) phenocrysts in a glass-rich groundmass with leucite and pyroxene microphenocrysts. Adjacent to the leucite phenocryst, groundmass has been removed leaving a gap now filled with thin-section glue. The garnet is melanite and compositional zoning is visible. Plane-polarized light, 2× magnification, field of view = 7 mm.

Foidite

Mineralogy: *Foids (>60%), ± Cpx, accessory Amp, Ap, Mll, Phl, Prv, Ttn, Wo* **SILICA: UNDERSATURATED**
IUGS/TAS classification: *Melano- to mesocratic/SiO_2 (<52 wt% at $Na_2O + K_2O = 14$ wt%)*
Texture: *Aphanitic to porphyritic*

The foidites are fine-grained, highly alkaline volcanic rocks with significant feldspathoids (>60%). The most common feldspathoid in foidite is nepheline followed by leucite, but other foids (analcime, nosean, haüyne, or sodalite) can be present. There are three subcategories of foidite based on the ratio of alkali feldspar/plagioclase: *Phonolitic foidite* (60–90% feldspathoid, alkali feldspar > plagioclase), *basanitic/tephritic foidite* (60–90% feldspathoid, plagioclase > alkali feldspar), and *foidite* (*sensu stricto*; 90–100% feldspathoid). The dominant feldspathoid mineral is added to the name of the rock (e.g. nepheline foidite, leucite foidite). Common accessory minerals include apatite, perovskite, titanite, melanite, wollastonite, amphiboles, and phlogopite.

Occurrence

Foidites are found in numerous tectonic settings and often associated with other alkaline rocks or carbonatites. Nepheline foidite is common in many igneous complexes associated with old shield areas that have thick lithosphere (e.g. the Kola Peninsula [Russia] or the carbonatites of the African Rift Valley).

Most petrogenetic models for these rocks require low-degree partial melting of heterogeneous asthenospheric mantle to explain their geochemical signatures and the link with alkaline lavas. After some process(es) induces mantle heterogeneity (tectonic extension, small-scale convection, mantle flow, etc.), the lithosphere is metasomatized to create amphibole- or phlogopite-rich veins that are later partially melted to generate low-Si alkali magmas. The presence or absence of amphibole, at the expense of phlogopite, is influenced by lithospheric thickness, with amphibole stable to 2.8–3.0 GPa and phlogopite to 8 GPa (the base of continental lithosphere). Foidites can also be generated via assimilation of limestone by basaltic magma – experiments at 1,050–1,150 °C and 0.1–500 MPa show that large amounts of limestone can be assimilated by basaltic magma and generate a CO_2-rich fluid phase that, combined with significant crystallization of Ca-rich clinopyroxene, produces strongly silica-undersaturated melts. The reaction is directly influenced by the MgO concentration of the magma:

olivine-saturated: $\underset{\text{calcite}}{2CaCO_3} + \underset{\text{melt}}{3SiO_2} + \underset{\text{olivine}}{Mg_2SiO_4} = \underset{\text{Cpx}}{2CaMgSi_2O_6} + \underset{\text{fluid}}{2CO_2}$

not olivine-saturated: $\underset{\text{calcite}}{CaCO_3} + \underset{\text{melt}}{MgO} + \underset{\text{melt}}{2SiO_2} \rightarrow \underset{\text{Cpx}}{CaMgSi_2O_6} + \underset{\text{fluid}}{CO_2}$

Leucite foidite from Rome (Italy). Whitish-green leucite phenocrysts nucleated on pyroxene are set in a fine-grained groundmass. Image is ~12 cm wide.
Image courtesy of J. St. John

Figure 2.65

A Nepheline foidite from Fogo (Cape Verde). Large, euhedral Ti-augite (brownish) and equant nepheline (colorless) in a fine-grained, glassy groundmass rich in microphenocrysts of both. Plane-polarized light, 2× magnification, field of view = 7 mm.

B Nepheline foidite from Fogo (Cape Verde). Same image as (A) but with crossed polarizers. Augite shows high birefringence and concentric compositional zoning. Nepheline shows characteristic first-order (gray) birefringence. Cross-polarized light, 2× magnification, field of view = 7 mm.

C Leucite foidite from Pitigliano (Italy). Seriate, equant, subhedral leucite phenocrysts in a brown, glassy, pyroxene-rich groundmass. Leucite has skeletal texture with radially arranged cavities (cartwheel-like) filled by dark glass. Plane-polarized light, 2× magnification, field of view = 7 mm.

D Leucite foidite from Pitigliano (Italy). Same image as (C) but with crossed polarizers. Leucite shows characteristic very low birefringence and pyroxene (higher birefringence) is easily distinguished. Cross-polarized light, 2× magnification, field of view = 7 mm.

Obsidian

Mineralogy: *Glass ± Qz, ± Pl, ± Kfs, ± Bt, ± Px, ± Hbl*

GLASSY ROCKS

IUGS classification: *Not applicable*

Texture: *Hypocrystalline to holohyaline*

Obsidian is glass produced from a rapidly cooled felsic lava. Obsidian is rich in SiO_2 (~65–80%) and often dark in color due to its amorphous form and the presence of iron and other transition elements. It typically has a water content $<1\%$, though at high pressures it can contain up to 10% H_2O. During extrusion, pressure may rapidly decrease, which increases viscosity and inhibits crystallization of the melt. It may contain gas bubbles which produce a golden sheen and may align with the flow direction. Obsidian is usually aphyric, although it may contain phenocrysts of feldspar, quartz, biotite, or pyroxene, but abundant crystallites (sub-micron crystals) are much more common. Being a glass, obsidian is susceptible to secondary hydration, devitrification, and other types of alteration, and this causes the formation of *spherulitic* or *perlitic* texture. The conchoidal fracture of obsidian is characteristic.

Occurrence

Obsidian is associated with the rapid eruption of higher-silica (dacitic to rhyolitic) melts mainly associated with convergent plate margins or regions of continental extension. Due to the high viscosity of silicic magma, obsidian commonly forms domes of modest size (< several kilometers in diameter). Domes are usually emplaced during the final stages of eruption, when the last, gas-depleted magma inflates in the central crater. Obsidian may form thick flows of limited extent (several hundred meters), the basal and/or upper *vitrophyre* of thicker rhyolitic lava flows, the chilled margins of rhyolitic dikes, or may be present as fragments in ash tuffs. Its glassy nature reflects very rapid cooling (e.g. on the margins or tops of rhyolitic flows); however, obsidian flows are too thick for their interiors to cool quickly and their formation is linked to inhibited diffusion, nucleation, and crystallization associated with highly polymerized and viscous silicic liquids. The 2008 eruptions of the Chaitén and Puyehue-Cordón Caulle volcanoes (Chile) documented an initial explosive phase, followed by the final effusion of obsidian flows. Classic examples of obsidian flows/domes include Glass Mountain (California) or Big Obsidian Flow (Oregon) in the United States, and those occurring on Lipari (Aeolian Islands, Italy).

Obsidian from Yellowstone (United States) showing typical conchoidal fracture (*top*) and partially devitrified obsidian in which the white spherulites are aligned with the flow direction (*bottom*).

Images courtesy of J. St. John

Figure 2.66

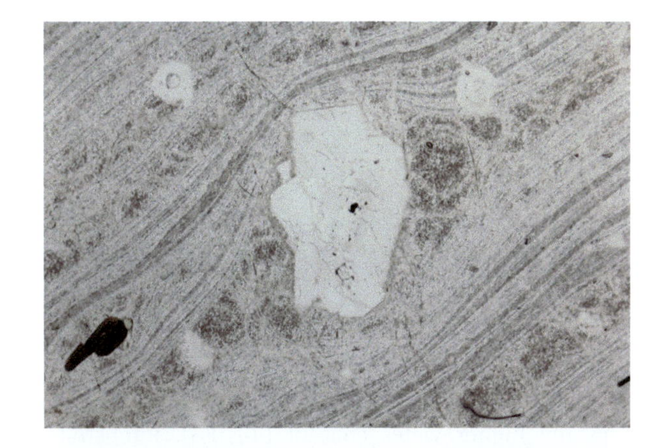

A Obsidian from Lipari (Italy). Feldspar crystals (colorless) and brown biotite (lower-left corner) embedded in a clear, glassy, crystallite-rich groundmass. Crystallite alignment defines a flow structure. Plane-polarized light, 2× magnification, field of view = 7 mm.

B Obsidian from Lipari (Italy). Same image as (A) but with crossed polarizers. The glassy groundmass is isotropic. Feldspar has gray interference colors, and biotite has brown birefringence. The small birefringent crystals in the isotropic groundmass are also feldspar. Cross-polarized light, 2× magnification, field of view = 7 mm.

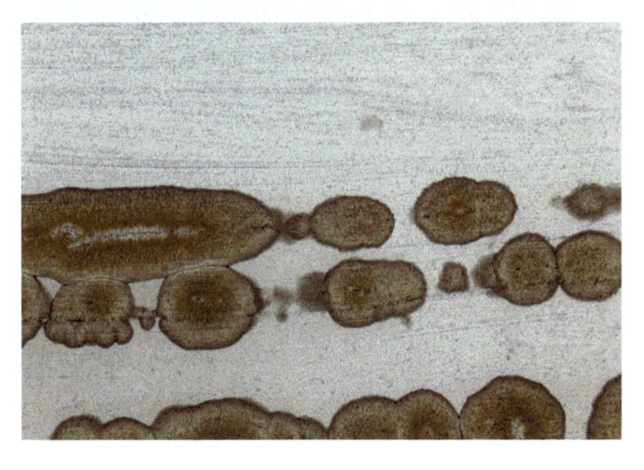

C Obsidian from Lipari (Italy). Partially devitrified obsidian with rounded, brown spherulites. Some spherulites have grown together. Plane-polarized light, 2× magnification, field of view = 7 mm.

Pitchstone

Mineralogy: *glass, ± Qz, ± Pl, ± Ksp, ± Px, ± Hbl, etc.*
IUGS classification: *Not relevant/glassy*
Texture: *Hypocrystalline to holohyaline*

GLASSY ROCKS

Pitchstone, a volcanic glass with conchoidal fracture, is distinguished from obsidian by its resinous luster, higher quantity of microlites and crystallites, and typically higher water content (4–10%). Pitchstone can be aphyric. Minerals, if present, depend on bulk composition and typically include quartz, alkali feldspar, plagioclase, pyroxene, hornblende, and fayalitic olivine. The composition of pitchstone is more variable than obsidian and ranges from andesitic–dacitic to rhyolitic. Pitchstone, like obsidian, is extremely susceptible to secondary alteration processes.

Occurrence

Pitchstone occurs mainly as dikes and sills, at the margins of acidic lava flows, or in thick lava flows. Pitchstone may grade to pitchstone porphyry (or vitrophyre) with abundant phenocrysts in a glassy groundmass which, like obsidian, commonly shows magmatic flow defined by the alignment of microlites; such microlites often have dendritic and skeletal morphologies. Pitchstone is widespread within the British Paleogene Igneous Province of England and in western Scotland and Ireland, where it occurs as minor intrusions and lava flows with wide textural and compositional variability. Elsewhere (Oregon, United States; Korea; New Zealand) it may be characterized by an anhydrous phenocryst assemblage distinct from the hydrous crystallite populations of the groundmass. The presence of hydrous groundmass phases such as hornblende and biotite with anhydrous phenocrysts such as ferrohedenbergite and ferrosilite may indicate increasing water saturation of the residual liquid due to the precipitation of anhydrous phases during magma ascent and emplacement. Pitchstone petrogenesis is mainly linked to magma mixing between basaltic magmas and hydrous crustal melts, as supported by disequilibrium textures in phenocrysts (spongy, sieve, and resorbed crystals), the presence of mafic glomeroporphyritic clots, and their often peraluminous chemistry, suggesting involvement of a crustal source. Other petrogenetic processes suggested include partial fusion of basaltic material and extreme fractional crystallization processes.

Pitchstone from Valles Caldera, New Mexico (United States). The sample show a black resinous luster with yellowish plagioclase crystals.
Image courtesy of J. St. John

Figure 2.67

A Pitchstone from Arran (Scotland). Resorbed quartz crystal (center) in a brown, glassy groundmass with small, dendritic hornblende crystallites. Plane-polarized light, 10× magnification, field of view = 2 mm.

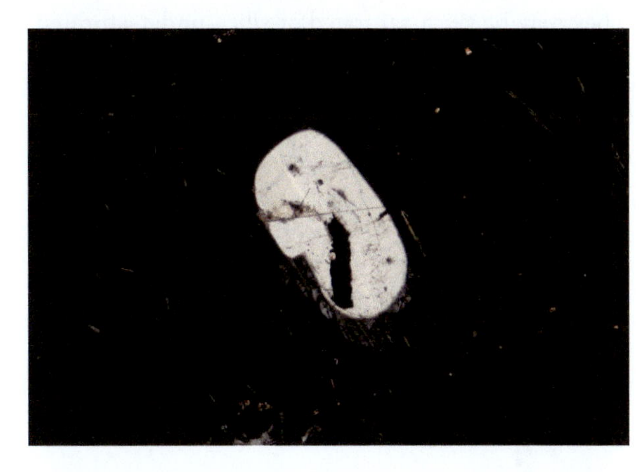

B Pitchstone from Arran (Scotland). Same image as (A) but with crossed polarizers. Glassy groundmass is mostly isotropic but larger crystallites of hornblende are birefringent. Resorbed quartz phenocryst has typical gray–white birefringence. Cross-polarized light, 10× magnification, field of view = 2 mm.

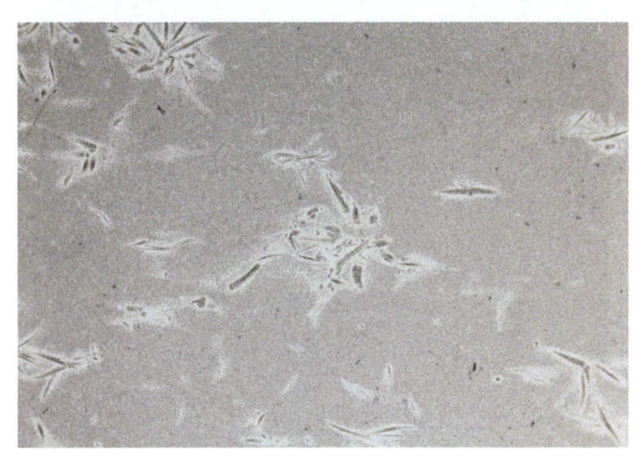

C Pitchstone from Arran (Scotland). The section is aphyric and contains only hornblende dendritic aggregates in a glassy, crystallite-rich groundmass. The growth of dendrites is diffusion-limited, generating an impoverished region (white) between the green crystallites and the grayish groundmass. Plane-polarized light, 2× magnification, field of view = 7 mm.

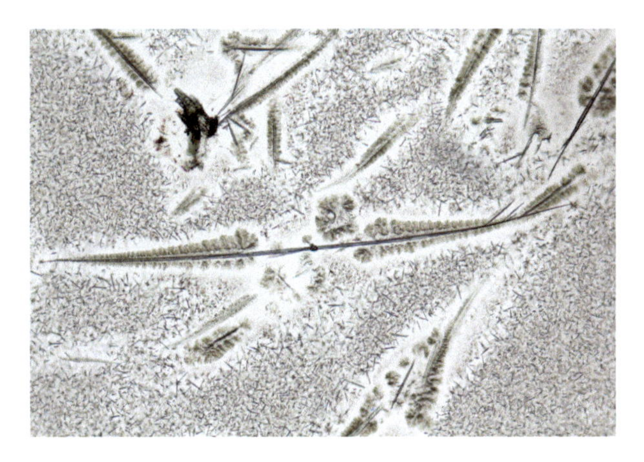

D Pitchstone from Arran (Scotland). Close-up of (C) showing hornblende dendrites. The border region (white) between the crystallites and the speckled groundmass is due to local diffusion. Plane-polarized light, 10× magnification, field of view = 2 mm.

TUFFACEOUS ROCKS, WELDED AND UNWELDED

Tuffs are pyroclastic rocks made of ejecta from volcanic eruptions. Tuffaceous rocks are classified on the basis of their pyroclastic fragments: *lithic tuff* (mostly rock fragments), *crystal tuff* (mostly crystals and crystal fragments), and *vitric tuff* (mostly pumice and glass fragments). Their matrix is often ash and these are known as ash-fall tuffs and ash-flow tuffs (also called *ignimbrites*). Very fine consolidated ash particles in tuffs may form a *cryptocrystalline* matrix. They may be silica-saturated or silica-undersaturated.

Tuffs are *welded* when sufficiently high temperatures are reached at the time of deposition – this allows glass shards and pumice fragments to fuse. The combined thickness (weight) of overlying material and its retained heat often result in deformation/flattening. Flattened, elongate pumice clasts are known as *fiamme* and are common in welded tuffs; their alignment produces the banded *eutaxitic* texture. Tuffaceous particles deposited far from the eruptive center are colder and consequently do not weld or compact as much.

Figure 2.68

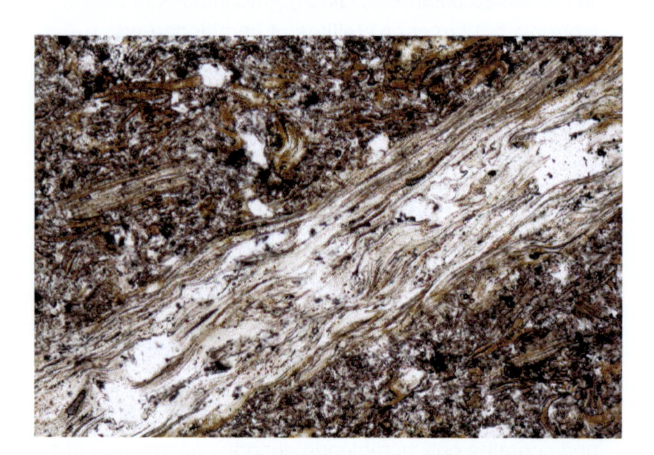

A Unwelded vitric tuff from Phlegraean Fields (Italy). A large glassy pumice fragment dominates the image and is surrounded by small pumice and shard fragments. These fragments are not welded or deformed. Plane-polarized light, 2× magnification, field of view = 7 mm.

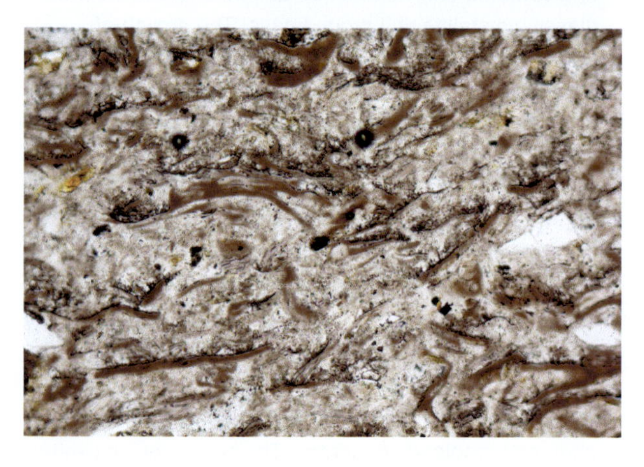

B Welded vitric tuff from Bingöl (Turkey). The variable colors (brown to beige) of these welded and flattened glassy fragments indicates their slightly different chemical composition. Note the typical Y-shaped glass shards still partly visible (lower-right, central, and upper-central part of image). Plane-polarized light, 10× magnification, field of view = 2 mm.

C Lithic-crystal tuff from Phlegraean Fields (Italy). Note the three lithic fragments (left half of image) set in a fine-grained groundmass together with feldspar (colorless) and opaque (black) crystals. The lithic fragments have different colors and textures indicating their different sources. Plane-polarized light, 2× magnification, field of view = 7 mm.

D Ignimbrite from Phlegraean Fields (Italy). A large, deformed, reddish-orange fiamme (pumice fragment) dominates the image. Note the flattening of the original gas bubbles inside the pumice fragment. The reddish color is due to oxidation. It is set in a fine-grained groundmass with feldspar (colorless) and lithic fragments (lower part of image). Plane-polarized light, 2× magnification, field of view = 7 mm.

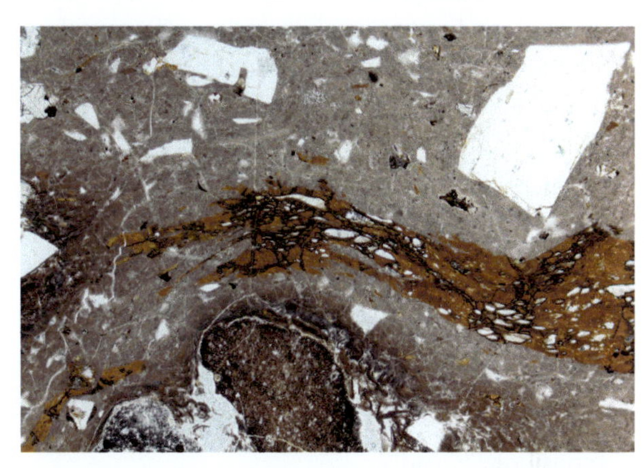

E Eutaxitic ignimbrite from Grantola (Italy). Euhedral plagioclase crystals (colorless) set in a flowing groundmass of welded and flattened pumice fragments. Color differences indicate slightly different chemical compositions. Plane-polarized light, 10× magnification, field of view = 2 mm.

APPLICATION 2.4

Crustal Anatexis

Partial melting of the crust (*crustal anatexis*) represents the interface between igneous and metamorphic petrology. Crustal anatexis is controlled by temperature (T), pressure (P), rock composition (X), and volatile content (X_v). When the physical state (T, P, X, X_v) of a rock changes, partial melting can occur. Such changes are common, occurring in both compressional and extensional tectonic settings, where deformation also facilitates melting.

As schematically illustrated below, the most effective ways to induce anatexis are to increase T, decrease P, or increase X_v. In this example, a sample exists in P/T space close to its dry solidus (i.e. the region of stability for being solid under dry conditions). To the left of the dry solidus (red line), the sample is solid (100% crystals). To the right of the dry solidus, it is solid (crystals) coexisting with melt, where the proportion of solid to liquid depends on the bulk composition, P, and T.

wet solidus
$+X_{H_2O}$
dry solidus
CRYSTALS
P
+T
-P
LIQUID + CRYSTALS
T

Increasing temperature. Increasing T can change the physical state of the sample (the green triangle) to the coexisting solid + liquid stability field. Increasing the local thermal gradient is the primary mechanism for achieving this. Increased temperatures associated with regions of active magmatism and magmatic underplating may induce partial melting of the country rock in this way.

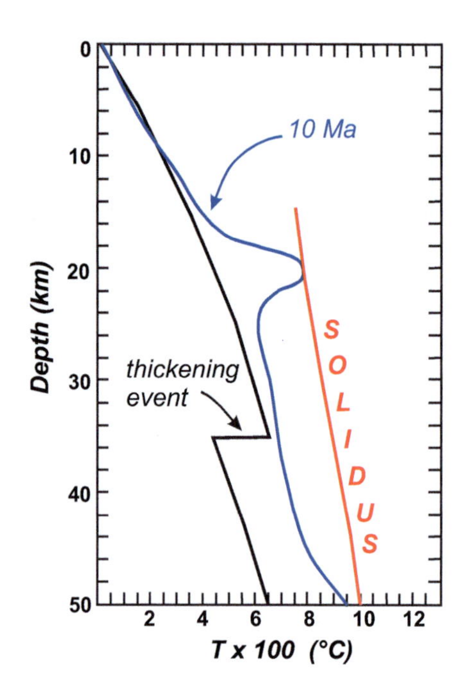

Thickening the crust during collision can also locally increase temperatures and lead to partial melting. This is especially true in crust with high concentrations of heat-producing (radioactive) elements, as in the Central Iberian Zone (Spain). Bea et al. (2003) developed a numerical thermal model for the Central Iberian Zone (CIZ) based on structural data indicating that the CIZ was thickened by underthrusting of lower crustal units (black line in the figure). These units are known to have high heat-producing element concentrations which contributed to the generation of transient geotherms. In their model (blue line in the figure), crustal thickening of rocks with high heat production produce an irregular geotherm that intersects the solidus. It is the intersection of the geotherm with the solidus at which partial melting can occur. In this scenario, crustal anatexis occurs at ~20 km depth, 10 Ma after thrusting.

Decreasing pressure. Decreasing pressure can also change the physical state of the sample. As shown in the *P*/*T* diagram, the effect of reducing *P* can effectively place a sample (the green triangle) within the stability field of solid + liquid. Localized *decompression melting* often follows crustal thickening where erosion, tectonic denudation, and/or thinning of the lithosphere occur. A classic example of decompression melts are the Himalayan leucogranites that formed ~30 Ma after the collision between Asia and India.

Changing composition/volatiles. The initial composition of the crust determines whether or not partial melting can occur. The melting temperature of a rock is influenced by the amount of volatiles present, especially water. As seen from the *P*/*T* diagram, an increase in water content (X_v) reduces the melting temperature of a rock and it is the abundance of water-bearing minerals (e.g. muscovite, biotite, and amphibole) that exert the principal control on whether or not partial melting can occur. Melt generation is limited by the amount and stability of these hydrous phases. Pore water from surrounding country rock may play an important role in anatexis, as does the abundance of heat-producing elements – both contributing to the overall physical state. Dry conditions (low water availability) suppress partial melting.

The melting of crustal rocks is seldom 100% efficient; instead, some components (minerals) of the rock will melt while other components remain unaffected – this produces a melt plus the residual rock; these "mixed" end-products are known as *migmatites*. On the Isle of Mull in Scotland (UK), a rare occurrence of incipient melting of Neoproterozoic Moine metasediment can be seen. The felsic partial melt follows S–C cleavage planes, presumably originally defined by hydrous micaceous minerals; these H_2O-bearing minerals would have provided the volatile content needed to reduce the melting temperature and thus induce melting of these metapelitic rocks. Such a process sits at the interface between igneous and metamorphic processes, and is discussed more in the next chapter.

Neoproterozoic Moine metasediment from the Isle of Mull (Scotland) with incipient melting due to regional metamorphism. The S–C fabric is now defined by felsic partial melt. The coin is ~2.5 cm diameter.

Image courtesy of M. Whitehouse

Bibliography

Textbooks

Frost, B.R., Frost, C.D., 2019. *Essentials of Igneous and Metamorphic Petrology* (2nd edn). Cambridge University Press.

Le Maitre, R.W. (ed.), 2002. *Igneous Rocks: IUGS Classification and Glossary of Terms* (2nd edn). Cambridge University Press.

MacKenzie, W.S., Donaldson, C.H., Guilford, C., 1982. *Atlas of Igneous Rocks and Their Textures*. Longman.

Philpotts, A.R., Ague, J.J., 2021. *Principles of Igneous and Metamorphic Petrology*. Cambridge University Press.

Sørensen, H., 1974. *The Alkaline Rocks*. Wiley.

Textures and Relations of Crystals

Bennett, E., Lissenberg, J., Cashman, K, 2019. The significance of plagioclase textures in mid-ocean ridge basalt (Gakkel Ridge, Arctic Ocean). *Contributions to Mineralogy and Petrology* 174, doi:10.1007/s00410-019-1587-1.

Boudreau, A., 2011. The evolution of texture and layering in layered intrusions. *International Geology Review* 53 (3–4), 330–353.

Donaldson, C.H., 1977. Laboratory duplication of comb layering in the Rhum pluton. *Mineralogical Magazine* 41(319), 323–336.

Godard, G., Martin, S., 2000. Petrogenesis of kelyphites in garnet peridotites: a case study from the Ulten zone, Italian Alps. *Journal of Geodynamics* 30(1–2), 117–145.

Le Bas, M., Maitre, R.L., Streckeisen, A., Zanettin, B., IUGS Subcommission on the Systematics of Igneous Rocks, 1986. A chemical classification of volcanic rocks based on the total alkali–silica diagram. *Journal of Petrology* 27(3), 745–750.

London, D., 2021. Pegmatites, in Alderton, D., Elias, S. (Eds.), *Encyclopedia of Geology* (2nd edn). Academic Press.

McBirney, A.R., Noyes, R.M., 1979. Crystallization and layering of the Skaergaard intrusion. *Journal of Petrology* 20(3), 487–554.

Mondal, S., Upadhyay, D., Banerjee, A., 2017. The origin of Rapakivi feldspar by a fluid-induced coupled dissolution–reprecipitation process. *Journal of Petrology* 58(7), 1393–1418.

Song, S., Cao, Y., 2021. Textures and structures of metamorphic rocks, in Alderton, D., Elias, S. (Eds.), *Encyclopedia of Geology* (2nd edn). Academic Press.

Streckeisen, A., 1976. To each plutonic rock its proper name. *Earth-Science Reviews* 12(1), 1–33.

Plutonic Rocks

Ultramafic Rocks

Aulbach, S., Lin, A.B., Weiss, Y., Yaxley, G.M., 2020. Wehrlites from continental mantle monitor the passage and degassing of carbonated melts. *Geochemical Perspectives Letters* 15, 30–34.

Barnes, C.G., Werts, K., Memeti, V., Paterson, S.R., Bremer, R., 2021. A tale of five enclaves: mineral perspectives on origins of mafic enclaves in the Tuolumne Intrusive Complex. *Geosphere* 17(2), 352–374.

Barrat, J.A., Bachèlery, P., 2019. La Réunion Island dunites as analogs of the Martian chassignites: tracking trapped melts with incompatible trace elements. *Lithos*, 344, 452–463.

Bedard, J.H., 1991. Cumulate recycling and crustal evolution in the Bay of Islands ophiolite. *Journal of Geology* 99, 225–249.

Chin, E.J., 2018. Deep crustal cumulates reflect patterns of continental rift volcanism beneath Tanzania. *Contributions to Mineralogy and Petrology* 173, 85.

He, Q., Zhang, S.-B., Zheng, Y.-F., Chen, R.-X., 2022. Peritectic minerals record partial melting of the deeply subducted continental crust in the Sulu orogen. *Journal of Metamorphic Geology* 40(1), 87–120.

Kepezhinskas, P., Berdnikov, N., Kepezhinskas, N., Konovalova, N., 2022. Metals in Avachinsky peridotite xenoliths with implications for redox heterogeneity and metal enrichment in the Kamchatka mantle wedge. *Lithos*, 106610.

Koepke, J., Feig, S.T., Berndt, J., Neave, D.A., 2021. Wet magmatic processes during the accretion of the deep crust of the Oman Ophiolite paleoridge: phase diagrams and petrological records. *Tectonophysics* 817, 229051.

Liu, J., Wang, J., Hattori, K., Wang, Z., 2022. Petrogenesis of garnet clinopyroxenite and associated dunite in

Hujialin, Sulu Orogenic Belt, Eastern China. *Minerals* 12(2), 162.

Ma, B., Qian, Z., Keays, R.R., et al., 2022. Petrogenesis of the Permian Luotuoshan sulfide-bearing mafic–ultramafic intrusion, Beishan Orogenic Belt, NW China: evidence from whole-rock Sr–Nd–Pb and zircon Hf isotopic geochemistry. *Journal of Geochemical Exploration* 233, 106920.

Maier, W.D., Halkoaho, T., Huhma, H., Hanski, E., Barnes, S.J., 2018. The Penikat intrusion, Finland: geochemistry, geochronology, and origin of platinum–palladium reefs. *Journal of Petrology* 59(5), 967–1006.

Martin, A.P., Cooper, A.F., Price, R.C., Doherty, C.L., Gamble, J.A., 2021. A review of mantle xenoliths in volcanic rocks from southern Victoria Land, Antarctica, in Martin, A.P., van der Wal, W. (Eds.), *The Geochemistry and Geophysics of the Antarctic Mantle.* Geological Society.

Milidragovic, D., Nixon, G.T., Scoates, J.S., Nott, J.A., Spence, D.W., 2021. Redox-controlled chalcophile element geochemistry of the Polaris Alaskan-type mafic–ultramafic complex, British Columbia, Canada. *The Canadian Mineralogist* 59(6), 1627–1660.

Montanini, A., Tribuzio, R., Thirlwall, M., 2012. Garnet clinopyroxenite layers from the mantle sequences of the Northern Apennine ophiolites (Italy): evidence for recycling of crustal material. *Earth and Planetary Science Letters* 351–352, 171–181.

Muroi, R., Arai, S., 2014. Formation process of olivine–clinopyroxene cumulates inferred from Takashima xenoliths, Southwest Japan arc. *Journal of Mineralogical and Petrological Sciences.* doi:10.2465/jmps.131003

O'Driscoll, B., Leuthold, J., Lenaz, D., et al., 2021. Melt percolation, melt–rock reaction and oxygen fugacity in supra-subduction zone mantle and lower crust from the Leka ophiolite complex, Norway. *Journal of Petrology*, 62(11).

Patkó, L., Liptai, N., Aradi, L.E., et al., 2020. Metasomatism-induced wehrlite formation in the upper mantle beneath the Nógrád-Gömör Volcanic Field (Northern Pannonian Basin): evidence from xenoliths. *Geoscience Frontiers* 11(3), 943–964.

Rezvukhin, D.I., Alifirova, T.A., Korsakov, A.V., Golovin, A.V., 2019. A new occurrence of yimengite-hawthorneite and crichtonite-group minerals in an orthopyroxenite from kimberlite: implications for mantle metasomatism. *American Mineralogist: Journal of Earth and Planetary Materials* 104(5), 761–774.

Score, R., Lindstrom, M., (Eds.), 1993. *Antarctic Meteorite Newsletter* 16(3).

Spengler, D., Alifirova, T., 2019. Formation of Siberian cratonic mantle websterites from high-Mg magmas. *Lithos* 326–327, 384–396.

Wang, C., Liang, Y., Dygert, N., Xu, W., 2016. Formation of orthopyroxenite by reaction between peridotite and hydrous basaltic melt: an experimental study. *Contributions to Mineralogy and Petrology* 171, 77.

Wang, K.Y., Song, X.Y., Yi, J.N., et al., 2019. Zoned orthopyroxenes in the Ni-Co sulfide ore-bearing Xiarihamu mafic–ultramafic intrusion in northern Tibetan Plateau, China: implications for multiple magma replenishments. *Ore Geology Reviews* 113, 103082.

Zhang, Z.W., Wang, Y.L., Qian, B., et al., 2018. Metallogeny and tectonomagmatic setting of Ni–Cu magmatic sulfide mineralization, number I Shitoukengde mafic–ultramafic complex, East Kunlun Orogenic Belt, NW China. *Ore Geology Reviews* 96, 236–246.

Gabbroic Rocks

Ashwal, L.D., 2010. The temporality of anorthosites. *The Canadian Mineralogist* 48(4), 711–728.

Barnes, C.G., Yoshinobu, A.S., Prestvik, T., et al., 2002. Mafic magma intraplating: anatexis and hybridization in arc crust, Bindal Batholith, Norway. *Journal of Petrology* 43(12), 2171–2190.

Cawthorn, R.G., Ashwal, L.D., 2009. Origin of anorthosite and magnetitite layers in the Bushveld Complex, constrained by major element compositions of plagioclase. *Journal of Petrology*, 50(9), 1607–1637.

Drouin, M., Godard, M., Ildefonse, B., Bruguier, O., Garrido, C.J., 2009. Geochemical and petrographic evidence for magmatic impregnation in the oceanic lithosphere at Atlantis Massif, Mid-Atlantic Ridge (IODP Hole U1309D, 30 N). *Chemical Geology*, 264 (1–4), 71–88.

Mathieu, L., MacDonald, F., 2022. Petrography and geochemistry of the intrusive rocks at the diorite-hosted Regnault Au Mineralization. *Minerals*, 12(2), 128.

O'Driscoll, B., Emeleus, C.H., Donaldson, C.H., Daly, J.S., 2010. Cr-spinel seam petrogenesis in the Rum Layered Suite, NW Scotland: cumulate assimilation and in situ crystallization in a deforming crystal mush. *Journal of Petrology*, 51(6), 1171–1201.

Otamendi, J.E., Ducea, M.N., Tibaldi, A.M., et al., 2009. Generation of tonalitic and dioritic magmas by coupled partial melting of gabbroic and metasedimentary rocks within the deep crust of the Famatinian magmatic arc, Argentina. *Journal of Petrology*, 50(5), 841–873.

Polat, A., Longstaffe, F.J., Frei, R., 2018. An overview of anorthosite-bearing layered intrusions in the Archaean craton of southern West Greenland and the Superior Province of Canada: implications for Archaean tectonics and the origin of megacrystic plagioclase. *Geodinamica Acta*, 30(1), 84–99.

Relvini, A., Martin, S., Carvalho, B.B., et al., 2021. Genesis of the Eastern Adamello plutons (northern Italy): inferences for the alpine geodynamics. *Geosciences* 12 (1), 13.

Renna, M.R., Tribuzio, R., 2011. Olivine-rich troctolites from Ligurian ophiolites (Italy): evidence for impregnation of replacive mantle conduits by MORB-type melts. *Journal of Petrology* 52(9), 1763–1790.

Sanfilippo, A., Tribuzio, R., 2013. Origin of olivine-rich troctolites from the oceanic lithosphere: a comparison between the Alpine Jurassic ophiolites and modern slow spreading ridges. *Ofioliti* 38(1), 89–99.

Granitic Rocks

Kemp, A., Wormald, R., Whitehouse, M., Price, R., 2005. Hf isotopes in zircon reveal contrasting sources and crystallization histories for alkaline to peralkaline granites of Temora, southeastern Australia. *Geology* 33 (10), 797–800.

Memeti, V., Paterson, S.R., Mundil, R., 2021. Coupled magmatic and host rock processes during the initiation of the Tuolumne Intrusive Complex, Sierra Nevada, California, USA: a transition from ephemeral sheets to long-lived, active magma mushes. *GSA Bulletin*, doi: 10.1130/B35871.1.

Moyen, J.F., Martin, H., 2012. Forty years of TTG research. *Lithos* 148, 12–336.

Zhao, F., Xue, S., Li, G., et al., 2022. Three types of Triassic granitoids in Changning-Menglian suture zone: petrological, geochemical, and geochronological constraints for source characteristics and petrogenesis. *Geological Journal*, doi: 10.1002/gj.4448.

Syenitic and Foid-Bearing Plutonic Rocks

Ashrafi, N., Jahangiri, A., Hasebe, N., Eby, G., 2018. Petrology, geochemistry and geodynamic setting of Eocene–Oligocene alkaline intrusions from the Alborz–Azerbaijan magmatic belt, NW Iran. *Geochemistry* 78(4), 432–461.

Bonin, B., 2007. A-type granites and related rocks: evolution of a concept, problems and prospects. *Lithos* 97, 1–29.

Cobbing, E.J., Pitcher, W.S., 1972. The coastal batholith of central Peru. *Journal of the Geological Society* 128(5), 421–454.

Çolakoğlu, A.R., Arehart, G.B., 2010. The petrogenesis of Sarıçimen (Çaldıran-Van) quartz monzodiorite: implication for initiation of magmatism (Late Medial Miocene) in the east Anatolian collision zone, Turkey. *Lithos* 119(3–4), 607–620.

Deniz, K., Kadioğlu, Y.K., 2016. Assimilation and fractional crystallization of foid-bearing alkaline rocks: Buzlukdağ intrusives, Central Anatolia, Turkey. *Turkish Journal of Earth Sciences* 25(4), 341–366.

Downes, H., Balaganskaya, E., Beard, A., Liferovich, R., Demaiffe, D., 2005. Petrogenetic processes in the ultramafic, alkaline and carbonatitic magmatism in the Kola Alkaline Province: a review. *Lithos* 85(1–4), 48–75.

Eyal, M., Litvinovsky, B., Jahn, B.M., Zanvilevich, A., Katzir, Y., 2010. Origin and evolution of post-collisional magmatism: coeval Neoproterozoic calc-alkaline and alkaline suites of the Sinai Peninsula. *Chemical Geology* 269, 153–179.

Ghanem, H., Jarrar, G.H., 2013. Geochemistry and petrogenesis of the 595 Ma shoshonitic Qunai monzogabbro, Jordan. *Journal of African Earth Sciences*, 88, 1–14.

Haloda, J., Rapprich, V., Holub, F.V., Halodova, P., Vaculovic, T., 2010. Crystallization history of Oligocene ijolitic rocks from the Doupovské hory Volcanic Complex (Czech Republic). *Journal of Geosciences* 55(3), 279–297.

Jung, S., Hoernes, S., Hoffer, E., 2005. Petrogenesis of cogenetic nepheline and quartz syenites and granites (Northern Damara Orogen, Namibia): enriched mantle versus crustal contamination. *The Journal of Geology* 113(6), 651–672.

Jung, S., Romer, R.L., Pfänder, J.A., Berndt, J., 2020. Petrogenesis of early syn-tectonic monzonite–granodiorite complexes: crustal reprocessing versus crustal growth. *Precambrian Research* 351, 105957.

Köksal, S., Toksoy-Köksal, F., Göncüoğlu, M.C., et al., 2013. Crustal source of the Late Cretaceous Satansarı monzonite stock (central Anatolia–Turkey) and its significance for the Alpine geodynamic evolution. *Journal of Geodynamics* 65, 82–93.

Litvinovsky, B.A., Zanvilevich, A.N., Wickham, S.M., Steele, I.M., 1999. Origin of syenite magmas in A-type granitoid series: syenite–granite series from Transbaikalia. *Petrology* 7, 483–508.

Litvinovsky, B.A., Jahn, B.M., Eyal, M., 2015. Mantle-derived sources of syenites from the A-type igneous suites: new approach to the provenance of alkaline silicic magmas. *Lithos* 232, 242–265.

Lynch, D.J., Musselman, T.E., Gutmann, J.T., Patchett, P.J., 1993. Isotopic evidence of Cenozoic volcanic rocks of the Pinacate volcanic field, Northwestern Mexico. *Lithos* 29, 295–302.

Rapela, C.W., Pankhurst, R.J., 1996. Monzonite suites: the innermost Cordilleran plutonism of Patagonia. *Earth and Environmental Science Transactions of The Royal Society of Edinburgh* 87(1–2), 193–203.

Riishuus, M.S., Peate, D.W., Tegner, C., Wilson, J.R., Brooks, C.K., 2008. Petrogenesis of cogenetic silica-oversaturated and -undersaturated syenites by periodic recharge in a crustally contaminated magma chamber: the Kangerlussuaq intrusion, East Greenland. *Journal of Petrology* 49(3), 493–522.

Temizel, I., 2014. Petrochemical evidence of magma mingling and mixing in the Tertiary monzogabbroic stocks around the Bafra (Samsun) area in Turkey: implications of coeval mafic and felsic magma interactions. *Mineralogy and Petrology* 108(3), 353–370.

Wenzel, T., Mertz, D.F., Oberhänsli, R., Becker, T., Renne, P.R., 1997. Age, geodynamic setting, and mantle enrichment processes of a K-rich intrusion from the Meissen massif (northern Bohemian massif) and implications for related occurrences from the mid-European Hercynian. *Geologische Rundschau* 86(3), 556–570.

Volcanic Rocks

Silica-Saturated Rocks

Avard, G., Whittington, A.G., 2012. Rheology of arc dacite lavas: experimental determination at low strain rates. *Bulletin of Volcanology* 74(5), 1039–1056.

Bachmann, O., Bergantz, G.W., 2008. Rhyolites and their source mushes across tectonic settings. *Journal of Petrology*, 49(12), 2277–2285.

Bryan, S.E., Ferrari, L., Reiners, P.W., et al., 2008. New insights into crustal contributions to large-volume rhyolite generation in the mid-Tertiary Sierra Madre Occidental province, Mexico, revealed by U–Pb geochronology. *Journal of Petrology* 49(1), 47–77.

Bullock, L.A., Gertisser, R., O'Driscoll, B., 2018. Emplacement of the Rocche Rosse rhyolite lava flow (Lipari, Aeolian Islands). *Bulletin of Volcanology* 80(5), 1–19.

Cabrera, A., Weinberg, R.F., Wright, H.M., 2015. Magma fracturing and degassing associated with obsidian formation: the explosive–effusive transition. *Journal of Volcanology and Geothermal Research* 298, 71–84.

Castro, J.M., Schipper, C.I., Mueller, S.P., et al., 2013. Storage and eruption of near-liquidus rhyolite magma at Cordón Caulle, Chile. *Bulletin of Volcanology* 75(4), 1–17.

Chen, L., Zhao, Z.F., 2017. Origin of continental arc andesites: the composition of source rocks is the key. *Journal of Asian Earth Sciences* 145, 217–232.

Costa, F., Singer, B., 2002. Evolution of Holocene dacite and compositionally zoned magma, Volcán San Pedro, southern volcanic zone, Chile. *Journal of Petrology* 43 (8), 1571–1593.

Cruz-Uribe, A.M., Marschall, H.R., Gaetani, G.A., Le Roux, V., 2018. Generation of alkaline magmas in subduction zones by partial melting of mélange diapirs: an experimental study. *Geology* 46(4), 343–346.

Geist, D., Howard, K.A., Larson, P., 1995. The generation of oceanic rhyolites by crystal fractionation: the basalt–rhyolite association at Volcán Alcedo, Galápagos Archipelago. *Journal of Petrology* 36(4), 965–982.

Grove, T.L., Kinzler, R.J., 1986. Petrogenesis of andesites. *Annual Review of Earth and Planetary Sciences* 14(1), 417–454.

Halder, M., Paul, D., Sensarma, S., 2021. Rhyolites in continental mafic large igneous provinces: petrology, geochemistry and petrogenesis. *Geoscience Frontiers* 12 (1), 53–80.

Holtz, F., Sato, H., Lewis, J., Behrens, H., Nakada, S., 2005. Experimental petrology of the 1991–1995 Unzen dacite, Japan. Part I: phase relations, phase composition and pre-eruptive conditions. *Journal of Petrology* 46(2), 319–337.

Kushiro, I., 2001. Partial melting experiments on peridotite and origin of mid-ocean ridge basalt. *Annual Review of Earth and Planetary Sciences* 29, 71–107.

Le Bas, M.J., Le Maitre, R.W., Streckeisen, A., Zanettin, B., 1986. IUGS Subcommission on the Systematics of Igneous Rocks: A chemical classification of volcanic rocks based on the total alkali-silica diagram. *Journal of Petrology* 27(3), 745–750.

Macdonald, R., 1974. Nomenclature and petrochemistry of the peralkaline oversaturated extrusive rocks. *Bulletin Volcanologique* 38(2), 498–516.

Macdonald, R., White, J.C., Belkin, H.E., 2021. Peralkaline silicic extrusive rocks: magma genesis, evolution, plumbing systems and eruption. *Comptes Rendus Géoscience* 353(S2), 1–53.

Perfit, M.R., Schmitt, A.K., Ridley, W.I., Rubin, K.H., Valley, J.W., 2008. Petrogenesis of dacites from the southern Juan de Fuca Ridge. *Geochimica et Cosmochimica Acta* 72(12), A736.

Preston, R.J., Hole, M.J., Still, J., Patton, H., 1998. The mineral chemistry and petrology of Tertiary pitchstones from Scotland. *Earth and Environmental Science Transactions of the Royal Society of Edinburgh* 89(2), 95–111.

Rooney, T.O., Sinha, A.K., Deering, C., Briggs, C., 2010. A model for the origin of rhyolites from South Mountain, Pennsylvania: implications for rhyolites associated with large igneous provinces. *Lithosphere* 2(4), 211–220.

Troll, V.R., Emeleus, C.H., Nicoll, G.R., et al., 2019. A large explosive silicic eruption in the British Palaeogene Igneous Province. *Scientific Reports* 9(1), 1–15.

Silica-Undersaturated Rocks

Ablay, G.J., Carroll, M.R., Palmer, M.R., Martí, J., Sparks, R.S.J., 1998. Basanite–phonolite lineages of the Teide–Pico Viejo volcanic complex, Tenerife, Canary Islands. *Journal of Petrology* 39(5), 905–936.

Brenna, M., Pontesilli, A., Mollo, S., et al., 2019. Intra-eruptive trachyte-phonolite transition: natural evidence and experimental constraints on the role of crystal mushes. *American Mineralogist: Journal of Earth and Planetary Materials* 104(12), 1750–1764.

Cooper, K.M., Kent, A.J., 2014. Rapid remobilization of magmatic crystals kept in cold storage. *Nature* 506, 480–483.

Costa, S., Masotta, M., Gioncada, A., Pistolesi, M., 2021. A crystal mush perspective explains magma variability at La Fossa Volcano (Vulcano, Italy). *Minerals*, 11(10), 1094.

Isakova, A.T., Panina, L.I., Stoppa, F., 2019. Formation conditions of leucite-bearing lavas in the Bolsena Complex (Vulsini, Italy): research data on melt inclusions in minerals. *Russian Geology and Geophysics* 60(2), 119–132.

Johansen, T.S., Hauff, F., Hoernle, K., Klugel, A., Kokfelt, T.F., 2005. Basanite to phonolite differentiation within 1550–1750 yr: U–Th–Ra isotopic evidence from the AD 1585 eruption on La Palma, Canary Islands. *Geology* 33(11), 897–900.

Konter, J.G., Jackson, M.G., 2012. Large volumes of rejuvenated volcanism in Samoa: evidence supporting a tectonic influence on late-stage volcanism. *Geochemistry, Geophysics, Geosystems* 13(6).

Laporte, D., Lambart, S., Schiano, P., Ottolini, L., 2014. Experimental derivation of nepheline syenite and phonolite liquids by partial melting of upper mantle peridotites. *Earth and Planetary Science Letters* 404, 319–331.

Lustrino, M., Fedele, L., Agostini, S., Prelević, D., Salari, G., 2019. Leucitites within and around the Mediterranean area. *Lithos* 324, 216–233.

Orlando, A., Conticelli, S., Armienti, P., Borrini, D., 2000. Experimental study on a basanite from the McMurdo Volcanic Group, Antarctica: inference on its mantle source. *Antarctic Science* 12(1), 105–116.

Rossi, S., Petrelli, M., Morgavi, D., et al., 2019. Role of magma mixing in the pre-eruptive dynamics of the Aeolian Islands volcanoes (Southern Tyrrhenian Sea, Italy). *Lithos* 324, 165–179.

Stoppa, F., Principe, C., Schiazza, M., et al., 2017. Magma evolution inside the 1631 Vesuvius magma chamber and eruption triggering. *Open Geosciences* 9(1), 24–52.

Suneson, N.H., Lucchitta, I., 1983. Origin of bimodal volcanism, southern Basin and Range province, west-central Arizona. *Geological Society of America Bulletin* 94(8), 1005–1019.

Trumbull, R.B., Bühn, B., Romer, R.L., Volker, F., 2003. The petrology of basanite–tephrite intrusions in the Erongo Complex and implications for a plume origin of Cretaceous alkaline complexes in Namibia. *Journal of Petrology* 44(1), 93–112.

Application Boxes

Bea, F., Montero, P., Zinger, T., 2003. The nature, origin, and thermal influence of the granite source layer of Central Iberia. *The Journal of Geology* 111, 579–595.

Davidson, J. Morgan, D., Charlier, B., Harlou, R., Hora, J., 2007. Microsampling and isotopic analysis of igneous rocks: implications for the study of magmatic systems. *Annual Reviews of Earth and Planetary Sciences* 35, 273–311.

Gaetani, G., Grove, T., 1998. The influence of water on melting of mantle peridotite. *Contributions to Mineralogy and Petrology* 131, 323–346.

Hirose, K., Kawamoto, T., 1995. Hydrous partial melting of lherzolite at 1 GPa: the effect of H_2O on the genesis of basaltic magmas. *Earth and Planetary Science Letters*, 133(3–4), 463–473.

Koga, K.T., Kelemen, P.B., Shimizu, N., 2001. Petrogenesis of the crust–mantle transition zone and the origin of lower crustal wehrlite in the Oman ophiolite. *Geochemistry, Geophysics, Geosystems* 2(9).

Piccardo, G.B., Zanetti, A., Müntener, O., 2007. Melt/peridotite interaction in the Southern Lanzo peridotite: field, textural and geochemical evidence. *Lithos* 94 (1–4), 181–209.

Rampone, E., Borghini, G., Basch, V., 2020. Melt migration and melt–rock reaction in the Alpine–Apennine peridotites: insights on mantle dynamics in extending lithosphere. *Geoscience Frontiers* 11, 151–166.

Renjith, M.L., 2014. Micro-textures in plagioclase from 1994–1995 eruption, Barren Island Volcano: evidence of dynamic magma plumbing system in the Andaman subduction zone. *Geoscience Frontiers*, 5(1), 113–126.

Salisbury, M., Bohrson, W., Clynne, M., Ramos, F., Hoskin, P., 2008. Multiple plagioclase crystal populations identified by crystal size distribution and in situ chemical data: implications for timescales of magma chamber processes associated with the 1915 eruption of Lassen Peak, CA. *Journal of Petrology* 49(10), 1755–1780.

3

METAMORPHIC ROCKS

3.1 Introduction

This chapter provides a visual basis for the identification of metamorphic minerals and rocks, and the recognition and meaning of their textures and fabrics under the microscope. We emphasize how what is seen in petrographic thin-section can be used to infer something about the metamorphic history preserved in a rock. This will enhance the reader's appreciation of metamorphic petrography and link this information to the many ways in which it can be applied. This book is not intended to replace a course or textbook on metamorphic petrology; to understand the mechanisms of metamorphism and fabric development, the reader is referred to the numerous fine metamorphic texts, such as Yardley and Warren (2021), Passchier and Trouw (2005), and Vernon (2018).

The textures and fabrics developed during metamorphism involve heat, deformation, and/or fluid flow, and reflect crystal nucleation, growth, and diffusion processes. Some textures, such as augen or granoblastic texture, describe the geometry of grains or grain aggregates, while features such as atolls, coronas, and symplectites reflect incomplete chemical/metamorphic reactions. Other textures and fabrics indicate deformation (e.g. cleavage, pressure shadows).

Temperature, pressure, and fluid compositions play major roles in enhancing or limiting the development of metamorphic textures and fabrics, while deformation controls their distribution in space. The mechanical properties of a rock and the deviatoric stress(es) acting on it during metamorphism can induce changes in mineral shapes, resulting in the development of planar fabrics or linear structures such as foliations, boudins, and lineations. In fact, deformation can facilitate the amount and direction of chemical exchange during metamorphism; for example, fluid infiltration along grain boundaries increases diffusivity at grain boundaries, assists grain-boundary sliding, and localizes deformation at these areas.

Metamorphic rocks, being derived from igneous and sedimentary protoliths, are complex – the number of minerals used to name them is greater than for most igneous rocks, but combining the knowledge of minerals from Chapter 1 with the information presented here, it is possible to begin recognizing, classifying, and understanding something about the formation and evolution of metamorphic rocks (Table 3.1).

Metamorphism affects the physical and/or chemical conditions (pressure, temperature, composition/

Table 3.1 Metamorphic rock names based or structure and protolith composition

Structural name
Slate – fine-grained, foliated pelite
Phyllite – coarser-grained, foliated pelite
Schist – coarse-grained, foliated metasediment or meta-igneous rock
Gneiss – coarse-grained, layered rock
Orthogneiss (meta-igneous)
Paragneiss (metasedimentary)
Granofels – high-T, coarse-grained, metasediment or meta-igneous rock

Protolith	Metamorphic name
Clay-rich sediment	Metapelite
Sand-rich sediment	Metapsammite
Clay + sand sediment	Metapelitic/metapsammitic
Quartz-rich sand	Quartzite
Marl	Calc-silicate
Limestone or dolostone	Marble
Basalt/gabbro	Metabasite/metagabbro
Granite	Metagranite

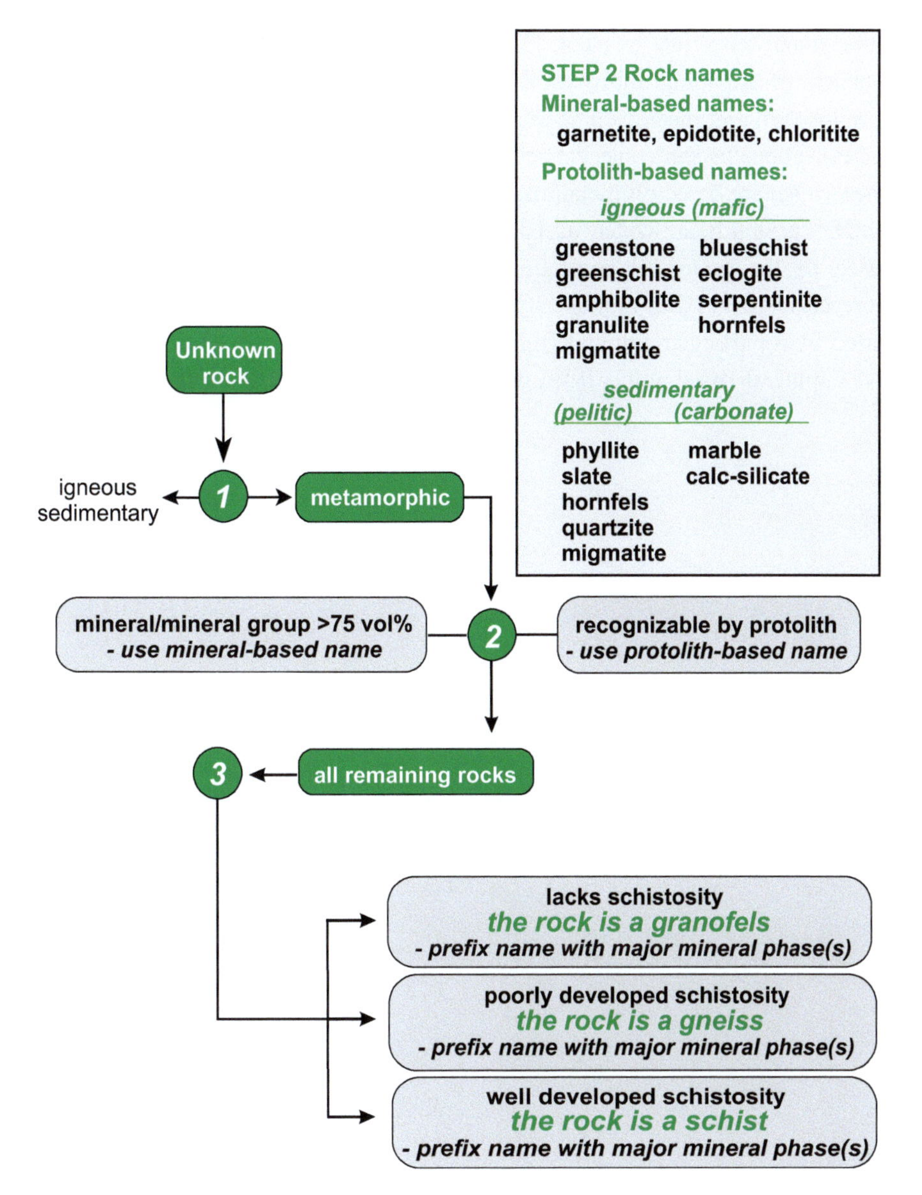

Figure 3.1 Flowchart for naming metamorphic rocks (following Fettes & Desmons, 2007). The specialized names at step 2 supersede the IUGS names of schist, gneiss, granofels. Only if a rock name from step 2 does not apply, should the reader proceed to step 3.

fluids) of a rock such that existing minerals may become unstable and break down, recrystallize, and/or grow new minerals – that is to say, it results in a change in form (*metamorphosis*). Unlike igneous processes, metamorphic processes modify the mineralogical, textural, and/or chemical constituents of the parent rock (the *protolith*) in the solid state (Table 3.1). The geometric relationship between primary and/or secondary minerals and matrix (or *host*) of a metamorphic rock defines its texture. Grain size and shape, the degree of recrystallization, porphyroclast/porphyroblast–matrix relationships, and the identification of fabrics related to deformation help us to understand metamorphic processes and evolution.

In naming metamorphic rocks, we adopt the IUGS recommendations of Fettes and Desmons (2007), in which specific mineral or rock names are applied if appropriate (see Fig. 3.1). If the specific names are inappropriate, then the basic structure (root name) is combined with the major minerals present. This approach has the benefit of providing information on fabric, protolith composition, and metamorphic grade. However, there are exceptions as some specific metamorphic rock names, which are based on metamorphism of a basaltic (mafic) protolith, are widely accepted and commonly used by the geological community. These names take precedence over IUGS nomenclature.

A rock dominated by a particular mineral (mode >75%) should have a mineral-based name. If features of the protolith can be determined, a protolith name should be applied. Except for these specialized names, metamorphic rocks should have a "root" name based on its fabric (Table 3.1), in which a schist has a well-developed, pervasive schistosity; a gneiss has a poorly developed or broadly spaced schistosity (>1 cm); and a granofels lacks fabric, with the minerals having a random orientation. This "root" name is then preceded by the major mineral(s) present (e.g. garnet gneiss or talc-bearing schist).

3.2 Metamorphic Textures

Grain Size and Shape

Acicular

A mineral's shape in metamorphic rocks is influenced by many factors, such as the mechanism for bonding atoms to the crystal surface. Many minerals have different atomic arrays on different crystallographic planes, and this promotes anisotropic growth – or a higher growth rate in a particular direction – to form acicular morphologies. *Acicular* is a term that describes the shape of a mineral, and it is applied to those minerals that develop a highly elongated shape (e.g. amphibole, sillimanite, kyanite). This morphology is typical of low-grade metamorphic rocks such as phyllite and schist, because as the metamorphic grade increases the shapes of grains and grain boundaries are modified due to recrystallization processes in order to reduce grain-boundary energy; this results in minerals becoming more equidimensional. The presence of acicular minerals gives the rock an anisotropic fabric (lineation). *Nematoblastic*, a term widely used in older literature and now obsolete, was used to describe acicular or prismatic minerals developed with strong preferred orientation and forming lineation.

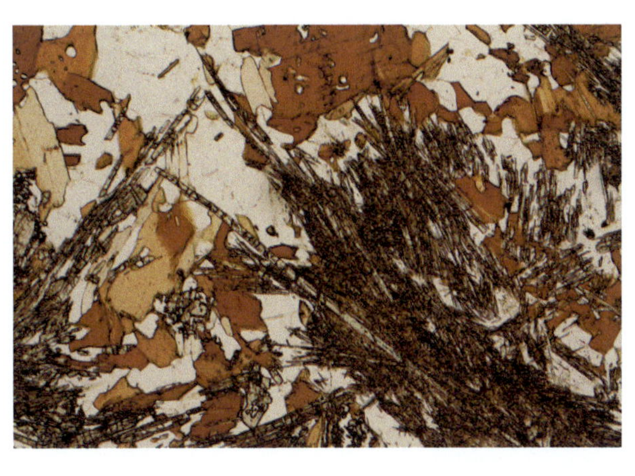

Figure 3.2

A Talc–tremolite schist from Val Chiavenna (Italy). Acicular tremolite (high relief and high birefringence in blue to yellow–green) surrounded by fibrous talc (with extremely high birefringence in pink and green). Cross-polarized light, 2× magnification, field of view = 7 mm.

B Sillimanite schist from Piona (Italy). Acicular to fibrous sillimanite (fibrolite) with pale brown colors, associated with biotite (orange–brown) and quartz (colorless). Plane-polarized light, 2× magnification, field of view = 7 mm.

Augen

Augen are large "eye-shaped" or ovoid crystals. Augen, often potassium feldspar, are usually set in a finer-grained, foliated matrix as in mylonite or gneiss. Although Kfs commonly forms augen, other minerals such as plagioclase, pyroxene, amphibole, or garnet may also form augen. Gneissic rocks with abundant augen are called *augen gneisses*. Augen are interpreted as either porphyroblasts grown during metamorphism, or as relict igneous phenocrysts (porphyroclasts) which are (at least partially) preserved because they are much stronger than the surrounding finer-grained quartz and mica that are more readily deformed and recrystallized. The latter, supported by microstructural, mineralogical, and chemical evidence, is widely accepted today.

Figure 3.3

A Augen gneiss from Valtellina (Italy). Augen are plagioclase porphyroclasts, some with Carlsbad twins (center-left crystal) and others with undulose extinction due to deformation (center-right crystal). Biotite and quartz define the matrix and a rough NE–SW foliation. Cross-polarized light, 2× magnification, field of view = 7 mm.

B Augen gneiss from Courmayeur (Italy). Augen are plagioclase porphyroclasts (with twinning) set in a fine-grained matrix of quartz, feldspar, and muscovite, defining a weak foliation. Note the muscovite layers flow around the larger porphyroclasts. Cross-polarized light, 2× magnification, field of view = 7 mm.

Granoblastic Textures: Decussate

Granoblastic decussate texture (also known as *diablastic* texture) is characterized by subhedral, interlocking, randomly oriented platy, tabular, or prismatic crystals, such as mica, amphibole, pyroxene, etc., with strong crystal anisotropy. The decussate texture minimizes a crystal's internal stress, and 120° grain boundaries between contiguous crystals might be present. Decussate texture generally refers to rocks with only one or two mineral species and it is common in hornfels and granofels.

Figure 3.4

A Decussate micaschist from Val Venosta (Italy). Note the interlocking, somewhat randomly oriented, biotite (brown) and muscovite (colorless) crystals. Plane-polarized light, 2× magnification, field of view = 7 mm.

B Decussate micaschist from Val Venosta (Italy). Same image as (A) but with crossed polarizers. Biotite and muscovite both show high birefringence, but muscovite is higher than the biotite and colorless in plane-polarized light. Cross-polarized light, 2× magnification, field of view = 7 mm.

Granoblastic Textures: Polygonal

Granoblastic polygonal texture defines a mosaic of approximately equidimensional subhedral or anhedral grains with straight grain boundaries and common 120° intersections that reflect equilibrium conditions. Polygonal texture is commonly developed in nonfoliated, high-grade metamorphic rocks such as marbles, quartzites, and granulites. The size of the grains depends mainly on the temperature and the presence of fluids, both of which promote larger crystal growth. Situations in which textural equilibrium is not fully achieved are indicated by somewhat irregular boundaries, known as *interlobate* texture. Monomineralic aggregates often produce more uniform polygonal textures. Polymineralic aggregates, on the other hand, result in grain boundaries with diverse energy states; thus, some mineral grain boundaries are minimized while others are maximized, but the net result is a lower total energy state for the rock.

Figure 3.5

A Polygonal texture in quartzite from Val Bognanco (Italy). These quartz crystals have straight grain boundaries that intersect at 120°. This texture documents a low-strain state for the sample. Cross-polarized light, 10× magnification, field of view = 2 mm.

B Polygonal marble from Carrara (Italy). Calcite crystals have mostly straight boundaries that intersect at 120°. Note the relatively uniform grain size across the thin-section. Cross-polarized light, 2× magnification, field of view = 7 mm.

Porphyroblasts

Porphyroblasts are large, euhedral to subhedral crystals that have grown during metamorphism. Porphyroblasts generally grow at the same time as the matrix, and the grain size difference between a porphyroblast and matrix is mainly due to different growth rates. Porphyroblasts are common in most metamorphic rocks, but are particularly widespread in metapelites (micaschist) and metabasites (greenschist/blueschist). A porphyroblast can be any metamorphic mineral from any metamorphic grade, but Al-silicates are the most common (e.g. garnet, biotite, staurolite, chloritoid, andalusite, kyanite, albite, epidote, hornblende).

Porphyroblasts are an important source of information for the reconstruction of the metamorphic history of a rock. As porphyroblasts grow, they can enclose adjacent minerals from the matrix and develop an internal inclusion pattern (or internal foliation). The mineral inclusions in a porphyroblast may preserve a different pressure–temperature (P–T) environment than the surrounding matrix. Furthermore, porphyroblasts like garnet with compositional solid solution are commonly chemically zoned and may record their metamorphic evolution from core to rim.

The relationship between the orientation of the internal foliation of the porphyroblast and the external matrix foliation can help determine the deformation history of a rock; porphyroblasts are therefore classified in relation to the foliation of the rock. Inclusion trails within porphyroblasts are referred to as S_i (i for internal) and the rock foliation outside the porphyroblasts is called S_e (e for external).

Based on such relationships, porphyroblasts can be divided into the following groups:

Pre-tectonic: Commonly associated with regions of contact metamorphism. Inclusions are randomly oriented and they can be surrounded by a matrix with a polyphase deformation history.

Inter-tectonic: These porphyroblasts have grown between two deformational phases and their internal inclusions are inherited from the earlier deformation phase while the matrix is affected by the later deformation phase.

Syntectonic: These are the most common and widespread type of porphyroblasts. Their growth occurs during a single deformation event. Internal inclusion trails are generally curved or oblique with respect to the external foliation. A very common characteristic, especially in inter- and syntectonic garnet porphyroblasts, is the spiral-shaped internal pattern (known as *snowball* structures) which is useful for shear sense analysis.

Post-tectonic: These porphyroblasts have grown after deformation has waned. They generally lack strain shadows, undulose extinction, and other evidence of deformation. If there are internal inclusion trails, they have continuity with the external foliation.

Figure 3.6

A Inter-tectonic garnet porphyroblast in garnet micaschist from Merano (Italy). The garnet porphyroblast (at extinction in the center of the image) shows an internal foliation consisting of rounded quartz inclusions. The internal foliation is at high angle to the external foliation of the rock and is inherited from an earlier deformation phase. Cross-polarized light, 10× magnification, field of view = 2 mm.

B Syntectonic "snowball" garnet porphyroblast in micaschist from Sondrio (Italy). This garnet shows an internal S-shaped foliation defining dextral rotation. The internal foliation is defined by quartz and small muscovite inclusions. Muscovite plus quartz define the external foliation, which wraps around the garnet porphyroblast. Cross-polarized light, 2× magnification, field of view = 7 mm.

C Post-tectonic biotite porphyroblast in micaschist from Resia (Italy). The sample is complexly folded (crenulated). The foliation is defined by layers of muscovite and opaque minerals. The foliation continues through the biotite in the center of the image. Plane-polarized light, 2× magnification, field of view = 7 mm.

Porphyroclasts

Porphyroclasts are relatively large single crystals in a fine-grained groundmass, common in deformed rocks such as mylonites, and represent remnants of resistant mineral grains derived from the protolith. Minerals that commonly form porphyroclasts are feldspar, kyanite, hornblende, garnet, and pyroxenes. A common feature of porphyroclasts are the "wings" or "tails" of dynamically recrystallized material. During deformation, the rigid porphyroclast rotates within the ductile matrix, perturbing the foliation and resulting in σ (sigma)- and δ (delta)-type porphyroclasts.

Figure 3.7

A σ-porphyroclast in mylonitic micaschist from Ivrea Verbano (Italy). The rounded garnet porphyroclast shows two mica-rich wings; the upper-right wing is chlorite-rich (pale green) while the lower-left wing is muscovite-rich (colorless). The garnet has small opaque inclusions and is surrounded by colorless quartz and biotite–muscovite-rich layers. The garnet indicates an apparent dextral sense of shear. Plane-polarized light, 10× magnification, field of view = 2 mm.

B σ-porphyroclast in mylonitic micaschist from Ivrea Verbano (Italy). Same image as (A) but with crossed polarizers. Quartz has interlobate grain boundaries; muscovite shows high-interference colors, and garnet is isotropic. Cross-polarized light, 10× magnification, field of view = 2 mm.

C δ-porphyroclast in mafic ultramylonite from the Tuscan Apennines (Italy). The two amphibole porphyroclasts are set in a very fine-grained matrix and both show delta-type wings. In both cases, the wings are derived by grain-size reduction of the porphyroclasts themselves. Plane-polarized light, 2× magnification, field of view = 7 mm.

Strain/Pressure Shadows

Strain shadows (also known as pressure shadows) form on both sides of a rigid crystal and are characterized by a different fabric than the rock matrix. Strain shadows may resemble porphyroclast systems, but the wings and tails of the latter have the same mineral composition as the rigid crystal and form by a different mechanism. Strain shadows develop from circulating fluid, which transports material from high-stress areas and reprecipitates it in low-stress areas. Commonly precipitated materials include quartz, carbonate, and chlorite, which may or may not be derived from the dissolution–reprecipitation of minerals from the surrounding wall-rock. Strain shadows usually have a gradual boundary with the surrounding matrix and a completely different structure than the surrounding matrix – for example, the foliation can be well developed in the rock but absent or very weakly developed in the strain shadow. Strain shadows are often accompanied by strain caps, which are areas rich in insoluble minerals and where the foliation is usually strongly developed. These occur on opposite sides of the rigid crystal, in the quarters orthogonal to the strain shadow. Sometimes, during prolonged deformation phases, strain shadows can assume a sigmoidal morphology and be useful as kinematic indicators.

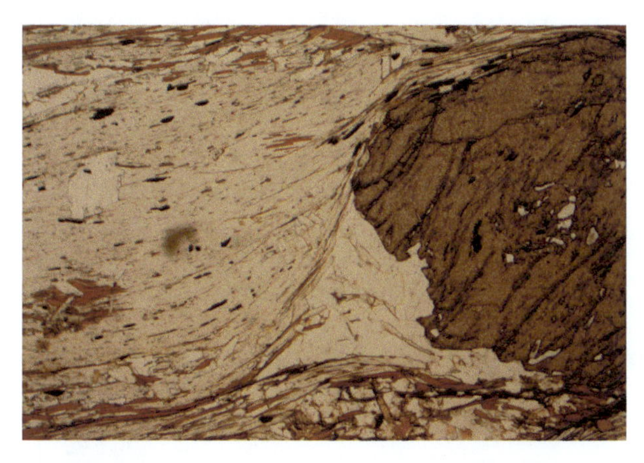

Figure 3.8

A Strain shadow in micaschist from Soazza (Switzerland). The colorless V-shaped strain shadow on the lower-left side of the brownish garnet porphyroblast is entirely composed of quartz. It has a different composition from the surrounding matrix which has a continuous foliation defined by muscovite, biotite (reddish), and magnetite (black). Plane-polarized light, 2× magnification, field of view = 7 mm.

B Strain shadow in micaschist from Soazza (Switzerland). Same image as (A) but with crossed polarizers. Quartz in the strain shadow has a granoblastic texture; the garnet is isotropic and muscovite has characteristic high birefringence. Cross-polarized light, 2× magnification, field of view = 7 mm.

BOX 3.1 Pseudomorphs

Pseudomorphs (literally "false form") are crystals completely or partially replaced by another mineral or mineral aggregate, but which retain the shape of the original mineral.

Common pseudomorph reactions	
Hbl ↔ Act + Chl	Lws ↔ Ep + Mrg + Qz+ Pl
St ↔ Chl + Ser	Ol ↔ Srp + Tlc + Mgs
Grt ↔ Chl + Bt	St ↔ Ser + Chl
Grt ↔ Sil + Bt	Pl ↔ Ser
	Bt ↔ Chl

Pseudomorphs can completely replace a mineral, leaving its shape unchanged, as in this altered basalt from Radicofani (Italy). The two olivine crystals are thoroughly replaced (pseudomorphed) by fibrous talc. Cross-polarized light, 10× magnification, field of view = 2 mm.

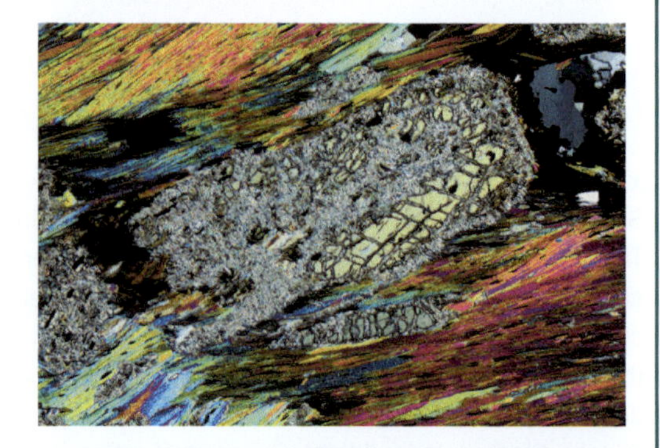

Pseudomorphs are very common in retrograde metamorphic rocks, as in this micaschist from Posada Asinara, Sardinia (Italy). The central staurolite crystal (yellow–green birefringence) is almost fully replaced by fine-grained sericite (blue–gray). The crystal is surrounded by high-birefringence muscovite. Cross-polarized light, 2× magnification, field of view = 7 mm.

Pseudomorphs can also occur in prograde metamorphism, as in this sillimanite–gneiss from Alpe Arami (Switzerland). The rounded garnet is pseudomorphed by an aggregate of sillimanite (fibrolite, light brown) and biotite (red–brown), surrounded by quartz (colorless). Plane-polarized light, 2× magnification, field of view = 7 mm.

Deformation Fabrics

Deformation Twins

Deformation twins (also known as mechanical twins or secondary twins) are distinct from growth twins (Table 3.2). The development of deformation twins is due to intracrystalline deformation, which involves shear of parts of the crystal structure parallel to a twin-glide plane. The development of deformation twins is influenced by the crystalline structure of a mineral, and consequently deformation twins occur in some minerals but not in others. Deformation twins are common in metamorphic rocks and may coexist with growth twins, often in the same crystal. Deformation twins can be distinguished from growth twins by the variable thickness along their length and their tapered terminations.

Deformation twins are particularly common in plagioclase, calcite, dolomite, barite, and many sulfide minerals. In plagioclase, deformation twins can develop according to two laws that may operate simultaneously, the albite law and the pericline law. The width and morphology of deformation twins in calcite can be used to determine the temperature conditions during deformation – Ferrill et al. (2004) showed how the mean calcite twin width correlates directly with temperature of deformation. Very small twins ($<1\ \mu m$) are common below 200 °C and dominate calcite crystals below 170 °C. Wider twins ($>1\ \mu m$) are typical of temperatures of 200–300 °C. At higher temperatures, dynamic recrystallization leads to the development of wide, bent, curved, and tapered twins that intersect each other.

Table 3.2 Differences between growth twins and deformation twins

Growth twins	Deformation twins
Simple or multiple (lamellar)	Multiple (never simple) and often in conjugate sets
Across the entire crystal	Heterogeneous distribution, often restricted to regions of a crystal
Uniform width along the entire twin lamellae with straight or stepped edges	Width changes gradually along the lamellae
Abrupt terminations of lamellae	Tapered terminations of lamellae

Figure 3.9 Plagioclase deformation twins in amphibolite from Valtellina, Italy (*left*) reflect the intersection of two different twin sets (both albite and pericline law twinning). Note how many twins terminate within the crystals (hornblende with brown birefringence and opaque magnetite also present). Deformation twins in marble from Val Venosta, Italy (*right*) are bent and show many tapered terminations. Cross-polarized light, 2× magnification, field of view = 7 mm.

BOX 3.2 Boudinage and Microboudinage

Boudinage, the layer-parallel extension of a stronger, more competent layer and a weaker, less competent layer involves stretching, necking, and finally segmentation and separation of the more competent layer(s). Boudinage can range from the macroscopic to microscopic scale (microboudinage). Microboudinage may affect single crystals or layers (e.g. clay-rich layers in marbles), or objects such as fossils and pebbles.

During boudinage, ductile layers flow into the spaces between "boudins" of the more competent material. This may form flanking structures or "neck" folds. In the above image, hornfelsed layers of a thermally metamorphosed limestone (dark layers) are boudinaged and slightly folded (top of image). Mineral aggregates often form in situ as individual boudins move apart; these can be used to establish metamorphic conditions during deformation, similar to strain shadows and extensional veins. Microboudinage can be used as a strain gauge to estimate extension: If the total length of a boudinaged grain or layer is divided by the sum of the length of each boudin, a minimum stretching value is obtained.

The thin-section images (below) of a boudinaged hornfels layer from the outcrop image above show fine-grained, slatey cleavage in the less competent layers. The more competent layer was stretched and extensional fractures between boudins filled with elongate calcite parallel to the extension direction (horizontal in this case).

(*Left*) Plane-polarized light; (*right*) cross-polarized light. Magnification 10×, field of view = 2 mm.

Foliation and Schistosity

Foliation is a nongenetic term used to describe any subparallel, planar feature that occurs *throughout* a rock (i.e. it is a *penetrative* fabric that may be primary or metamorphic in origin). Such planar features are defined by variations in mineral composition or grain size, preferred orientation of elongate or platy minerals, planar microfractures, or any combination of these. Foliations are either *spaced* or *continuous*.

Spaced foliation is characterized by distinct domains of microlithon and cleavage. The microlithon represents the region of rock between adjacent cleavage domains. Cleavage domains are defined by minerals with a preferred orientation, such as mica-rich, phyllosilicate-rich, or quartz + feldspar-rich domains. In metapelites, the cleavage domains are usually rich in phyllosilicate minerals and other minerals like ilmenite, graphite, rutile, apatite, and zircon.

Continuous foliation (also known as penetrative or pervasive foliation) is characterized by equally spaced planar surfaces throughout the rock defined by the preferred orientation of platy minerals (e.g. phyllosilicates, amphiboles, and/or flattened crystals such as quartz or calcite).

Schistosity is the general term for foliation planes rich in muscovite and/or biotite, visible without the aid of magnification. Schistosity tends to be wavy and discontinuous due to its coarser grain size.

Figure 3.10

A Spaced foliation in schist from Levigliani (Italy). The rock is characterized by parallel cleavage domains rich in muscovite (high birefringence) alternating with microlithons richer in quartz plus some muscovite. Cross-polarized light, 2× magnification, field of view = 7 mm.

B Spaced foliation in schist from Levigliani (Italy). This magnified view of (A) shows the difference between the cleavage domains and microlithons. Cleavage domains are characterized by the planar arrangement of muscovite, while the muscovite in the microlithon domains has a weaker preferred orientation oblique to the cleavage domains. Cross-polarized light, 10× magnification, field of view = 2 mm.

C Continuous foliation in micaschist from Alpe Serra (Italy). The sample has a parallel arrangement of biotite (brown) and quartz (colorless) which defines a continuous schistosity. Plane-polarized light, 2× magnification, field of view = 7 mm.

D Continuous foliation in micaschist from Alpe Serra (Italy). Same image as (C) but with crossed polarizers. Biotite shows characteristic medium–high birefringence while quartz shows its lower (gray) birefringence. Cross-polarized light, 2× magnification, field of view = 7 mm.

E Schistosity in micaschist from Val di Fosse (Italy). The sample is characterized by a continuous schistosity consisting of biotite-rich layers (brown) alternating with muscovite-rich layers (colorless). The schistosity is weakly bent. Plane-polarized light, 2× magnification, field of view = 7 mm.

Crenulation Cleavage

Crenulation cleavage is a secondary spaced foliation formed by the reorientation of a pre-existing foliation, i.e. an older foliation (commonly a slaty cleavage or a schistosity) folded by a new foliation. The new foliation typically develops at a high angle to the earlier foliation. Crenulation cleavage is a common fabric in phyllosilicate-rich rocks such as slates, phyllites, and schists, and found in all metamorphic grades.

Crenulation cleavage initially develops by kinking of the pre-existing foliation, with bending and rotation of pre-existing phyllosilicate minerals. As deformation proceeds, the hinge angle of the crenulation progressively tightens and pressure solution, recrystallization, and grain growth dominate since significant strain and chemical gradients are established between the hinge and limb regions of the crenulation folds. Limbs become sites of higher strain (relative to the hinges), resulting in solution and removal of minerals such as quartz, feldspar, and calcite; these minerals are removed from limb zones and precipitated in the hinge regions, thus the limbs are enriched in minerals such as biotite, white mica, chlorite, or graphite. Although the development of crenulation cleavage may progressively erase the pre-existing foliation, it might be preserved in microlithons or low-strain sites (adjacent to porphyroblasts/porphyroclasts). Crenulation cleavage may change its morphology or even disappear completely in the same thin-section due to the presence of phyllosilicate-rich and phyllosilicate-poor layers.

Figure 3.11

A Incipient crenulation cleavage in micaschist from Merano (Italy). The rock shows a continuous schistosity (oriented roughly vertical) affected by an incipient crenulation (horizontal bands) which bends the more birefringent mica crystals. Note the extinction of muscovite in the developing crenulation. Quartz (gray birefringence) is also present. Cross-polarized light, 10× magnification, field of view = 2 mm.

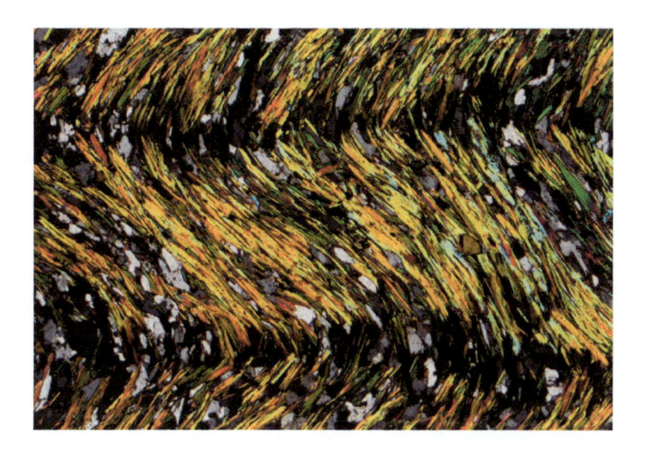

B Crenulation cleavage in micaschist from Maine (United States). The rock is characterized by a well-developed crenulation cleavage which defines the new foliation (oriented diagonally). The old foliation is preserved within the microlithons as folds. Plane-polarized light, 2× magnification, field of view = 7 mm.

C Crenulation cleavage in micaschist from Maine (United States). Same image as (B) but with crossed polarizers. Muscovite (high birefringence) defines the new foliation, while quartz (gray birefringence) is folded within the microlithons. Cross-polarized light, 2× magnification, field of view = 7 mm.

D Crenulation cleavage in phyllite from Sardinia (Italy). The lower-right part of the image has a well-developed crenulation cleavage in the phyllosilicate-rich zone; the upper-left part of the image has less phyllosilicate minerals, which prevents crenulation cleavage development. The white layer that separates the two regions is a folded calcite vein. Plane-polarized light, 2× magnification, field of view = 7 mm.

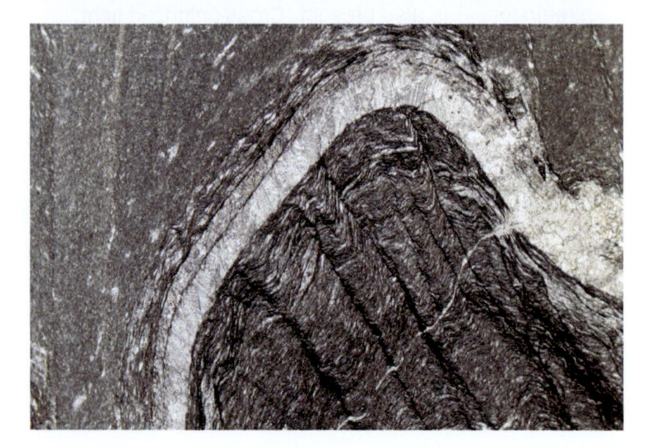

BOX 3.3 Foliation Terminology

Continuous Foliation

Continuous cleavage	the planar arrangement of fine-grained minerals
Continuous schistosity	the planar arrangement of coarser-grained minerals

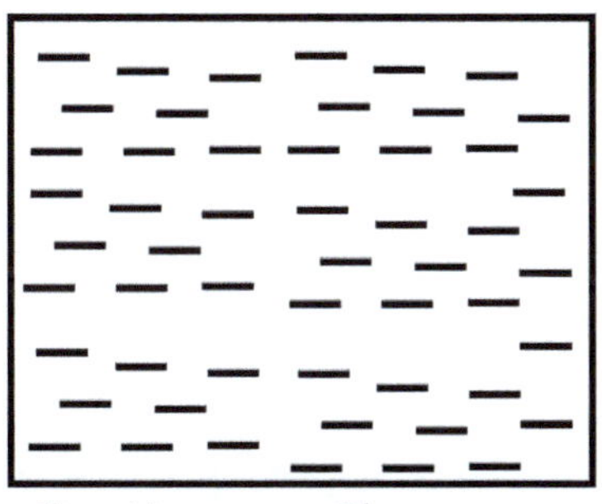

Spaced Foliation

Disjunctive cleavage/schistosity	interrupted at the mesoscopic scale
Crenulation cleavage	folding of an earlier foliation

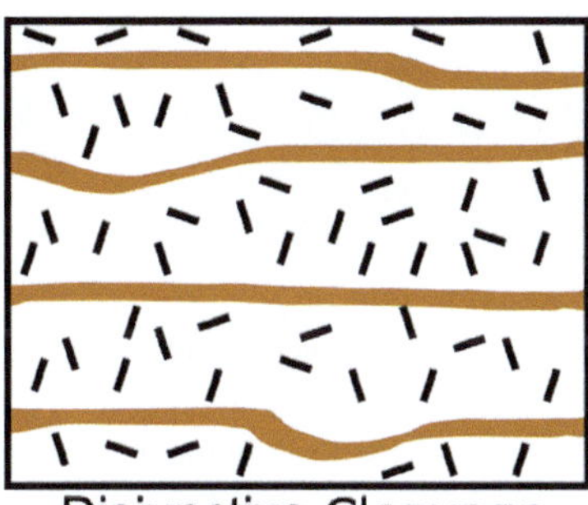

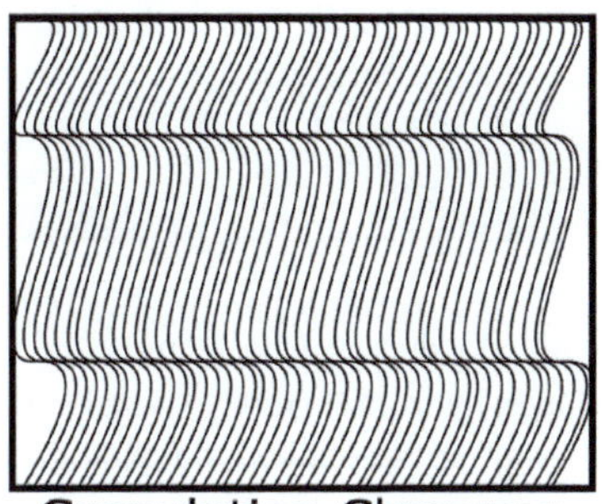

Microlithon rock between adjacent cleavage domains

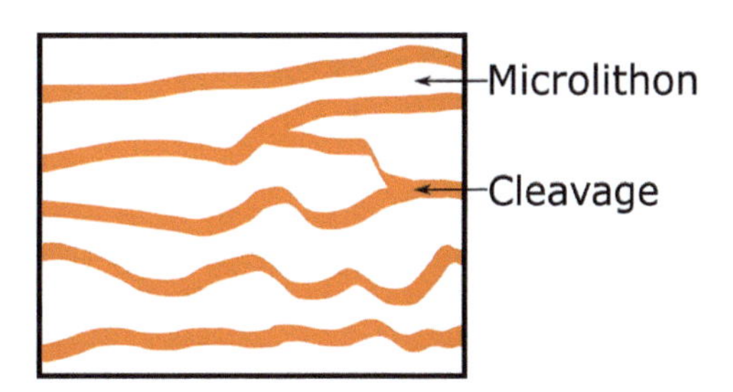

Spaced foliation shapes

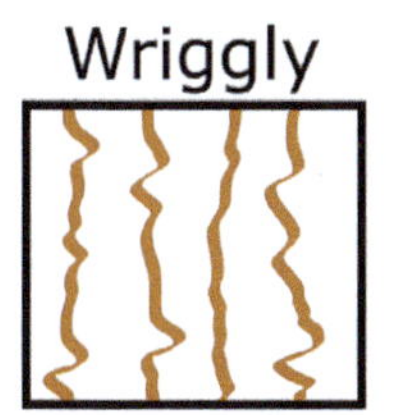

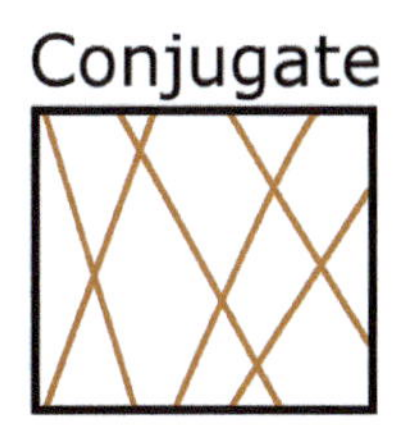

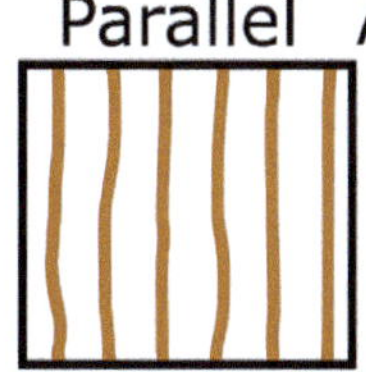

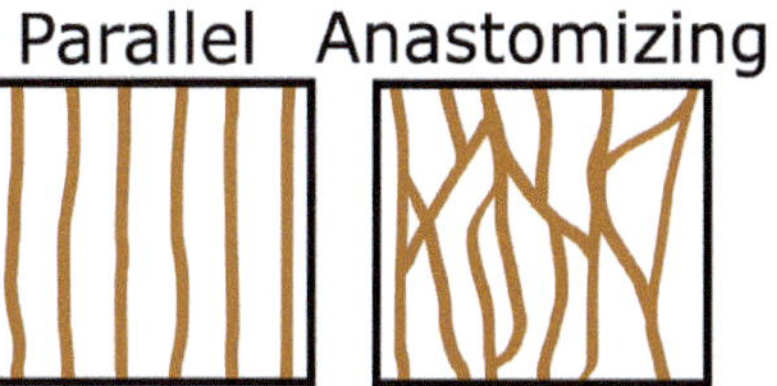

Kink Bands

Kink bands are sites of local deformation within a crystal. They are commonly wedge-shaped, tapered, or tabular, and reflect an abrupt change in the crystallographic orientation of part of a crystal. Kink bands can sometimes be confused with undulose extinction, but kink bands have sharply defined, high-angle boundaries that are visible in both plane- and cross-polarized light. Kink bands, together with deformation lamellae and deformation twins, represent the microstructural evidence of crystal plastic deformation in single crystals. Kink bands are common in minerals with strongly anisotropic crystal structures that have only one slip plane (mica, kyanite, and pyroxene), but can be found in crystals of any symmetry class such as quartz, feldspar, amphibole, and olivine. Kink bands develop during deformation to accommodate shortening parallel to the maximum compressive stress direction – a single slip plane is often inadequate for maintaining homogeneous deformation, which consequently localizes into sharp bends to accommodate crystal shortening.

Figure 3.12

A Kink bands in kyanite gneiss from Alpe Arami (Switzerland). The large, horizontal kyanite crystal shows multiple kink bands with wedge-shaped morphology. A smaller kyanite crystal (bottom, left of center) shows similar features. Other minerals present include quartz (gray birefringence), biotite (high birefringence), and other kyanite porphyroblasts. Cross-polarized light, 2× magnification, field of view = 7 mm.

B Kink bands in weakly deformed granite from Adamello (Italy). The large biotite crystal has kink bands with wedge-shaped morphology. Note how the kinks deflect the biotite cleavage. Cross-polarized light, 2× magnification, field of view = 7 mm.

Poikiloblasts and Poikiloclasts

A poikiloblast is a porphyroblast with fine-grained, randomly oriented inclusions of other minerals which give the crystal a spongy aspect. Poikiloblastic texture is the metamorphic analog of poikilitic texture in igneous rocks. In the latter, however, the inclusions pre-date their host, whereas in a poikiloblast the host and its inclusions form synchronously. If the host is a relict crystal of the protolith or derived from a previous metamorphic event, then the term poikilo*clast* is used. An extreme example of poikiloblastic texture is *web texture*, in which the host crystal grows interstitially, forming a crystallographically continuous network. Web texture often forms at the margins of porphyroblasts as they grow into the matrix and reflects rapid growth in the presence of intergranular fluid(s).

Poikiloblastic texture is particularly common in garnet, staurolite, cordierite, and hornblende. Poikiloblast inclusions can be of two types: (1) inert phases not involved in the porphyroblast-forming reaction and subsequently enveloped as the porphyroblast grows, or (2) residual phases involved in the porphyroblast-forming reaction or a reactant that wasn't completely consumed. For example, the abundant quartz inclusions in andalusite or kyanite often result from the pyrophyllite decomposition reaction:

$$Al_2Si_4O_{10}(OH)_2 = Al_2SiO_5 + 3SiO_2 + H_2O$$

The quartz produced by this reaction is not readily removed and is therefore incorporated in the porphyroblast. Poikiloblasts represent an important source of information for reconstructing the metamorphic evolution of rocks:

- They can contain different types of inclusions from core to rim and can therefore be used to establish the P–T evolution of the rock. Furthermore, some inclusions can be used as geothermobarometers or can be dated isotopically to determine pressure–temperature–time (P–T–t) paths.
- The inclusion trails within a porphyroblast may record the deformation history of a rock. Through analysis of the relationships between internal (S_i) and external (S_e) foliation(s), the deformation history may be determined.

Figure 3.13

A Andalusite poikiloblast in staurolite–schist from Val Varrone (Italy). The andalusite poikiloblast (gray birefringence) contains randomly oriented quartz inclusions and fractures perpendicular to the long axis. Other minerals present are biotite (high birefringence) and quartz (gray–white birefringence). Cross-polarized light, 10× magnification, field of view = 2 mm.

B Kyanite poikiloblast in kyanite–schist from Resia (Italy). The kyanite crystal (NW–SE in the center of the image) has oriented quartz and muscovite inclusions that result in an internal foliation. Kyanite cleavage is distinctive. Other minerals present include muscovite (blue birefringence), biotite (orange birefringence, bottom-left), and quartz. Cross-polarized light, 2× magnification, field of view = 7 mm.

C Garnet poikiloblast in garnet micaschist from Valtellina (Italy). This is an extreme example of poikiloblastic texture known as *web texture*. Garnet (brown) has grown along the boundaries between the crystals of quartz (colorless), resulting in a "web-like" structure. Plane-polarized light, 10× magnification, field of view = 2 mm.

BOX 3.4 Lineations

Lineation is a general term to describe any penetrative, repeating linear feature that occurs within a rock. Lineations are represented by minerals, mineral aggregates, crenulations, or intersections, and are often associated with planar foliations.

Mineral lineations. The parallel arrangement of elongate or stretched mineral grains. In this example, black tourmaline is aligned in a metamorphosed granitic dike.

Aggregate lineations. The parallel arrangement of elongate aggregates or deformed objects, such as pebbles and fossils. In this example, elongate andalusite porphyroblasts in a micaschist define an aggregate lineation.

Crenulation lineations. The aligned hinges of microfolds associated with crenulation cleavage (here, subhorizontal lines) define a lineation and are a common feature in many low- to medium-grade schists.

Intersection lineations. The intersection between any two planar fabrics (foliation and bedding, two sets of cleavage, etc.) defines a lineation. During folding, many rocks develop an axial planar foliation that intersects bedding and generates small elongated rods known as "pencils" (as shown; image ~1 m wide).

Image courtesy of S. Papeschi

Ribbons

Ribbons are highly elongate single crystals or polymineralic aggregates with disc- or lens-shaped morphology. They are common in highly deformed metamorphic rocks such as mylonites and gneisses. Monocrystalline or polycrystalline ribbons define, or highlight, the foliation of a rock. Most ribbons consist of quartz, but although less common, mica, feldspar, and orthopyroxene ribbons are also known. Ribbons form by intense intracrystalline plastic deformation, which is a function of the metamorphic conditions (pressure and temperature) on their crystalline structure. An essential factor for ductile crystal deformation is the presence of slip systems, which allows dislocation along a slip plane. In low-grade metamorphic rocks, feldspar deforms by brittle fracturing and may develop winged porphyroclasts, while quartz deforms ductilely and generates elongate ribbons often with undulose extinction and extinction bands parallel to their long axis, or recrystallizes into polycrystalline quartz aggregates. Quartz ribbons usually wrap around feldspar. At medium- to high-grade conditions, both feldspar and quartz deform ductilely, and may form monomineralic and polymineralic ribbons – in this context quartz ribbons usually lack intracrystalline deformation.

Figure 3.14

A Quartz ribbons in mylonitic granite from Rio de Janeiro (Brazil). Elongated and flattened quartz ribbons (arranged E–W) wrapping completely sericitized feldspar (with dusty appearance). The brown birefringence crystals are biotite. Cross-polarized light, 2× magnification, field of view = 7 mm.

B Plagioclase ribbons in a mylonitic granulite from Ivrea Verbano (Italy). Plagioclase has been deformed and squeezed between pyroxene porphyroclasts (most at extinction). The plagioclase ribbon shows folded deformation twins. Cross-polarized light, 2× magnification, field of view = 7 mm.

APPLICATION 3.1

Shear Sense Indicators

Numerous textures and fabrics seen in thin-section can be used to determine the direction of shear during deformation. Common shear sense indicators include S–C fabrics, porphyroclasts and porphyroblasts, and so-called "mica fish."

S–C fabrics consist of a foliation or shear plane (S) transected by planar shear (C) bands. S–C fabrics are most common in granular rocks such as porphyroclastic granitoids, but may develop in all types of rocks. S-surfaces may form (1) synchronously with the shear bands; (2) in the late stages of the same deformation that results in the shear bands; or (3) as an earlier fabric later deformed by the shear bands. Regardless, S–C fabrics develop parallel to shear zone margins and provide important information about the direction of shear.

At the beginning of deformation, S-surfaces develop at an angle of about 45° to C-surfaces; as the shearing progresses, the angle between S–C planes reduces and approaches parallelism in highly sheared mylonites. S-surfaces "lean" in the direction of motion, thus S–C fabrics are used to determine the direction of shear.

S–C fabric in eclogitic micaschist from the Sesia Lanzo zone (Italy). Both images are the same. In the lower image the surfaces C and S have been highlighted and the sense of shear is indicated by the arrows. Planar C-surfaces (bold dashed lines in lower image) are roughly parallel to each other, while curved S-surfaces (fine dotted lines in the lower image) are defined by muscovite (high birefringence) and quartz (low birefringence). The combined S–C fabric indicates top-to-the-left (sinistral) shear.

Image courtesy of M. Zucali

Porphyroclasts/porphyroblasts, with their deformation tails or "wings," can also be used to determine shear sense. These structures form in response to differential stress at subsolidus conditions and are particularly common in high-strain zones.

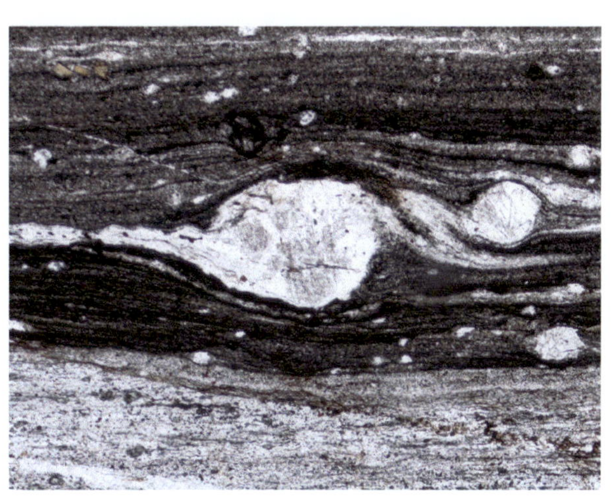

Secret Pass mylonite, Ruby Mountains, Nevada (United States). The large feldspar porphyroclast in this mylonite is a δ-type with top-to-the-left (sinistral) shear sense. A lithological contrast is visible in the lower part of the image. Plane-polarized light, magnification 5×, field of view = 2.45 mm.

Image courtesy of J. Platt

Mica fish also make excellent shear sense indicators. Due to the mineral structure of muscovite, it is easily sheared along its cleavage planes, which produces sigmoidal, lenticular, or parallelogram shapes that record the shear direction. Here, muscovite documents top-to-the-left (sinistral) shear.

Muscovite in a mylonitic micaschist from Bianzone (Italy). The sample has a well-developed foliation (oriented diagonally) and muscovite ("fish") seem to "swim" in a fine-grained matrix of phyllosilicate and graphite, defining a top-to-the-left (sinistral) shear sense. Cross-polarized light, magnification 2×, field of view = 7 mm.

Image courtesy of M. Zucali

Stylolites

Pressure solution (stress-induced solution transfer) is a common process both during sediment diagenesis and at low metamorphic grades, and occurs when grains are dissolved at intergranular or intercrystalline contacts. This process can lead to the development of numerous structures such as truncated detrital grains, dislocation of veins or linear structures, and *stylolitic* surfaces. Stylolites are irregular, curved, or toothed surfaces that form during chemical compaction. Their morphology is variable, but generally consist of interlocking columns capped by subplanar concentrations of insoluble material. As shown, the columns are typically parallel to the direction of maximum compressive stress.

Stylolites can be divided into two main types. *Sedimentary* (or diagenetic) stylolites are linked to sediment compaction during diagenesis and are parallel to bedding. *Tectonic* stylolites are formed by tectonic compression – their orientation depends on the orientation of the stress field, which is often transverse to bedding. Stylolites are more common in carbonates than in other rocks due to the high susceptibility of carbonate minerals to dissolution; for example, below 250 °C carbonate minerals are more soluble than quartz, while above 250 °C quartz is more soluble than carbonate. Stylolites are commonly rich in insoluble material due to the local removal of dissolved material via pressure solution. The insoluble material consists of clay minerals, oxides, bituminous material, sulfides, heavy minerals, etc.

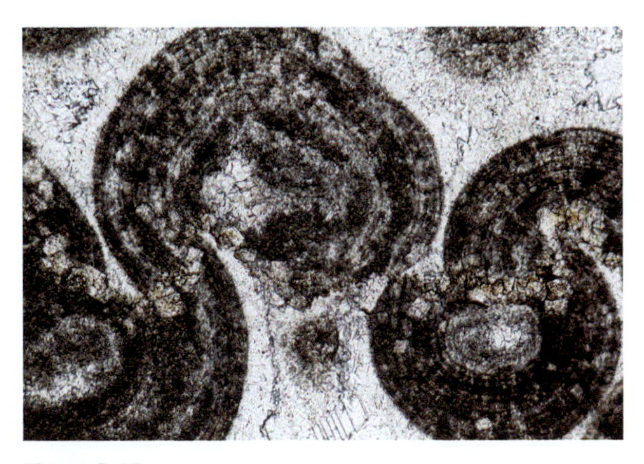

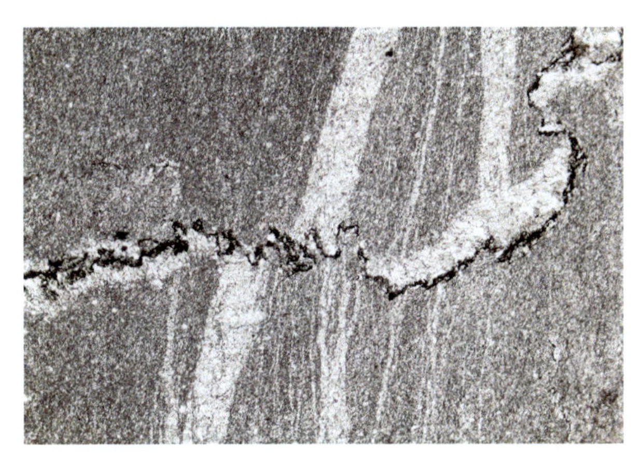

Figure 3.15

A Sedimentary stylolite in oolitic limestone from Sardinia (Italy). The stylolite (blocky, light brown) is roughly horizontal across the center of the image and made of insoluble oxidized material. Plane-polarized light, 10× magnification, field of view = 7 mm.

B Tectonic stylolite in a contact-metamorphosed limestone from Bazena (Italy). The stylolite is marked by black insoluble material and cross-cuts white calcite veins. Plane-polarized light, 2× magnification, field of view = 7 mm.

APPLICATION 3.2

Stylolites as Strain Markers and Conduits for Diagenetic Fluids

Stylolites can provide information about the rocks in which they are formed. They provide insight to various geological processes controlling the formation of sedimentary reservoirs or deformation zones. Stylolites play an important role during diagenesis and may reduce the thickness of a sedimentary body, which in turn can impact the rheological properties of the upper crust. In addition, the orientation of sedimentary and tectonic stylolites can be used to determine the maximum compressive stress direction.

Volume change. Stylolites are generated by pressure solution, which imparts a volume change to the rock. This can be estimated using the geometric method of Gratier et al. (2013). During stylolite genesis, progressive dissolution produces an "apparent" shift of the vein; the dissolved rock width (X) can be estimated by measuring the distance of the apparent shift (d_s) and the angle (α) between the vein and the stylolite using the equation:

$$X = d_s \tan \alpha$$

Diagenesis. Sedimentary and tectonic stylolites may have important implications for diagenetic reactions and the circulation of fluids in sedimentary basins. Anastomosing stylolites parallel to sedimentary bedding may control the geometry and extent of large-scale (meters to kilometers) diagenetic alteration and fluid circulation. In an analysis of the Lower Cretaceous Benicàssim carbonate platform in the Maestrat Basin (eastern Spain), Gomez-Rivas et al. (2022) showed that diagenetic reactions involving carbonate and fluid circulation were controlled by stylolite networks. Their detailed study documented how diagenetic processes, such as dolomitization, are influenced and controlled by the presence of complex networks of stylolites.

Stylolites are baffles for diagenetic fluids

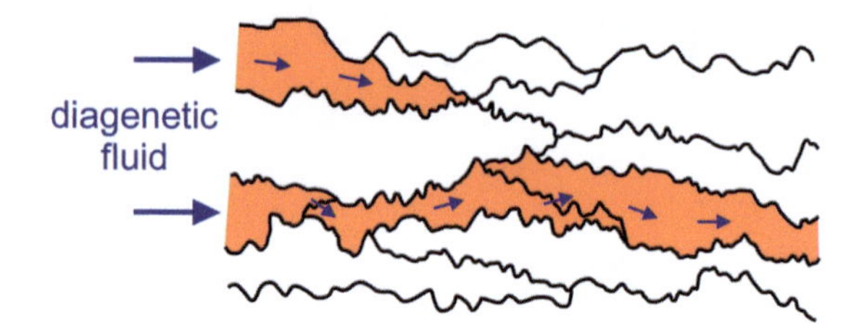

For example, the sketch above illustrates the potential control of stylolites on diagenetic fluid flow. If the fluid reactivity and environmental conditions permit, diagenetic fluid (purple arrows) is channeled along stylolite networks (wavy lines), facilitating diagenetic

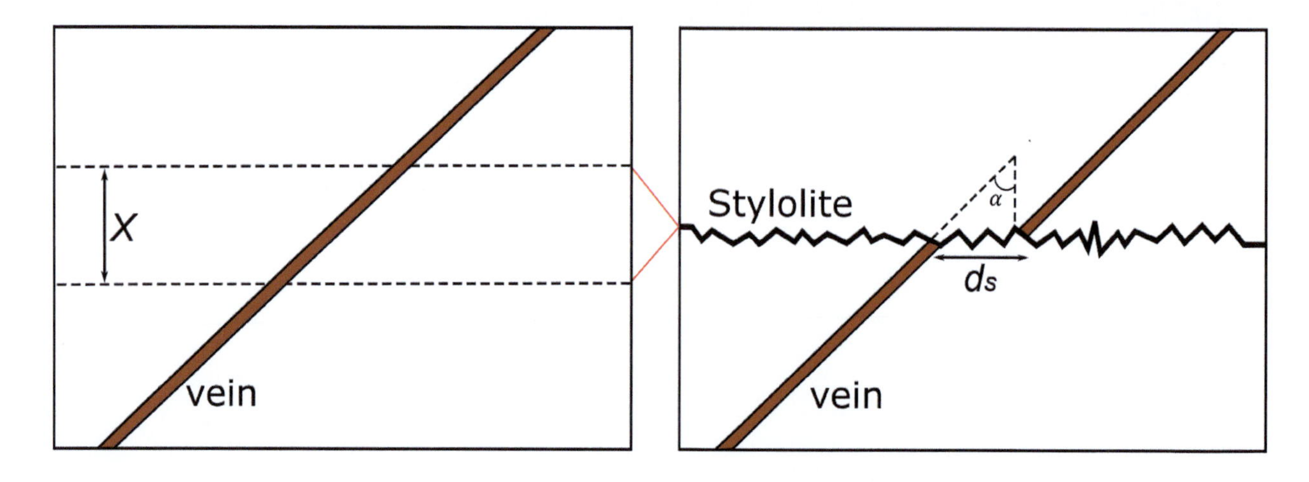

Image courtesy of E. Gomez-Rivas

reactions (orange zones). Once diagenetic fluids get through the barrier of an individual stylolite, they can pervade the nearby rock volume. Stylolite morphology and its sealing properties determine whether fluid flow is discontinuous (upper pink region) or continuous (lower pink region).

This scenario is evident in the Benicàssim carbonate platform. As shown in the outcrop photo above, Gomez-Rivas et al. (2022) identified a "dolomitization front" that follows an undulose stylolite network. The "dolomitization front" is very clear – it weaves up and down, following the wave-like stylolite network.

Tectonics. Stylolites may also record fluid flow in changing tectonic conditions. Stylolite networks close to faults in the Benicàssim carbonate platform apparently acted as conduits to fluids released during high pressure, generating hydraulic breccia and deposition of high-temperature calcite. This is seen in outcrop (below), where the hydraulic breccia occurs with stylolites containing high-temperature calcite cement (white).

Image courtesy of E. Gomez-Rivas

Veins and Strain Fringes

Veins are dilated fractures filled with polycrystalline minerals. They document the circulation of fluids. They are ubiquitous microstructures in sedimentary, igneous, and metamorphic rocks. The fluids may be derived from metamorphic reactions (such as devolatilization, dehydration, pressure solution) or be from an external source(s). Since veins represent sites of dilation, circulating fluids undergo a pressure decrease that may cause saturation of chemical components dissolved in solution and result in their consequent precipitation and crystallization. Vein-filling minerals can be variable and do not always relate to the mineralogy of the surrounding rock. The most common vein minerals are quartz, calcite, epidote, and sulfides; mica and feldspar may also occur, as well as andalusite or sillimanite in high-grade metamorphic veins. Veins can provide information about a rock or a region as, for example, their internal structures (the shape and orientation of vein-filling minerals) can be used to determine paleo-stress fields, deformation kinematics, and fluid pressure. The geochemistry of vein-filling minerals and their fluid inclusions can constrain metamorphic conditions, fluid characteristics, and fluid origin. In addition, veins often contain datable minerals and economically important ore deposits (such as gold, silver, or metal sulfides). Their internal morphology is extremely variable and complex, resulting from different combinations of crystal shape, crystal growth directions, and stress regimes. A common mechanism that leads to the formation of veins is called *crack and seal*: repeated hydraulic fracturing followed by mineral precipitation from the fluid which seals the fracture. Veins can be classified according to various factors, such as crystal morphology (blocky, fibrous, stretched, etc.) and vein-filling minerals:

Syntaxial veins: Vein-filling minerals grow from the wall toward the center of the vein in optical continuity with mineral grains of the same composition as in the host rock (e.g. quartz veins in sandstones and calcite veins in limestones or marbles). As crystals grow in the same direction as vein opening, they become elongate in the growth direction, forming blade-like crystals with *comb structure*. Crystals meet at a medial line (or medial suture), which represents the original fracture and, if deformation accompanies crystal growth, they may assume a curved fibrous morphology.

Antitaxial veins: Vein-filling minerals grow toward the vein walls away from the center of the vein. Antitaxial veins consist of minerals that are absent or uncommon in the host rock (e.g. calcite fibers in quartzite). Single fibrous crystals show crystallographic continuity from one wall to the other. If present, the median line is commonly marked by small wall-rock fragments. The morphology of the crystals will be fibrous if deformation accompanies the crystals' growth.

Fringe structure (or pressure fringes): These represent a particular type of site dilation that develops on the two low-pressure sides of rigid objects (usually ore minerals) in a ductilely deforming rock. These rigid objects cause a local perturbation of the stress field and fluids deposit new material at these low-pressure sites. Pressure fringes are characterized by the presence of fibrous crystals and differ from pressure shadows in that the latter have diffuse boundaries and lack a fibrous internal structure. Since these structures develop during deformation, their internal and external shape can be used as kinematic indicators.

Figure 3.16

A Syntaxial vein in limestone from La Spezia (Italy). The vein developed in a fine-grained limestone. It is composed of calcite, which grows from the vein walls and meets in the vein center along the median line. The calcite crystals are elongate in the growth direction. Cross-polarized light, 10× magnification, field of view = 2 mm.

B Antitaxial veins in slate from Pisa (Italy). The vein consists of elongate quartz (not present in the surrounding rock) oriented perpendicular to the vein walls with crystallographic continuity from one wall to the other. Cross-polarized light, 2× magnification, field of view = 7 mm.

C Fringe structure in pyrite-rich slate from Casaio (Spain). The pyrite aggregate (opaque) is surrounded by a pressure fringe of quartz with an elongate morphology. In the upper part of the fringe, adjacent to the pyrite aggregate, a thin calcite fringe is also visible (distinctive birefringence). Cross-polarized light, 10× magnification, field of view = 2 mm.

Reaction and Disequilibrium Textures

Atoll

The term "atoll crystals" or "atoll structure" is a descriptive term used for minerals characterized by a ring-shaped morphology in which a core mineral(s) is surrounded by a rim of another mineral. This structure is very common in sulfides but is also found in other minerals, particularly garnets from a wide range of metamorphic environments around the world. This is a reaction texture and a special form of corona texture – its origin invokes numerous processes, such as preferential dissolution of mineral cores by fluid infiltration, polymetamorphism, or coalescence of small crystals that may envelop other phases and form a ring-shaped morphology. Atoll garnet occurs in low- to high-grade metamorphic rocks of differing composition and are often associated with whole (non-atoll) garnets.

A common characteristic of atoll garnet worldwide is that they are of different composition with respect to whole garnets they may occur with and also show compositional differences between the core and rim of a single atoll garnet. For example, atoll garnet in eclogites from the Tso Morari complex (India) have a higher concentration of Ca + Mn in their cores than in their rims, and their formation is inferred to be due to fluid infiltration along pre-existing fractures during eclogite exhumation. These fluids would have favored the breakdown of the original garnet cores. In many cases, such as the Bohemian Massif, there is also evidence of element exchange between garnet cores and rims, in which the elements released by the breakdown of early-formed garnet cores are incorporated into newly grown garnet rims.

Figure 3.17

A Atoll garnet in micaschist from Val Venosta (Italy). The central part of this garnet has a polycrystalline core of biotite (brown) and quartz (colorless). Biotite growth seems to perfectly follow the outer edge of the crystal (top-left and bottom-right), while garnet (brownish) defines the remaining crystal rim. The garnet crystal is wrapped by a continuous foliation defined by muscovite. Plane-polarized light, 10× magnification, field of view = 2 mm.

B Atoll garnet in eclogite from Mucrone (Italy). The garnet crystal (isotropic) shows a core nearly completely replaced by muscovite (different orientations but high-interference colors). Some relic garnet forms "islands" within the crystal. Cross-polarized light, 10× magnification, field of view = 2 mm.

BOX 3.5 Reaction Rims and Epitaxy

Changing metamorphic conditions, during either prograde or retrograde processes, can lead to the development of *reaction rims*. Reaction rims can be complex and have highly variable morphology. Reaction rims that form around grains are called *coronas*. A corona results from the incomplete reaction between the core mineral and neighboring phases. Coronas are often composed of symplectic intergrowths or a new stable mineral can preferentially nucleate on a crystallographically similar pre-existing phase with a systematic relationship between the two crystal structures (e.g. sillimanite on muscovite, amphibole on pyroxenes, staurolite on kyanite). *Epitaxy* describes a type of mineral growth in which one mineral utilizes aspects of an existing mineral's crystal lattice.

Coronas are particularly common in regionally metamorphosed metabasites. In this example from Sondalo (Italy), an olivine (yellow–green birefringence) is surrounded by a double reaction rim of (1) orthopyroxene and (2) a fine-grained intergrowth of amphibole and spinel (adjacent to the host plagioclase). Cross-polarized light, 2× magnification, field of view = 7 mm.

Coronas are widespread in retrograde high-pressure rocks such as granulites or eclogites. In this example from Isla de Margarita (Venezuela), rutile (brown–red) is mantled by titanite (light brown). The host mineral is phengite (colorless). Plane-polarized light, 20× magnification, field of view = 1 mm.

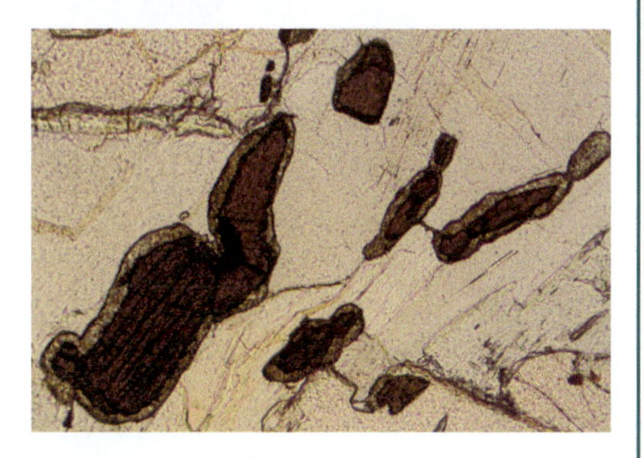

This example of epitaxy from Sponda (Switzerland) shows a large kyanite crystal (strong cleavage) interleaved with staurolite (pale yellow and perpendicular fractures). Plane-polarized light, 2× magnification, field of view = 7 mm.

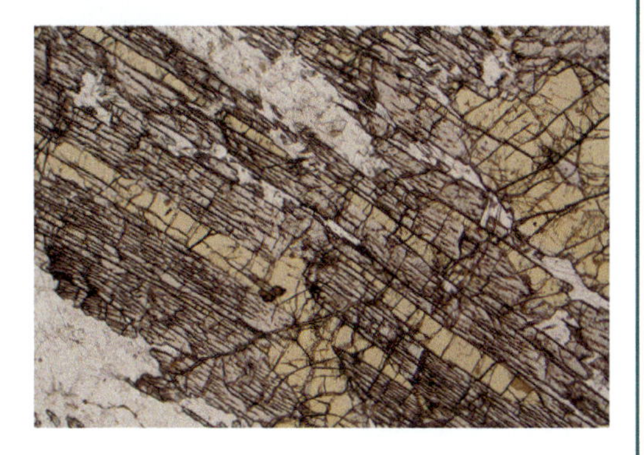

BOX 3.6 Symplectites

Symplectites are vermicular intergrowths of two minerals that grow simultaneously in a solid state reaction by breakdown of an earlier unstable solid solution phase. Symplectites are common in high-grade metamorphic rocks (e.g. granulite, eclogite), especially in those that have experienced decompression during uplift or exhumation processes. Symplectite-forming reactions commonly are incomplete so that reactants and products can be observed together. This may be due to various factors: The symplectite reaction zone expands with time, implying a substantial reduction of chemical potential gradients in the symplectite zone, or a slowing of diffusion rates and thus reaction rates. Furthermore, symplectites form in uplift contexts, where the temperature decreases over time causes a further slowing of diffusion rates. Symplectites can be a valuable tool for determining tectono-thermal histories of metamorphic terranes.

Common symplectites in high-grade rocks include Grt = Crd + Qz ± Opx, as in this example, where a large isotropic garnet (bottom) is surrounded by a symplectic intergrowth of orthopyroxene (with yellow-orange birefringence), quartz (gray birefringence) and cordierite (white–yellow birefringence). The crystals with blue birefringence near the garnet are sapphirine. Granulite from Val Codera (Italy), cross-polarized light 2× magnification, field of view = 2 mm.

Other common symplectites are Omp = Pl + Px ± Hbl and Grt = Hbl/Px + Pl ± Ilm in eclogitic rocks, as shown here, in which both types can be seen. Garnet is surrounded by a symplectic intergrowth of plagioclase (colorless), hornblende (green), and ilmenite (black). The surrounding omphacite is now replaced by a symplectic intergrowth of diopside and albite. Eclogite from Sardinia (Italy). Plane-polarized light 2× magnification, field of view = 2 mm.

Zoning

Compositional zoning in metamorphic minerals can form in two different ways: (1) During crystal growth, and (2) post-solidification. Compositional zoning reflects changes in P–T conditions. Metamorphic growth zoning is very similar to magmatic growth zoning, and is directly related to the availability of elements suitable for the growth of the crystals.

Post-solidification, as metamorphic conditions change, compositional zoning can be affected by ion-exchange reactions between the outer parts of a mineral and the rock matrix. These ion-exchange reactions can affect the whole crystal, especially at high temperature or in the presence of intergranular fluids that facilitate diffusion of chemical components, or can affect only the outer part of the crystals at lower temperatures or if intergranular fluids are absent. Zoning in metamorphic minerals can be concentric, patchy, or oscillatory:

Concentric compositional zoning commonly occurs during mineral growth, is usually less complex than in igneous rocks, and can be marked by inclusions. It can be linked to the change of chemical components as a result of metamorphic reactions involving other minerals in the rock, or due to change in intergranular fluid composition. Concentric compositional zoning is common in many minerals such as plagioclase, in which it is usually reversed (Ca-rich rim and Ca-poor core), developing with increasing temperature as Ca-plagioclase becomes progressively more stable at higher temperatures. Other commonly concentrically zoned minerals are epidote, allanite, corundum, tourmaline, mica, amphiboles, zircon, and garnet.

Although its zoning is difficult to observe under a polarizing microscope, one of the most commonly zoned minerals, and one of the most studied, is garnet. Being isotropic, its zoning can be observed through other tools such as an electron microprobe (EMP), SEM backscatter imaging, or X-ray element mapping. Mn-rich garnets are stable at lower temperatures than more Fe–Mg-rich ones, so garnet cores typically have much higher Mn than the rims.

Patchy zoning is less common than concentric compositional zoning and is thought to be due to dissolution and overgrowth as metamorphic and deformation conditions change. Patchy zoning is quite common in plagioclase, epidote, corundum, tourmaline, amphiboles, and garnet, as well as in many sulfide and metallic oxide minerals.

Oscillatory zoning is the rarest and least common type of zoning, and is generally believed to develop during open-system growth conditions involving changes in fluid composition. One of the minerals that most commonly shows oscillatory zoning is hydrogrossular (grossular–andradite) garnet in skarns. Its oscillatory zoning is revealed as anomalous birefringence.

At high temperatures, compositional zoning can be obscured or completely obliterated by diffusion processes since intracrystalline diffusion can lead to homogenization of earlier growth-zoning profiles. For example, in upper amphibolite facies, garnet shows “flat” chemical profiles which are related to these processes.

Figure 3.18

A Concentric compositional zoning in hornblende gneiss from Valtellina (Italy). The picture is dominated by a large plagioclase crystal, with sharp compositional zoning (dark to light gray). The orange birefringent crystals are hornblende, with small rectangular inclusions of biotite. Cross-polarized light, 2× magnification, field of view = 7 mm.

B Patchy zoning in zoisite skarn from Adamello (Italy). The large zoisite crystal shows an irregular patchy zoning, highlighted by birefringence variations. The gray birefringent crystals surrounding the zoisite are quartz. Cross-polarized light, 2× magnification, field of view = 7 mm.

C Oscillatory zoning in garnet skarn from Haytor (England). The entire picture is dominated by hydrogrossular garnets with spectacular oscillatory zoning and anomalous birefringence. Cross-polarized light, 2× magnification, field of view = 7 mm.

BOX 3.7 Exsolution in Metamorphic Rocks

Exsolution features are common in minerals from magmatic and metamorphic environments. In metamorphic rocks, exsolution textures can be induced by P–T variations and help to determine the metamorphic history of a rock.

Pressure-controlled exsolution textures are common in rocks from continental collisional belts and from xenoliths in kimberlites. Exsolution lamellae occur particularly in minerals from high- and ultra high-pressure environments (e.g. garnet–peridotite from the deeper mantle, exhumed deeply subducted oceanic crust, thick continental granitic crust). Pressure–temperature-controlled exsolution induced by cooling and decompression are common in garnet, clinopyroxene, and Cr-spinel.

Exsolution lamellae associated with decreasing pressure
Omp exsolves Qz
Grt exsolves Px, Rt/Na-Amp
Ol exsolves Ilm, Cr–Ti-Mag
Opx exsolves Cr-Spl
Cpx, Grt, Ttn, or Chr exsolves Qz/Coe
Cpx exsolves Kfs/Grt/Phl/Ph

This ultra high-pressure calc-silicate rock is from the Kokchetay Massif (Kazakhstan). The central crystal is diopsidic clinopyroxene (red birefringence) that has exsolved potassium feldspar (parallel alignment of white to yellow birefringence). Garnet (isotropic) and calcite (note distinctive twinning) are also present. Cross-polarized light, magnification = 20× 0.8 mm wide.

This ultra high-pressure clinopyroxenite is from Horní Bory (Czechia). The central large clinopyroxene crystal (yellow birefringence) has crystallographically controlled exsolution of garnet (aligned isotropic blebs). Cross-polarized light, 10× magnification, 3.5 mm wide.

Images courtesy of H.-P. Schertl

3.3 Metamorphic Facies

Metamorphic changes occur when a rock is heated (prograde metamorphism) or cooled (retrograde metamorphism). The type of metamorphism that occurs depends on temperature, confining pressure, directed pressure (deformation), chemical activity or fugacity of water, and their variation through time. The presence of fluid facilitates metamorphic reactions. For rocks metamorphosed under the same conditions, different bulk-rock compositions will produce different mineral assemblages. The corollary to this statement is also true – for a particular bulk-rock composition, different mineral assemblages must reflect different metamorphic conditions. The relationship between metamorphism, rock composition, and pressure and temperature of metamorphism is represented by the concept of *metamorphic facies*. Metamorphic facies were developed largely from investigations of metamorphosed basalts and are therefore linked to mafic protolith composition (Eskola, 1920).

The eight metamorphic facies (shown below) cover a wide range of pressures and temperatures. Today experimental data are used to define these facies, which extend to much higher pressures (the so-called ultra high-pressure facies rocks) than shown here. The position of the fields depends on bulk composition, metamorphic reaction kinetics, and the activity of fluid(s); consequently, the facies and their boundaries (shown in white) vary by ±25 °C and ±1 MPa.

Low-grade metamorphism of a mafic protolith is marked by the appearance of minerals such as zeolite, or at ~200 °C by prehnite and pumpellyite. Higher-grade facies may be divided (e.g. upper- and lower-greenschist facies, amphibolite and garnet–amphibolite facies, medium- and high-temperature eclogite facies, or low-, medium-, and high-pressure granulite facies). The dashed lines in the facies diagram approximate P–T gradients associated with the following metamorphic environments: contact metamorphism (1), volcanic arcs (2), collisional mountain belts – 30 °C/km (3), stable continents (4), and accretionary prisms (5). The general conditions for anatexis (melting H_2O-saturated crustal rocks) is represented by the dashed and dotted black line – conditions to the right of this line are where partial melts may be generated.

With increasing temperature, under the lowest pressure conditions (contact metamorphism), additional distinctions between albite–epidote, hornblende, pyroxene, and sanidinite (san.) hornfels are made. Metamorphism of a basaltic protolith will result in systematically different mineral assemblages under the pressure and temperature conditions defined by each metamorphic facies.

Metamorphic facies of a basaltic protolith

Metamorphic facies	Mineral assemblage	Other associations
Zeolite	Lmt + Anl + Hul + Wrk	± Relict igneous grains
Prehnite–pumpellyite	Prh + Pmp + Chl ± Ab ± Act ± Ep	± Relict igneous grains
Greenschist	Act + Ab + Chl ± Ep (low grade), Hbl + Act ± Ab (high grade)	
Amphibolite	Hbl + Pl ± Ep (low grade), Hbl + Pl ± Ep ± Grt ± Px (high grade)	
Granulite	Opx + Cpx + Pl ± Hbl ± Grt ± Ol	
Blueschist	Gln + Ab + ± Lws	± Relict igneous grains
Eclogite	Omp + Grt – Pl	
Granulite	Cpx + Opx +Pl	

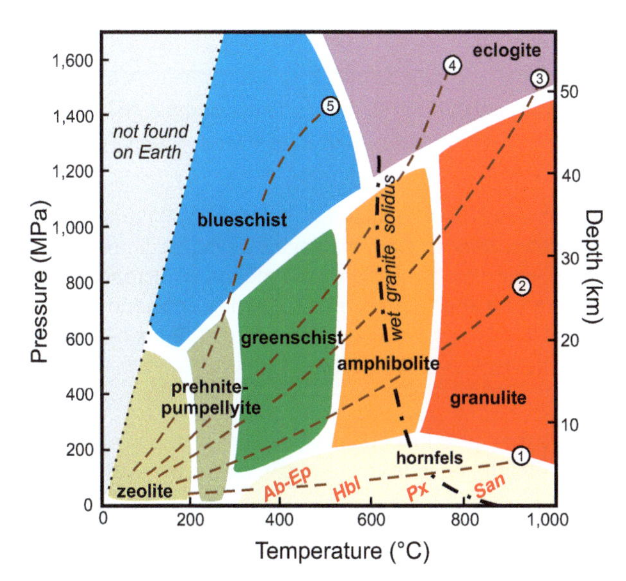

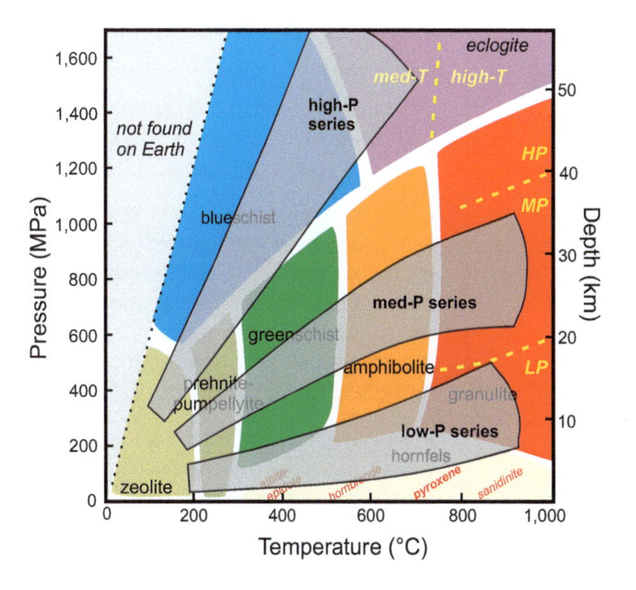

Several metamorphic facies may occur within a single region; together these facies represent a *metamorphic belt.* A metamorphic belt usually defines a *metamorphic series:* a group of metamorphic facies related by a common tectonic process or setting. Three metamorphic series reflect the main thermo-tectonic settings operating on Earth:

- the low-pressure series (>30 °C/km)
- the medium-pressure series (10–30 °C/km)
- the high-pressure series (5–10 °C/km)

The greenschist, amphibolite, and granulite facies of the low-pressure and medium-pressure series dominate the metamorphic record on Earth. The low-pressure series represents metamorphic belts associated with high temperatures (e.g. hydrothermal or contact metamorphism, rift zones, and high heat flow orogens). The medium-pressure series occurs over a similar temperature range but under conditions of higher pressure, as are typical of orogenic belts and regional metamorphism associated with continents. There is some overlap between the low- and medium-pressure metamorphic series, but if metapelitic rocks are present then distinct metamorphic zones will form, permitting the discrimination of these two series. The high-pressure series (which continues to pressures >1,600 MPa, not shown here) is characteristic of subduction zones and dominated by rocks of the blueschist and eclogite facies. Such rocks are often overprinted by lower pressure/temperature facies when they later return to the surface.

BOX 3.8 Metamorphic Zones

The intensity or degree of metamorphism may generally be referred to as the "metamorphic grade," and is a convenient way to refer to the general degree of metamorphism of a rock without specifying the exact relationship between temperature and pressure.

In the Scottish Highlands in the late 1800s, George Barrow documented changes in mineralogy associated with regional metamorphism of Neoproterozoic- to Cambrian-aged pelitic sediments. He recognized that a similar protolith produced specific new minerals as a function of increasing metamorphic grade. The first appearance of each *index* mineral defined a metamorphic zone and the identification of six metamorphic zones documented a regime of increasing pressure or temperature (as shown below). Metamorphic zones documenting regional (Barrovian) metamorphism are now recognized around the world.

However, it is important to realize that *index* minerals form as a function of whole-rock chemistry (i.e. when compatible with the bulk composition of the protolith); no index mineral can form if the specific chemical constituents required for it are absent.

Barrovian zones

Metamorphic zone	Mineral assemblage	Other associations
Chlorite	Chl + Ms + Qz + Ab	± Cal ± Stp ± Pg
Biotite	Bt + Chl + Ms + Qz + Ab	± Cal
Garnet	Grt + Bt + Chl + Ms + Qz + Ab	± Cld ± Pl
Staurolite	St + Grt + Bt + Chl + Ms + Qz + Ab	± Mnz
Kyanite	Ky ± St + Grt + Bt + Ms + Qz + Pl	
Sillimanite	Si ± St + Grt + Bt + Ms + Qz + Pl	± Ky relics

Metamorphic zone	Mineral reaction(s)	Condition
Biotite	Ms(Si) +Chl → Bt + Ms(Al) + Qz + H_2O	Moderate P/T
Garnet	Chl + Ms → Grt + Bt + Qtz + H_2O	Moderate P/
Staurolite	Grt + Ms + Chl → St + Bt + Qz + H_2O	increasing T
	Chl + Ms → St + Bt + Qz + H_2O	↓
	Cld + Ms → St + Grt + Bt + Qz + H_2O	
Kyanite	Ms + St + Chl → Bt + Ky + Qz + H_2O	
	St + Ms + Qz → Al_2SiO_5 + Bt + H_2O	
Sillimanite	Ky → Sil	
	St + Ms + Qz → Grt + Bt + Sil + H_2O	
	St + Qz → Grt + Sil + H_2O	
	Ms + Qz → Al_2SiO_5 + Kfs + H_2O	
Partial melting	Kfs + Qz + Al_2SiO_5 + H_2O → melt	Moderate P/High T
	Ms + Qz + H_2O → Sil + melt	
	Ms + Qz → Kfs + Sil + melt	
	Ms + Pl(Na) + Kfs + Qz + H_2O → melt	
	Bt + Sil + Qz → Kfs + Crd + melt	
	Bt + Sil + Qz → Kfs + Grt + melt	
	Bt + Sil + Qz → Crd + Grt + Kfs + melt	
Ultra high temperature	Crd + Grt →Sil + Opx	High T (900 °C)
	Opx + Sil → Spr + Qz	High T (1,000 °C)

Zeolite Facies

Metamorphic series: *very low pressure, very low temperature*

Zeolite facies metamorphism occurs at $<200\,^{\circ}C$ and $<400\,MPa$; it documents the transition from diagenesis to low-grade (sub-greenschist) metamorphism. In rocks of basic to intermediate composition, zeolite formation is attributed to alteration at low temperatures in the presence of CO_2-poor or CO_2-absent aqueous fluids. Rocks of the zeolite facies include any in which zeolite and quartz are present, whether diagenetic or metamorphic. Zeolites are low-density hydrated silicates that are rarely stable at temperatures above 300 °C. Low- and high-temperature zeolite facies are recognized, with the former represented by heulandite (or stilbite) + analcime + quartz, and the latter by laumontite + albite + quartz (± pumpellyite, prehnite, or epidote). The most abundant zeolite facies minerals include the heulandite group, analcime, and laumontite. The formation of epidote + chlorite + quartz marks the upper limit of the zeolite facies.

Metasomatism (alteration due to fluids) is the process by which rocks are altered at zeolite facies. Under zeolite facies conditions, primary minerals and textures of the protolith (either igneous or sedimentary) are commonly preserved and the rocks are generally undeformed. Alteration occurs due to porosity and/or the presence of highly reactive minerals and glass. For example, volcanic glass is easily altered to zeolite and high-temperature volcanic minerals (olivine, pyroxene) are particularly susceptible to hydration. Under zeolite facies conditions, pelitic assemblages low in iron, magnesium, and calcium often produce clay minerals such as illite. Progressive changes in the zeolite assemblage are sensitive to bulk-rock composition.

Zeolite facies is a product of burial metamorphism, hydrothermal alteration, and thermal metamorphism. Therefore, it is linked to thicker accumulations of volcaniclastic sediments in subsiding basins, to subduction zones, and to geothermal fields associated with volcanic centers. The latter includes parts of ocean spreading centers where circulating hydrothermal fluids have lower temperatures.

Zeolite facies as a function of protolith composition

Protolith	Rock type	Mineral parageneses
Ultramafic	Peridotite, serpentinite	Serpentine group minerals, brucite, dolomite, magnesite
Mafic	Basalt, andesite, gabbro, diorite	Chlorite, serpentine, zeolites (+ analcime and wairakite), quartz, albite, prehnite, pumpellyite, calcite, dolomite, ± laumontite
Pelitic	Shale, mudstone	Chlorite, illite, quartz, albite, calcite, dolomite
Quartzofeldspathic	Sandstone, rhyolite, granite, chert	Quartz, plagioclase, potassium feldspar, garnet, biotite, muscovite, ± hornblende, ± clinopyroxene
Calcareous	Limestone, dolomite, marl	Calcite, dolomite, quartz, chlorite, illite, albite

Figure 3.19

A Zeolite facies basalt from Bergamo (Italy). A green mixture of chlorite + celadonite (part of the mica group) + palagonite forming small spheroidal aggregates and replacing the original volcanic glass. Fibrous zeolites (left) are mixed with brown calcite. Two amygdales filled with zeolite are also present (upper-right quadrant). Plane-polarized light, 2× magnification, field of view = 7 mm.

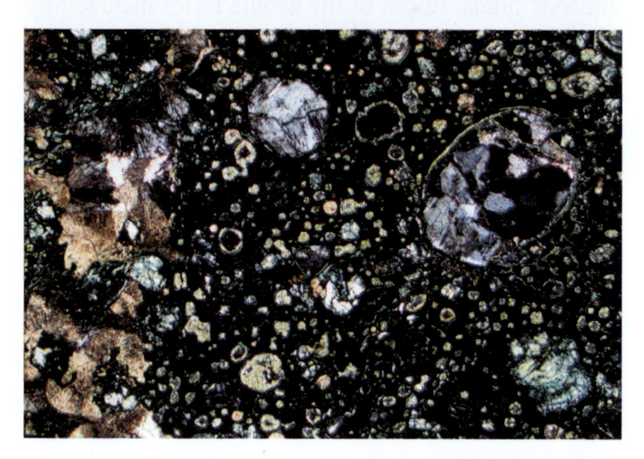

B Zeolite facies basalt from Bergamo (Italy). Same image as (A) but with crossed polarizers. The chlorite + celadonite have very low birefringence (almost isotropic), but along rims of the small spheroidal aggregates their birefringence is visible. The vesicle-filling zeolite shows typical gray birefringence. Cross-polarized light, 2× magnification, field of view = 7 mm.

C Zeolite facies pelite from Sardinia (Italy). A fine-grained mixture of illite + chlorite (brownish-green), and quartz (clear). The small black "dots" are magnetite. Plane-polarized light, 2× magnification, field of view = 7 mm.

D Zeolite facies pelite from Sardinia (Italy). Same image as (C) but with crossed polarizers. The fine-grained mixture of illite and chlorite is characterized by the high birefringence of illite; quartz has low birefringence. Cross-polarized light, 2× magnification, field of view = 7 mm.

Prehnite–Pumpellyite Facies

Metamorphic series: *low pressure, low temperature*

Prehnite–pumpellyite facies metamorphism occurs at ~200–300 °C and <650 MPa, is transitional between the very low-grade zeolite facies and greenschist facies, and is sometimes referred to as *sub-greenschist facies*. Prehnite can form within the stability fields of heulandite, laumontite, and wairakite when associated with CO_2-poor or CO_2-absent aqueous fluids at very low temperatures; thus, recognition of prehnite–pumpellyite facies requires both the *presence* of prehnite and/or pumpellyite, and the *absence* of zeolites, actinolite, lawsonite, or jadeite. Due to prehnite–pumpellyite facies mineralogy, these rocks are often blue–green in color.

Primary protolith minerals and textures may be preserved, which in mafic igneous precursors includes relict calcic-plagioclase, augite, hornblende, biotite, and orthoclase. In psammitic rocks, prehnite and pumpellyite are sparse or absent. Spongy prehnite enclosing quartz is interpreted to document the breakdown of laumontite or heulandite. Mineralogical changes in argillite and limestone are less obvious, although clay largely disappears.

The prehnite–pumpellyite facies is a hallmark of hydrothermal alteration associated with mid-ocean ridges, burial metamorphism, and the thickened prisms of sedimentary and volcanic rocks accreted at continental margins. The upper portions of ophiolites may show similar alteration, where it may be related to either hydrothermal or thermal metamorphism. The prehnite–pumpellyite facies tends to be poorly developed in on-shore hydrothermal systems, where pumpellyite is commonly absent.

In the low-pressure series, with increasing metamorphic grade zeolite facies (often with wairakite) transitions to prehnite–pumpellyite facies, and then to greenschist facies, as seen in the Tanzawa Mountains of Japan. In the medium-pressure series, with increasing metamorphic grade the prehnite–pumpellyite facies transitions to a pumpellyite–actinolite assemblage, and then to greenschist facies (e.g. similar to regional metamorphism in the Wakatipu district of southern New Zealand). In the high-pressure series, with increasing metamorphic grade the prehnite–pumpellyite facies passes into the lawsonite–albite–chlorite facies (e.g. transitional to blueschist facies, as also seen in southern New Zealand).

Prehnite–pumpellyite facies as a function of protolith composition

Protolith	Rock type	Mineral parageneses
Ultramafic	Peridotite, serpentinite	Serpentine group minerals, brucite, dolomite, magnesite
Mafic	Basalt, andesite, gabbro, diorite	Chlorite, serpentine, prehnite, pumpellyite, quartz, albite, calcite, dolomite
Pelitic	Shale, mudstone	Chlorite, muscovite, clay minerals, quartz, albite, calcite, dolomite
Quartzofeldspathic	Sandstone, rhyolite, granite, chert	Quartz, albite, prehnite, pumpellyite, epidote, chlorite, illite, kaolinite
Calcareous	Limestone, dolomite, marl	Calcite, dolomite, quartz, clay minerals, albite

Figure 3.20

A Prehnite–pumpellyite facies gabbro from Livorno (Italy). The original igneous texture is completely erased by the new growth of prehnite and pumpellyite. Prehnite (colorless) occupies the space between pumpellyite (fractured, gray, high relief). Plane-polarized light, 2× magnification, field of view = 7 mm.

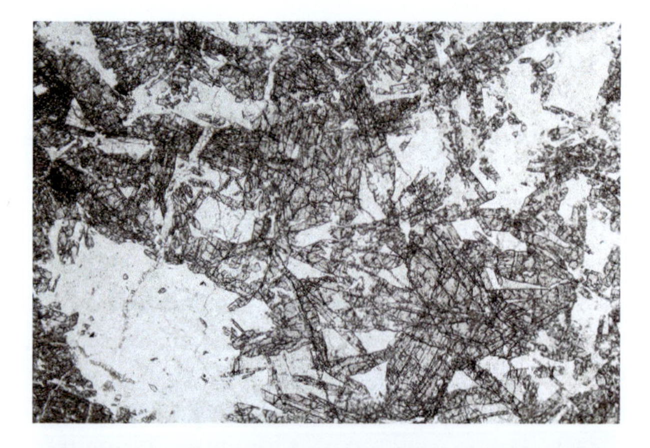

B Prehnite–pumpellyite facies gabbro from Livorno (Italy). Image same as (A) but with crossed polarizers. Pumpellyite has blue birefringence, while prehnite shows low-order greenish-gray colors. Cross-polarized light, 2× magnification, field of view = 7 mm.

C Prehnite–pumpellyite facies pelite from Sardinia (Italy). Epidote (euhedral, lozenge-shape, high relief, light brown) in a fine-grained, clay-rich matrix (bottom half of image) occurs with green chlorite and black magnetite (top of image). The chlorite may indicate near-greenschist facies conditions or may reflect a mafic sedimentary protolith. Plane-polarized light, 2× magnification, field of view = 7 mm.

D Prehnite–pumpellyite facies pelite from Sardinia (Italy). Same image as (C) but with crossed polarizers. Epidote shows characteristic nonuniform birefringence, chlorite is purple–brown (indicating high Fe content). Cross-polarized light, 2× magnification, field of view = 7 mm.

Greenschist Facies

Metamorphic series: *predominantly medium-pressure associated with regional metamorphism*

Greenschist facies metamorphism occurs at about 300 °C and 100 MPa up to ~500 °C and 1,000 MPa, and forms part of both the low-pressure and medium-pressure metamorphic series. In both series it is bound by the lower-temperature prehnite–pumpellyite facies and at higher temperature transitions into the amphibolite facies. It is generally absent from the high-temperature hornfels metamorphic series.

If the protolith is ultramafic, principal greenschist facies minerals may include serpentine, brucite (Mg-hydroxide), olivine (forsterite), and the calcic-amphibole tremolite. When the protolith is mafic, common metamorphic minerals include chlorite, epidote, and actinolite. In an igneous or sedimentary protolith with increasing quartz and feldspar, common alteration reactions may produce muscovite and related sheet silicates via reactions (1) and (2).

$$\underset{\textit{K-feldspar}}{3KAlSi_3O_8} + 2H^+ = \underset{\textit{muscovite}}{KAl_3Si_3O_{10}(OH)_2} + 6SiO_2 + 2K^+ \quad (1)$$

$$\underset{\textit{anorthite}}{CaAl_2Si_2O_8} + 2K^+ + 4H^+ = \underset{\textit{muscovite}}{2KAl_3Si_3O_{10}(OH)_2} + 3Ca^{2+} \quad (2)$$

Consequently, quartzofeldspathic sediments and their metamorphosed equivalents are often richer in SiO_2 and have a higher ratio of Al_2O_3 to alkalis than their protolith, whereas immature sediments from more mafic sources tend to retain higher FeO and MgO.

If significant non-carbonate minerals are present in the protolith, calcium, magnesium, or calcium–magnesium silicates may form (e.g. diopside, forsterite, tremolite), along with Ca-rich garnet and phlogopite. For example, a protolith containing potassium feldspar may generate phlogopite during metamorphism, a protolith containing clay can produce aluminous minerals, or a protolith containing iron oxides may produce Fe-bearing minerals. Calcite, dolomite, potassium feldspar, and quartz may be present in rocks of any grade, if the protolith bulk-rock composition is suitable.

Metamorphism of carbonate rocks generally results in grain-coarsening, but this depends strongly on the composition of H_2O–CO_2 fluids as *decarbonation*, *dehydration*, and *mixed-volatile reactions* may entirely consume calcite and dolomite. Reactions associated with the higher temperatures of contact metamorphism (sandinite hornfels facies) generate more unusual minerals, such as åkermanite, tilleyite, larnite, etc. Calc-silicate rocks known as skarns are typically the result of metasomatism along the contact of intrusions and their country rock, and they may host economic ore deposits.

Greenschist facies as a function of protolith composition

Protolith	Rock type	Mineral parageneses
Ultramafic	Peridotite, serpentinite	Serpentine group, brucite, forsterite, tremolite
Mafic	Basalt, andesite, gabbro, diorite	Chlorite, actinolite, epidote, albite, quartz
Pelitic	Shale, mudstone	Chlorite, muscovite, albite, biotite, garnet, quartz
Quartzofeldspathic	Sandstone, rhyolite, granite, chert	Muscovite, plagioclase, quartz, titanite
Calcareous	Limestone, dolomite, marl	Calcite, dolomite, muscovite, quartz

Figure 3.21

A Greenschist facies metapelite from Sardinia (Italy). The garnet porphyroclast (center) contains quartz inclusions (clear) and is fractured. The foliation wraps around the garnet and is defined by muscovite (colorless), biotite (brown), and elongate ilmenite (black). The colorless crystals in the matrix are quartz. Plane-polarized light, 2× magnification, field of view = 7 mm.

B Greenschist facies metapelite from Sardinia (Italy). Same image as (A) but under crossed polarizers. The isotropic garnet porphyroclast has quartz inclusions at a high angle to the foliation. Muscovite has high birefringence and quartz low birefringence. Cross-polarized light, 2× magnification, field of view = 7 mm.

C Calc-schist from Merano (Italy). The rock is dominated by calcite with typical cleavage. Muscovite (colorless) and aligned, prismatic epidote (high relief and brown color) define the foliation (best seen in XPL). Plane-polarized light, 2× magnification, field of view = 7 mm.

D Calc-schist from Merano (Italy). Same image as (C) but under crossed polarizers. Calcite shows its characteristic high birefringence, muscovite has blue colors, and epidote shows a high but variable birefringence. The foliation is defined by the alignment of muscovite and epidote. Cross-polarized light, 2× magnification, field of view = 7 mm.

Amphibolite Facies

Metamorphic series: *medium-pressure series associated with regional metamorphism*

Amphibolite facies metamorphism is represented by a range of pressures and temperatures most often associated with regional metamorphism (the medium-pressure series). In P–T space amphibolite facies extends from ~500 °C at 200 MPa to ~700 °C at 1,100 GPa. At lower temperatures it is bound by the greenschist facies and at higher temperature by granulite facies. At higher pressures and temperatures amphibolite facies gives way to the medium-temperature eclogite facies by the disappearance of hornblende and the appearance of omphacitic pyroxene. The transition to amphibolite facies from greenschist facies associated with increasing temperature is typically accompanied by the appearance of calcic-plagioclase in place of albite and the disappearance of epidote. With increasing pressure muscovite forms at the expense of biotite and K-feldspar.

While remnant protolith mineralogy at amphibolite facies is rare, it is more common to find phenocrysts of olivine, pyroxene, plagioclase, or even magmatic amphibole, pseudomorphed by hornblende. Original magmatic textures, especially crude magmatic layering in layered intrusions, may also be preserved.

Amphibolite facies metamorphic terranes are associated with a variety of tectonic regimes, oceanic subduction to continental subduction to orogenic rifting. Amphibolite facies rocks can be produced by hydration of pre-existing rocks at high temperatures and crustal depths of <80 km during different types of orogenesis. Amphibolite facies rocks were originally considered a prograde metamorphic product of greenschist facies rocks under conditions of elevated pressure and temperature. Today there are numerous studies linking amphibolite facies rocks with retrograde metamorphism of granulite facies and eclogite facies rocks. During collisional orogeny, prograde metamorphism may be indicated by a series of processes related to increasing temperature, whereas retrograde metamorphism is indicated by processes associated with decreasing temperature (regardless of changes in pressure). On the other hand, prograde metamorphism may be characterized by increasing pressure, with retrograde metamorphism reflecting decreasing pressure (regardless of changes in temperature). Ultra high-pressure terranes often experience an increase in temperature during decompressional exhumation, resulting in a specific style of retrograde metamorphism associated with ultra high-pressure assemblages: granulite facies dehydration occurs at depth, while amphibolite facies hydration occurs at shallower crustal levels.

Amphibolite facies as a function of protolith composition

Protolith	Rock type	Mineral parageneses
Ultramafic	Peridotite, serpentinite	Talc, forsterite, anthophyllite, tremolite, orthopyroxene
Mafic	Basalt, andesite, gabbro, diorite	Hornblende, plagioclase, quartz, garnet
Pelitic	Shale, mudstone	Muscovite, biotite, quartz, garnet, plagioclase, staurolite, kyanite, sillimanite
Quartzofeldspathic	Sandstone, rhyolite, granite, chert	Quartz, plagioclase, potassium feldspar, garnet, biotite, muscovite, ± hornblende, ± clinopyroxene
Calcareous	Limestone, dolomite, marl	Calcite, dolomite, quartz, biotite, amphibole, diopside, potassium feldspar, wollastonite

Figure 3.22

A Amphibolite facies peridotite, Sondrio (Italy). Predominantly olivine (colorless to "dusty" from alteration and tiny black spinel inclusions) with granoblastic texture. Note the straight grain boundaries right of center. Anthophyllite (Mg-amphibole) is prismatic to acicular, colorless, and fractured perpendicular to its long axis. Plane-polarized light, 2× magnification, field of view = 7 mm.

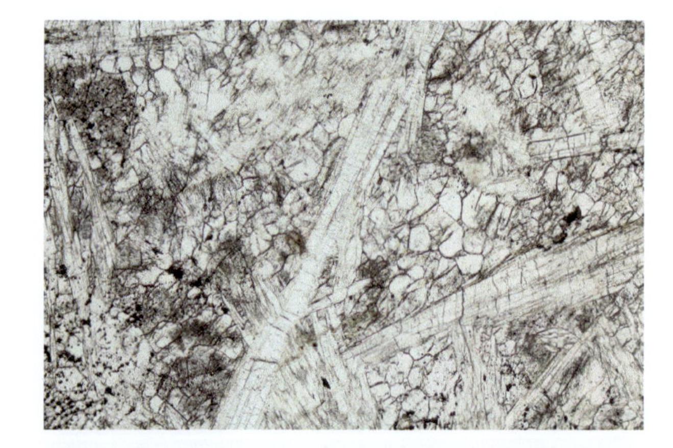

B Amphibolite facies peridotite, Sondrio (Italy). Image same as (A) but with crossed polarizers. Both olivine and anthophyllite have high birefringence but their crystal habits are distinct. Cross-polarized light, 2× magnification, field of view = 7 mm.

C Amphibolite facies quartzofeldspathic schist, Darjeeling (India). The image contains a large, pale brown sillimanite fractured perpendicular to its long axis (diagonal crystal, lower-right), biotite (red–brown), quartz and feldspar (both colorless). Plane-polarized light, 2× magnification, field of view = 7 mm.

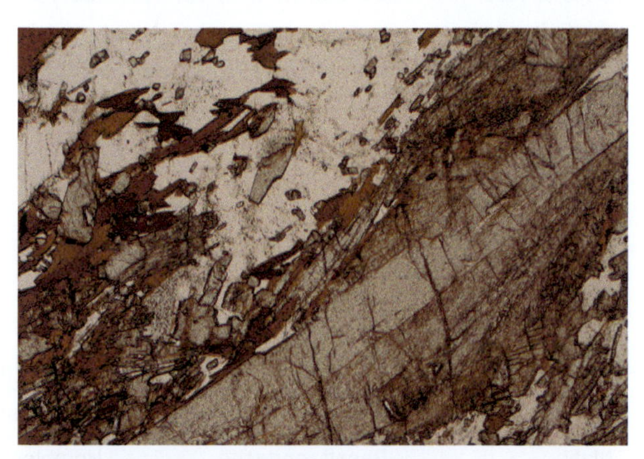

D Amphibolite facies quartzofeldspathic rock, Darjeeling (India). Same image as (C) but with crossed polarizers. Sillimanite shows orange–yellow birefringence and biotite a deeper red birefringence. Plagioclase shows typical gray birefringence. Cross-polarized light, 2× magnification, field of view = 7 mm.

BOX 3.9 Transport and Diffusion

Intergranular fluids are chemical transport agents that facilitate the metamorphic reactions taking place. Fluids not only host dissolved material, but as transport agents, move the atoms, ions, and molecules through the system and thereby speed up chemical reactions. These chemical constituents (1) move from minerals that are breaking down into the fluid, or (2) move from the fluid to the new metamorphic minerals that are growing. This generally describes prograde dehydration and retrograde rehydration metamorphic reactions, respectively.

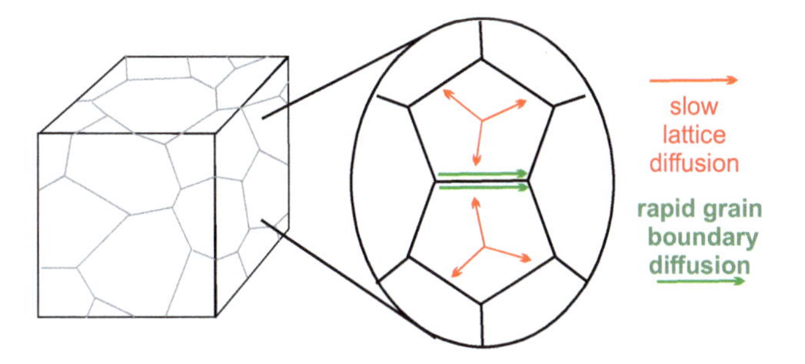

During dehydration or rehydration reactions, atoms or molecules are transported via lattice diffusion (also called volume or intragranular diffusion) or grain boundary diffusion. While lattice diffusion is more effective at higher temperatures, grain boundary diffusion is orders of magnitude faster than lattice diffusion. This is shown schematically in the above image – physically moving atoms across point defects *within* the crystal lattice of a mineral (lattice diffusion, in red) is slower and less efficient than transporting atoms in a fluid migrating along grain boundaries (in green). In this context, grain boundaries act as "highways" for fluid motion and chemical exchange, promoting the rapid transfer of chemical constituents from mineral to fluid (or vice versa).

It is difficult to initiate chemical change without a liquid transporting agent since lattice diffusion is a slow and inefficient process. Minerals reacting with fluids often have irregular, corroded boundaries, while new growing minerals have "convex outward" planar faces. In this olivine marble, grain boundary diffusion is occurring between carbonate (grayish) and forsteritic olivine (central crystal with high birefringence), producing an irregular rim of variably developed serpentine (fibrous, blue–gray overgrowth). Sample from Calabria (Italy), 10× magnification, field of view 2 mm.

BOX 3.10 Fluids, Temperature, and Metamorphism

Fluid plays an important role in metamorphism via reactions associated with hydration/dehydration, carbonation/decarbonation, infiltration, and/or dissolution/reprecipitation.

- Hydration/dehydration: consumes (hydration) or liberates (dehydration) H_2O, associated with decreasing and increasing temperature, respectively.
- Carbonation/decarbonation: consumes (carbonation) or liberates (decarbonation) CO_2 associated with decreasing and increasing temperature, respectively.
- Infiltration: introduction of a fluid from an external source – for example, contact metamorphism.
- Solution/reprecipitation: reacting minerals dissolve into a fluid at one location, while simultaneously precipitating from the fluid at another.

The presence of fluid, which may or may not be water, facilitates element mobility, deformation, and metamorphism. Most rocks contain some water/fluid, either in the minerals themselves, in the pore space that exists between grains, or in the tiny fractures present in the rock. Such fluid typically consists of water with CO_2 (or other dissolved gases) and dissolved ions, and is influenced by the composition of the mineral grains with which it is in direct contact.

As the metamorphic temperature and pressure of a rock changes, the composition of the fluid must also change. As a rock is buried and pressure and temperature increase (prograde metamorphism), hydrous minerals such as amphibole change to anhydrous minerals such as pyroxene. It follows that higher-grade metamorphic rocks have fewer hydrous minerals. As deeply buried rocks are exhumed to shallower crustal levels, anhydrous minerals like pyroxene can be hydrated to form amphibole; this is consistent with the link between hydrous mineralogies in medium- to lower-grade metamorphic rocks. The image below illustrates fluid infiltration.

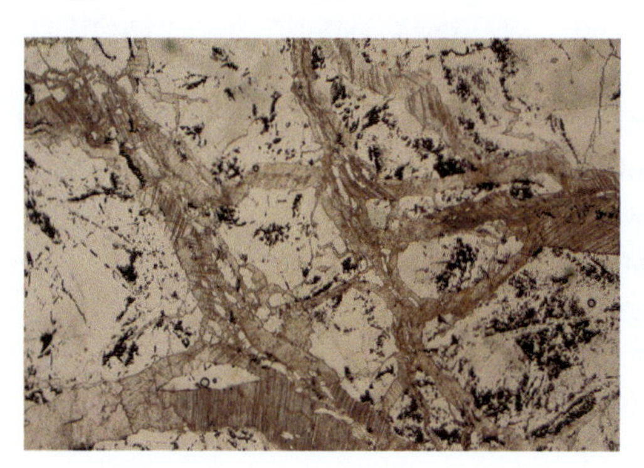

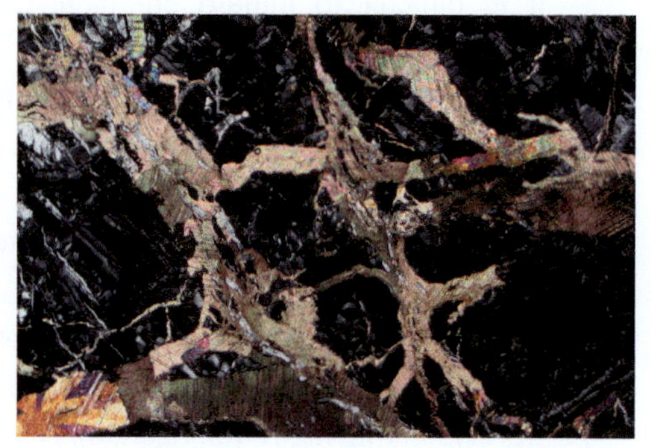

Ophicarbonate from the Votri complex, Liguria (Italy). These images show a complex fracture network in serpentinite. (*Left*, plane-polarized light) Secondary carbonate (pale brown, moderate relief) reflects extensive circulation of hydrothermal fluids; serpentine is colorless and the opaque material is magnetite. (*Right*, cross-polarized light) The same image showing the very high birefringence (pastel colors) of carbonate-filling veins, while the host serpentine has low, first-order birefringence. Both images 2× magnification, field of view = 7 mm.

Granulite Facies

Metamorphic series: *low- and medium-pressure series*

Granulite facies metamorphism occurs at high temperature (>700 °C) and a wide range of pressures associated with contact (low-pressure series) and regional (medium-pressure series) metamorphism (<50 km depth). In P–T space the granulite facies field extends from approximately 700 °C at 200 MPa to ~1,000 °C at 1,400 GPa. At lower temperature it is bound by amphibolite facies conditions and at lower pressure by hornfels facies conditions. At higher pressures, granulite facies gives way to the high-temperature eclogite facies. Mafic granulites can be subdivided into low-pressure (olivine, plagioclase), medium-pressure (orthopyroxene, clinopyroxene, plagioclase, garnet, hornblende), and high-pressure (garnet, clinopyroxene, plagioclase, hornblende) types.

During regional metamorphism of a basaltic protolith, the transition from amphibolite facies to granulite facies is represented by the breakdown of amphibole and the appearance of pyroxene (a mineralogy similar to the mafic igneous protolith). At higher temperatures, dehydration melting may generate an anhydrous (granulitic) residue after removal of the melt. In quartzofeldspathic rocks this might involve the dehydration of biotite via reactions such as:

Bt + Pl + Qz = Opx + Crd + Kfs + liquid (~ 800–900 °C, 300 MPa)

Bt + Pl + Qz = Opx + Grt + Kfs + liquid (~890–975 °C, 1,000 MPa)

The role of fluids (H_2O, CO_2-rich fluids, or brines) can be locally important. At granulite facies the infiltration of fluids (or external melt) can produce highly migmatized quartzofeldspathic gneisses with abundant leucosome, whereas dehydration melting of a quartzofeldspathic protolith produces anhydrous minerals overprinting the original rock fabric.

Granulite facies dehydration occurs at depth under ultra high-pressure conditions. In the Gneiss–Eclogite Unit of the Erzgebirge (Saxony, Germany), early high-pressure metamorphism (≤830 °C and 2.1 GPa) was followed by decompression, deformation, and channelized fluid flow that resulted in hydration and partial melting (migmatite formation at 600–700 °C and 0.7–1.0 GPa). This metamorphic sequence is attributed to continental collision followed by orogenic collapse. Granulite facies rocks may also be overprinted under eclogite facies conditions during prograde processes.

Granulite facies as a function of protolith composition

Protolith	Rock type	Mineral parageneses
Ultramafic	Peridotite, serpentinite	Forsterite, orthopyroxene, clinopyroxene
Mafic	Basalt, andesite, gabbro, diorite	Clinopyroxene, orthopyroxene, plagioclase, quartz, garnet
Pelitic	Shale, mudstone	Quartz, potassium feldspar, plagioclase, sillimanite, garnet, biotite, orthopyroxene, cordierite (UHT = sapphirine + quartz)
Quartzofeldspathic	Sandstone, rhyolite, granite, chert	Orthopyroxene, clinopyroxene, garnet, potassium feldspar, cordierite (low-pressure), sillimanite or kyanite (high-pressure)
Calcareous	Limestone, dolomite, marl	Calcite, dolomite, quartz, diopside, potassium feldspar, wollastonite, forsterite

Figure 3.23

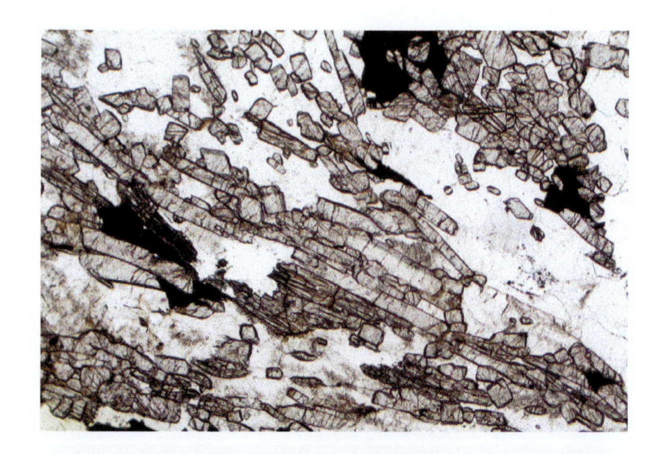

A Granulite facies pelitic protolith from Monterosso Calabro (Italy). Sillimanite (prismatic, high relief, fractures perpendicular to elongation) defines a coarse foliation. The square crystals are basal sections of sillimanite. Quartz (colorless), K-feldspar ("dusty" appearance), and magnetite (black) are also present. Plane-polarized light, 2× magnification, field of view = 7 mm.

B Granulite facies pelitic protolith from Monterosso Calabro (Italy). Same image as (A) but with crossed polarizers. Sillimanite shows medium–high birefringence (note basal sections show lower birefringence). Quartz (right) shows gray–white birefringence, while K-feldspar (lower-left) has lower birefringence than quartz. Plagioclase shows typical gray birefringence. Cross-polarized light, 2× magnification, field of view = 7 mm.

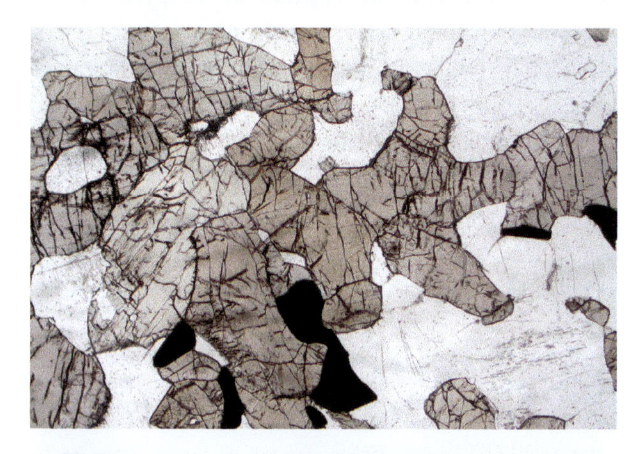

C Granulite facies quartzofeldspathic protolith from Madras (India). The sample is defined by orthopyroxene (weak pleochroism in pale browns, fractured), plagioclase (colorless), and quartz (indistinguishable in PPL). Plane-polarized light, 2× magnification, field of view = 7 mm.

D Granulite facies quartzofeldspathic protolith from Madras (India). Same image as (C) but with crossed polarizers. Orthopyroxene has low birefringence (yellow to gray), plagioclase is gray with polysynthetic twins, and untwinned quartz with gray–white birefringence (bottom center). Cross-polarized light, 2× magnification, field of view = 7 mm.

Blueschist Facies

Metamorphic series: *high-pressure/low-temperature series associated with subduction*

Blueschist facies metamorphism occurs over a broad P–T environment and is linked to the high-pressure metamorphic series. In P–T space the blueschist facies extends from ~100 °C at 600 MPa to ~550 °C at 1,500 GPa, and to 400 °C at 2,000 MPa. The blueschist facies is bound by the lower-pressure zeolite, subgreenschist (pumpellyite–actinolite), and greenschist facies, by the higher-temperature amphibolite facies, and at higher pressures and temperatures by the eclogite facies. With increasing *temperature*, glaucophane breaks down to form chlorite, lawsonite to epidote, and glaucophane reacts with lawsonite to form actinolite; thus, blueschist facies is replaced by greenschist facies. With increasing *pressure*, glaucophane reacts with albite to form omphacite, and chlorite breaks down to form garnet; thus, blueschist facies is transformed to eclogite facies.

Ultramafic rocks with typically anhydrous mineral assemblages must be hydrated (e.g. during ocean-floor metamorphism) for metamorphic reactions to occur. Blueschist facies metamorphism of an ultramafic protolith will produce serpentine minerals, brucite, dolomite, and magnesite, whereas a basaltic protolith may generate glaucophane, lawsonite, or epidote. A pelitic protolith will produce glaucophane, phengite, lawsonite, or epidote, whereas a quartzofeldspathic protolith is likely to generate lawsonite and muscovite. A carbonate protolith will produce glaucophane, aragonite, dolomite, phengite, and epidote. Most protoliths produce chlorite and albite at lower pressures, and sodic clinopyroxene at higher pressures. The dehydration of lawsonite commonly generates a pseudomorph of epidote, muscovite, and albite.

Blueschist facies high-pressure–low-temperature environments are represented by subduction zones, which are dominated by the low geothermal gradients (4–14 °C km^{-1}) needed to produce blueschist facies minerals. Blueschist facies rocks are stable in subduction zones at depths of 25–60 km; at greater depths/higher pressures blueschist facies rocks are transformed to eclogite facies rocks. The preservation of blueschist facies rocks requires rapid exhumation, otherwise the rocks retrogress as they pass from high- to low- pressure, or from lower- to higher-temperature environments; the latter often results in greenschist facies overprinting during metamorphic retrogression.

Blueschist facies as a function of protolith composition

Protolith	Rock type	Mineral parageneses
Ultramafic	Peridotite, serpentinite	Serpentine group, mica group, talc, epidote, iron oxides
Mafic	Basalt, andesite, gabbro, diorite	Alkali-amphibole (mostly glaucophane), lawsonite, epidote, jadeite, phengite, chlorite, garnet, quartz
Pelitic	Shale, mudstone	Alkali-amphibole, lawsonite, epidote, jadeite, carpholite, chloritoid, talc, muscovite, chlorite, garnet, albite, aragonite, quartz
Quartzofeldspathic	Sandstone, rhyolite, granite, chert	Glaucophane, lawsonite, muscovite, jadeite, chlorite, paragonite, garnet, feldspar, quartz
Calcareous	Limestone, dolomite, marl	Aragonite/calcite, dolomite, muscovite

Figure 3.24

A Blueschist from Cervinia (Italy). Large glaucophane (dark blue) with fractures perpendicular to the long axis and epidote (fractured, brownish), set in a schistose matrix of muscovite (transparent) and glaucophane (gray–blue). Minor chlorite also present. Muscovite content is consistent with a pelitic protolith. Plane-polarized light, 2× magnification, field of view = 7 mm.

B Blueschist from Ghinivert (Italy). Prismatic glaucophane (deep blue) is aligned horizontally and defines a coarse foliation alternating with quartz-rich (colorless) layers. Minor green chlorite is present; the brown crystals in the upper part of the image are stilpnomelane, which indicates a protolith with high iron content. The quartz layers suggest a quartzofeldspathic protolith. Plane-polarized light, 10× magnification, field of view = 2 mm.

C Blueschist from Ghinivert (Italy). Same image as (B) but with crossed polarizers. Almost all glaucophane is at extinction; some crystals are oriented diagonally and show low birefringence. Quartz is granoblastic with interlobate grain boundaries. Cross-polarized light, 10× magnification, field of view = 2 mm.

Eclogite Facies

Metamorphic series: *the high-pressure series*

Eclogite facies metamorphism requires pressures in excess of ~1,200 MPa and a low (5–10 °C km^{-1}) geothermal gradient. At high pressures, plagioclase breaks down completely: The albite component enters into omphacite and the anorthite component contributes to epidote or garnet, the latter being Mg-rich (unlike garnet in other facies). In quartzofeldspathic rocks, the sheet silicates biotite and chlorite are replaced by phengitic white mica and sometimes talc, but in water-absent conditions such transformations may be absent.

At lower temperatures eclogite facies is bound by blueschist facies, and at lower pressures by amphibolite and granulite facies. Medium- and high-temperature eclogite facies are informally recognized at a temperature of ~700 °C. A close spatial association of eclogite and blueschist or amphibolite facies rocks is not uncommon, but it is only recently that the significance of fluid(s) in this context has been understood. The retrogression of metabasic eclogite requires the addition of water, whereas metafelsic eclogite is susceptible to spontaneous devolatilization reactions of micaceous minerals on decompression. This may explain why some quartzofeldspathic rocks are "metastable" during high-pressure/ultra high-pressure metamorphism.

The eclogite stability field can exceed pressures of 2,000 MPa, conditions of the so-called *ultra high-pressure* rocks. Inclusions of ultra high-pressure minerals such as coesite (>2,600 MPa) or microdiamond (>3,500 MPa) imply much higher pressures than previously understood. During ultra high-pressure metamorphism at depths of >100 km, quartz converts to its high-pressure polymorph coesite (e.g. the Western Gneiss Region [Norway]). The reverse transformation is rapid and coesite is only preserved if protected as an inclusion within a strong mineral host (i.e. garnet or zircon). Though ultra high-pressure minerals like coesite are relatively rare, evidence of their previous existence may be preserved: The conversion of coesite to quartz during decompression involves volume expansion which results in radial fractures forming in the host mineral around the ultra high-pressure mineral inclusion. In addition, during retrograde metamorphism coesite quickly inverts back to quartz, often generating a distinctive palisade texture. Both radial fractures in the host mineral and the palisade texture of the coesite pseudomorph provide evidence of ultra high-pressure processes.

Eclogite facies as a function of protolith composition

Protolith	Rock type	Mineral parageneses
Ultramafic	Peridotite, serpentinite	Olivine, pyroxene, garnet
Mafic	Basalt, andesite, gabbro, diorite	Garnet, omphacite, quartz, kyanite, rutile, coesite (in UHP rocks)
Pelitic	Shale, mudstone	Phengite, quartz, omphacite, garnet, kyanite, rutile, coesite/diamond (in UHP rocks)
Quartzofeldspathic	Sandstone, rhyolite, granite, chert	Jadeite, muscovite, garnet, phengite, glaucophane, potassium feldspar, quartz
Calcareous	Limestone, dolomite, marl	Aragonite, dolomite, omphacite, epidote, quartz, phengite, garnet, coesite/diamond (in UHP rocks)

Figure 3.25

A Eclogitic granite from Monte Mucrone (Italy). Igneous plagioclase is replaced by a fine-grained, brownish aggregate of jadeite, zoisite, and quartz. This is surrounded by a narrow corona of jadeite (light gray), followed by a corona of garnet (darker outline). Biotite (orange–red) also has coronas of garnet and is replaced in its central parts by muscovite (colorless). The surrounding colorless region is mostly quartz. Plane-polarized light, 10× magnification, field of view = 2 mm.

B Eclogitic granite from Monte Mucrone (Italy). Same image as (A) but with crossed polarizers. The minerals replacing igneous plagioclase are not distinguishable, but the coronas of jadeite and garnet (isotropic) are. Biotite shows high birefringence. Quartz is recrystallized with 120° grain boundaries. Cross-polarized light, 10× magnification, field of view = 2 mm.

C Eclogite facies whiteschist from Dora Maira (Italy). In this ultra high-pressure rock a large pyrope hosts numerous coesite + quartz inclusions. The higher relief of coesite is distinguishable next to quartz. The fracture in the host garnet results from the volume expansion of the coesite to quartz transition during decompression. Kyanite is also present (small crystal with good cleavage, center-left). Plane-polarized light, 10× magnification, field of view = 2 mm.

D Eclogite facies whiteschist from Dora Maira (Italy). Same image as (C) but with crossed polarizers. The pyrope is isotropic. During retrogression, coesite quickly inverts, forming a radiating rim of palisade quartz – easily seen as coesite has lower birefringence (and higher relief) than quartz. Kyanite shows brown–orange birefringence. Cross-polarized light, 10× magnification, field of view = 2 mm.

Hornfels Facies

Metamorphic **series:** *low pressure, high temperature*

Hornfels facies reflects metamorphism at low pressure (approximately <200 MPa) and high temperature (up to >1,000 °C) – that is, contact metamorphism. Rocks of the hornfels facies are commonly hard, dark, and often fine-grained, although "spotted" slates and schists are occasionally present. The high temperatures associated with hornfels facies promote recrystallization of existing minerals and crystallization of new minerals, and this process commonly results in porphyroblasts or relict phenocrysts set in a finer-grained granoblastic matrix. The pre-existing structure of the parent rock is only preserved at lower temperatures.

The mineralogy associated with hornfels facies is a function of temperature and protolith composition. Contact metamorphism of a mafic protolith (mafic hornfels) may contain feldspar, hornblende, ± pyroxene. Pelites and clay-rich rocks yield biotite hornfels. Limestone and dolomite generate calc-silicate hornfels (marbles), and quartz-rich sandstones produce quartzites. Cordierite and andalusite are stable hornfels facies minerals that are absent from higher-pressure settings.

The uppermost part of continental crust is dominated by (meta)sediment, so metapelite hornfels is common where plutons intrude shallow crustal levels. The highest temperatures are recorded adjacent to the pluton and the metamorphic halo (aureole) extends several meters to a few kilometers from the intrusion, decreasing in temperature with distance. Several thermal regimes are defined in association with contact metamorphism of mafic country rocks that may involve partial melting reactions at the highest temperatures (sanidinite hornfels facies).

Hornfels facies as a function of protolith composition

Protolith	Rock type	Mineral parageneses
Ultramafic	Peridotite, serpentinite	Serpentine group minerals, talc, tremolite, chlorite, hornblende, forsterite, pyroxene
Mafic	Basalt, andesite, gabbro, diorite	Feldspar, epidote, hornblende, pyroxene, garnet, titanite, biotite
Pelitic	Shale, mudstone	Quartz, feldspar, biotite, andalusite, sillimanite, kyanite, cordierite, pyroxene, sapphirine
Quartzofeldspathic	Sandstone, rhyolite, granite, chert	Quartz, plagioclase, potassium feldspar, garnet, biotite, muscovite
Calcareous	Limestone, dolomite, marl	Calcite, dolomite, quartz, tremolite, talc, forsterite, diopside, wollastonite, monticellite, åkermanite

Thermal regimes and diagnostic minerals of hornfels facies

Regime	Temperature	Facies	Pelitic protolith
Low-temperature	300–500 °C	Albite, epidote	Biotite, muscovite, quartz
Medium-temperature	500–650 °C	Hornblende	Cordierite, andalusite, biotite
High-temperature	650–800 °C	Pyroxene	Potassium feldspar, cordierite, orthopyroxene
Ultra high-temperature	>800 °C	Sanidinite	Cordierite, sanidine, tridymite, corundum, spinel, orthopyroxene

Figure 3.26

A Albite–epidote hornfels (pelitic protolith) from Val Lanterna (Italy). The sample is characterized by epidote (high relief, pale brown, fractured), quartz (colorless), and hornblende (prismatic, green). The rock has a massive structure (lacking a preferred orientation). Plane-polarized light, 2× magnification, field of view = 7 mm.

B Albite–epidote hornfels (pelitic protolith) from Val Lanterna (Italy). Same image as (A) but with crossed polarizers. Epidote shows characteristic high birefringence, quartz has interlobate grain boundaries and white–gray birefringence, while hornblende shows low colors. Cross-polarized light, 2× magnification, field of view = 7 mm.

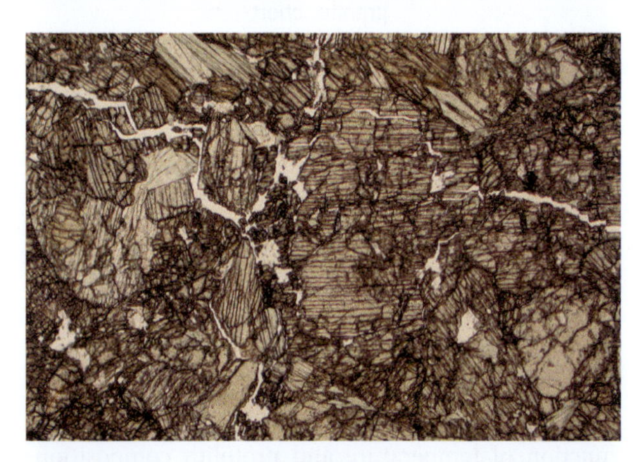

C Pyroxene hornfels (mafic-ultramafic protolith) from Siberia (Russia). The sample is dominated by clinopyroxene (brownish, pronounced cleavage, fracture), with minor calcite (pale, lower relief, typical cleavage, center-left) and minor biotite (pale green, bladed habit, upper-left). The white stripes are fractures generated during thin-section preparation. Plane-polarized light, 2× magnification, field of view = 7 mm.

D Pyroxene hornfels (mafic-ultramafic protolith) from Siberia (Russia). Same image as (C) but with crossed polarizers. Clinopyroxene and biotite show high birefringence, while calcite (center-left) shows very high (pastel) birefringence. Cross-polarized light, 2× magnification, field of view = 7 mm.

APPLICATION 3.3

Pressure and Temperature Paths

Thermobarometry is the estimation of the temperature (T) and pressure (P) recorded in a rock. Metamorphic changes often record the P–T environment in which these changes occurred (i.e. increasing [prograde] or decreasing [retrograde] P–T paths), and this is especially pronounced during orogenesis.

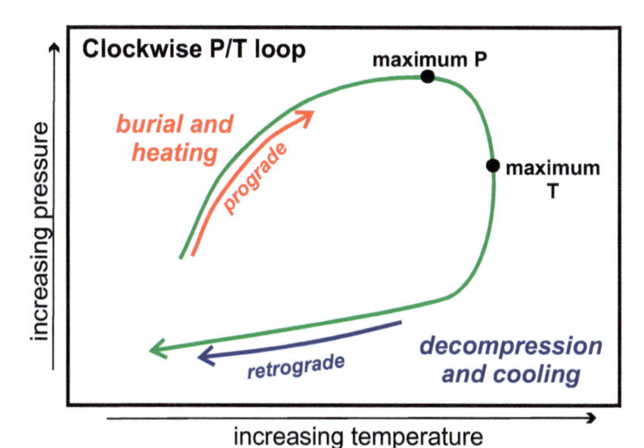

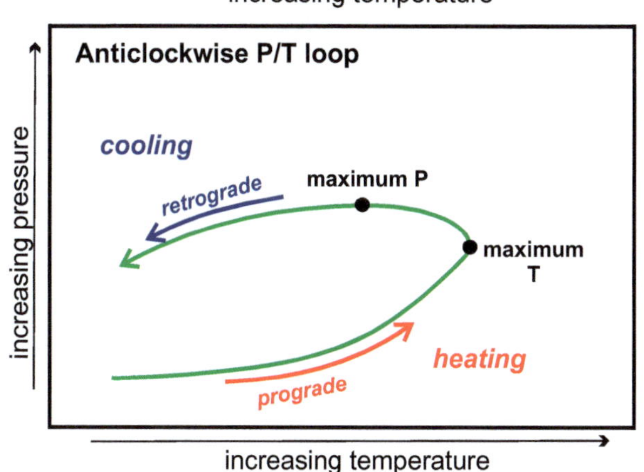

P–T loops may be considered clockwise or anticlockwise, and depend on the relative influence and timing of crustal thickening, heating, and cooling. Metamorphic conditions may reach a peak in P before T (*clockwise*), or they may record a peak in T before P (*anticlockwise*). In a clockwise P–T path, pressure and temperature increase during prograde metamorphism until a maximum P is reached; due to the poor thermal conductivity of rocks, changes in T typically lag behind changes in P. After prograde metamorphism, clockwise P–T paths are generally associated with *isothermal decompression,* followed by slower decompression and cooling along the retrograde path. Clockwise P–T loops are associated with crustal thickening. Anticlockwise P–T paths generally record a prograde path of heating during which the peak thermal maximum is achieved prior to maximum pressure. This is followed by *isobaric cooling* along the retrograde path. Anticlockwise paths are associated with igneous intrusions and environments of crustal thinning (e.g. hotspots and rift zones).

Careful petrographic analysis can be used to determine the minerals and textures that represent the different parts of the P–T path. Solid solution minerals (e.g. feldspar and garnet) are particularly useful if their compositional zoning preserves the different/changing P–T conditions in which they grew. Such chemical zoning can document the relative timing of mineral growth (earlier older cores followed by later younger rims). Given the refractory nature of garnet and its wide stability field, it is commonly used to document evolving P–T conditions.

Different tectonic environments are typified by distinct P–T paths. A recent study by Copley and Weller (2022) linked P–T paths with the thermal evolution of a thickening crust. They reproduced distinct thermal styles on the basis of P–T path geometry and concluded that thickening of the mid- to upper crust above a rigid lower crustal substrate could account for most orogenic scenarios. In addition, they determined that the curvature of the P–T loop as maximum temperature is approached is the most diagnostic feature to allow the thermal and tectonic history of a mountain belt to be estimated.

In the below figure, three P–T scenarios are summarized and each scenario correlates with distinct

Petrographic evidence for the P/T path

P/T path	Evidence
Prograde (pre-peak)	mineral inclusions, poikiloblastic textures – an earlier mineral formed at a lower P/T is trapped within a later mineral formed at a higher P/T; may be erased at high-T (granulite facies)
Peak	porphyroblastic textures – large euhedral crystals within a fine-grained matrix both formed during peak metamorphism
Retrograde (post-peak)	coronas, symplectites, cross-cutting relationships – minerals formed at lower P/T mantled by higher grade minerals; intergrowth of peak and retrograde minerals; peak minerals cut by retrograde minerals

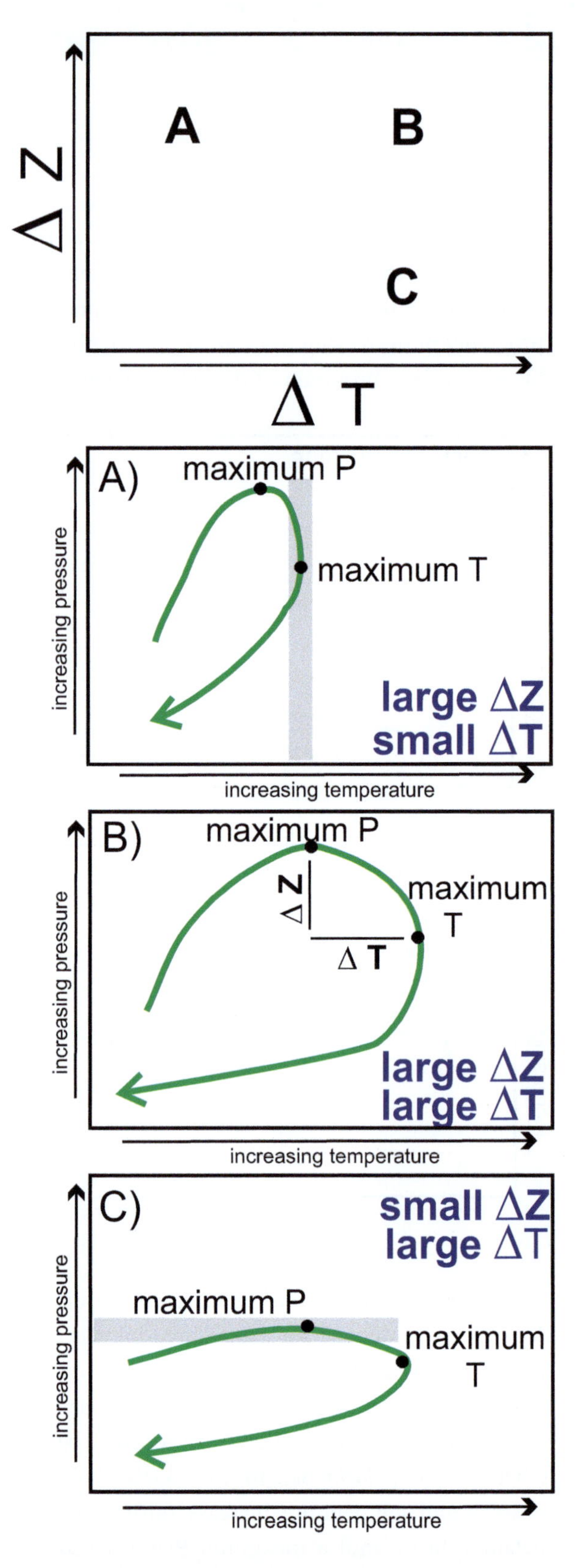

differences in depth/pressure (ΔZ) and temperature (ΔT). The temperature difference between the maximum pressure and maximum temperature points of each P–T loop is ΔT; ΔZ is the depth difference between these two points. All three P–T trajectories represent clockwise paths. Each style of thermal evolution documents distinct relative magnitudes of ΔT and ΔZ (as shown in B).

The first pane illustrates where the three characteristic P–T loops occur relative to each other in P–T space. The curvature of the P–T loop places direct constraints on the rates of radiogenic heating, thickening, erosion, and the initial crustal thickness associated with orogeny. Pane (A) represents isothermal decompression (highlighted by the gray bar). Such loops require the crust to have reached high temperatures while thick, with limited temperature change due to rapid exhumation. Pane (B) represents a broad P–T loop. Broad P–T loops reflect scenarios involving rocks that begin at deeper crustal levels prior to rapid thickening. Pane (C) represents isobaric heating (highlighted by the gray bar). This thermal style, also known as a hairpin P–T loop, is mainly related to local heating adjacent to intrusions.

The preserved metamorphic assemblage in a rock commonly relates to its peak P–T conditions because prograde metamorphism generally involves dehydration reactions. Retrograde metamorphism involves rehydration/revolatization reactions under decreasing

P–T which, in theory, allows prograde mineral assemblages to revert to those more stable at less extreme conditions. This is relatively uncommon, however, since volatiles released during prograde metamorphism usually migrate out of the rock and are no longer available to recombine with the rock during uplift/exhumation. Localized retrograde metamorphism usually occurs along fractures, for example, which form during exhumation/uplift and provide a pathway for fluids to re-enter the system.

3.4 Metamorphic Rocks

Schist

Mineralogy: Bt, Mu, Qz, Amp, Grt, Sil, St, Ky, ± accessories **ROOT**
IUGS classification: Schist (root name)
Texture: Strongly foliated and often crenulated

A schist is a strongly foliated and commonly porphyroclastic metamorphic rock. The term *schist* is a textural term that also encompasses slates and phyllites ("schistose rocks"). By definition, a schist is composed of >50% tabular, lamellar, or prismatic minerals coarse enough (>0.25 mm) to be seen in a hand specimen without a magnifying lens. A schist has a well-developed schistosity formed during dynamic metamorphism. It commonly has compositional layering which may have originated by metamorphic differentiation processes or been inherited from original sedimentary structures (i.e. bedding). The term schist does not imply any particular mineral composition as a schist may form from any protolith at any metamorphic grade. However, pelitic and quartzofeldspathic schists are among the most common and abundant types. The mineralogy of a schist depends largely on the protolith composition and the metamorphic grade, but the most common minerals are phyllosilicates (biotite, muscovite, and chlorite) and quartz; feldspar may be present in small amounts. The presence of index minerals, or characteristic minor constituents, should be added to the rock name (e.g. garnet micaschist, staurolite micaschist, hornblende schist, talc schist). As metamorphic grade increases, dehydration reactions convert the phyllosilicates in a schist to granular minerals, destroying the schistosity and generating a gneissic texture.

Occurrence

Schist (especially pelitic schist) forms due to regional metamorphism in convergent plate boundary environments or in contact metamorphic zones around intrusions. They are widespread in numerous orogens, and classic examples include the schists of the Barrovian metamorphic zones and the Buchan zones of the Scottish Highlands.

Garnet micaschist from Bhutan. The rock is characterized by a well-developed schistosity defined by white mica (grayish) and biotite (black). Lens-shaped quartz (vitreous) and feldspar (white) are present. The rounded reddish crystals are garnet porphyroclasts. Sample is ~10 cm wide.

Image courtesy of A. Tindle

Figure 3.27

A Quartz micaschist from Resia (Italy). The sample is folded and characterized by a muscovite-rich zone (lower half) which shows a well-developed crenulation cleavage, and a more quartz-rich region (upper half) with poorly developed crenulation cleavage. Quartz shows normal low birefringence and muscovite higher birefringence. Cross-polarized light, 2× magnification, field of view = 7 mm.

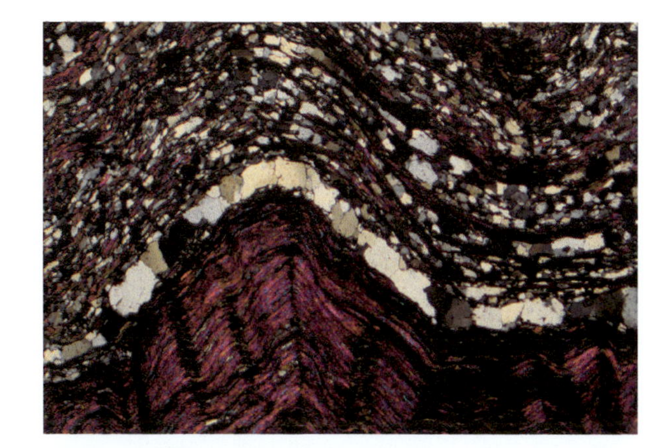

B Garnet micaschist from Dora Maira (Italy). The rounded central crystal is an isotropic garnet porphyroblast, wrapped by a coarse-grained schistosity defined by muscovite (high birefringence) and quartz (gray birefringence, with granoblastic texture). Some quartz inclusions in the garnet are visible. Cross-polarized light, 10× magnification, field of view = 2 mm.

C Chloritoid micaschist from Calabria (Italy). Muscovite-rich layers are interspersed with quartz-rich layers. The muscovite layers are crenulated. Chloritoid (prismatic, low birefringence) is also present. Cross-polarized light, 2× magnification, field of view = 7 mm.

D Calcite micaschist from Tyrol (Italy). This sample contains idioblastic calcite (gray–pink birefringence) wrapped by a coarse-grained schistosity defined by muscovite (high birefringence). Cross-polarized light, 2× magnification, field of view = 7 mm.

Gneiss

Mineralogy: Qz, Fsp, Bt, Crd, Mu, Amp, Px, Grt, Sil, St, Ky, ± accessories **ROOT**
IUGS classification: Gneiss (root name), poor or widely spaced schistosity
Fabric: Foliated (banded), granoblastic

Gneiss is a rock name based on texture. It describes an igneous (*orthogneiss*) or sedimentary (*paragneiss*) protolith that is metamorphosed at higher pressures and temperatures than a schist. A gneiss is a recrystallized, medium- to coarse-grained metamorphic rock that displays poor fissility, but has well-developed foliation in the form of compositional banding (*gneissic banding*) or augen (*augen gneiss*). There are diverse compositions of gneiss, which mirror the diverse composition of protoliths; gneiss mineralogy is equally varied, reflecting the diversity of protolith compositions. Orthogneiss is typically dominated by quartz and feldspar, whereas paragneiss often contains garnet and biotite. The formation of silicate minerals such as biotite, cordierite, sillimanite, kyanite, staurolite, andalusite, and garnet in a gneiss are consistent with an aluminous protolith. Such minerals are used to preface the name of the rock (e.g. biotite gneiss or garnet–staurolite gneiss).

The layering in a banded gneiss may reflect original compositonal layering of the protolith, but more commonly it reflects the segregation of quartz and feldspar from more mafic minerals such as garnet, pyroxene, amphibole, and mica during deformation at high temperature. An augen gneiss is a metamorphosed and deformed porphyritic magmatic rock in which original igneous phenocrysts (often feldspar) are deformed into lenticular or elliptical shapes (augen or porphyroclasts) within a finer-grained matrix of quartz and mica. During deformation and metamorphism, the stronger augen are less deformed than the weaker matrix.

At the highest metamorphic grades, gneiss formation approaches the zone of partial melting. Given the right protolith composition, migmatite formation can occur (i.e. a mixed rock with both a relict metamorphic structure and a newly crystallized melt).

Occurrence

Gneiss forms due to regional metamorphism near convergent plate boundaries where amphibolite facies to granulite facies conditions exist (>600 °C at 200–1,400 MPa) and can also occur under high-pressure conditions such as eclogite gneiss (e.g. Western Gneiss region [Norway]). Gneiss forms the core of stable continental shields, especially older cratons that are now exposed at the surface. Gneiss formation may reflect multiple orogenic events in which the first event produces granitic basement and the second deforms and melts this basement, producing a gneiss dome.

Banded, garnet-bearing gneiss with felsic (quartz and feldspar) and mafic (biotite-rich) layers. Sample is ~11 cm wide.

Image courtesy of J. St. John

Figure 3.28

A Orthogneiss from Ivrea Verbano (Italy). Gneissic banding is defined by the alternation of felsic (quartz and plagioclase) and mafic (brown biotite and colorless muscovite) layers. Two rounded garnets are present (one in the upper-right and one in the upper part of the central biotite layer). Plane-polarized light, 2× magnification, field of view = 7 mm.

B Orthogneiss from Ivrea Verbano (Italy). Same image as (A) but with crossed polarizers. Quartz shows gray birefringence and interlobate grain boundaries. Muscovite shows blue and biotite pinkish birefringence. Plagioclase is granoblastic and garnet is isotropic. Cross-polarized light, 2× magnification, field of view = 7 mm.

C Paragneiss from Val di Pejo (Italy). A coarse foliation is defined by layers of biotite + epidote and quartz + feldspar. Biotite is golden-brown and epidote has high relief. Quartz is colorless, while feldspar appears "dusty" due to alteration. Plane-polarized light, 2× magnification, field of view = 7 mm.

D Paragneiss from Val di Pejo (Italy). Same image as (C) but with crossed polarizers. Epidote (zoisite) has diagnostic, deep blue anomalous birefringence, quartz has higher birefringence (whitish) than feldspar (grays), and biotite has high birefringence (pinks). Cross-polarized light, 2× magnification, field of view = 7 mm.

Granofels

Mineralogy: *Qz, Fsp, Grt, Px, Amp, ± Ph, ± Zo*

ROOT

IUGS classification: *Granofels (root name), lacks fabric*

Fabric: *Granular*

Granofels is the name used to describe a granular metamorphic rock with a granoblastic texture, devoid of schistosity. More specific rock names always take precedence and should be used instead of granofels whenever applicable. For example, a marble is also a granofels, but the name "marble" provides more information and is therefore preferred.

Granofels can be massive or layered (e.g. banded granofels). Protolith composition determines granofels mineralogy, but quartz, feldspar, pyroxene, and/or garnet are typical and visible without magnification. Granofels lack schistosity due to equant (rather than elongate) minerals or the random arrangement of inequant minerals, and in thin-section grain boundaries may intersect at 120° angles. The mineral mode is used as a prefix to the rock name (e.g. amphibole–plagioclase granofels).

Occurrence

Granofels rocks can occur in a wide variety of tectonic settings. Mafic granofels can be associated with high-pressure settings related to subduction (the Zermatt-Saas, Western Alps), high-temperature contact metamorphism in the mid-crust (Arthur River complex in Fiordland, New Zealand), or high-pressure granulite facies conditions associated with continent–continent collision (Dronning Maud Land, East Antarctica).

Granofels from Blankaholm (Sweden). The sample is composed of quartz (white), feldspar (pale pink), and biotite (black–brown).

Image courtesy of www.skan-kristallin.de

Figure 3.29

A Granofels from Valle Bognanco (Italy). This quartz-rich granofels has a weak foliation defined by quartz (clear) and biotite (green–brown). Allanite (the large central crystal) has its characteristic brown color with patchy zoning along the right rim, and the small, dark, rounded crystals are titanite. Plane-polarized light, 10× magnification, field of view = 2 mm.

B Granofels from Valle Bognanco (Italy). Same image as (A) but with crossed polarizers. Allanite shows anomalous birefringence masked by the crystal color. Biotite shows higher birefringence. Quartz has a granoblastic texture with gray birefringence. Cross-polarized light, 10× magnification, field of view = 2 mm.

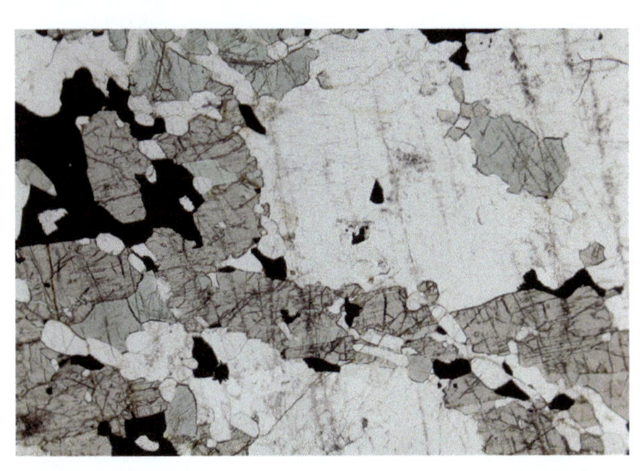

C Granofels from Rogaland (Sweden). This granulite facies charnockite has a massive (non-schistose) structure composed of orthopyroxene (pale brown), clinopyroxene (pale green, e.g. crystal in the upper-right quadrant), and plagioclase (clear, with some sericitized fractures). Plane-polarized light, 2× magnification, field of view = 7 mm.

D Granofels from Rogaland (Sweden). Same image as (C) but with crossed polarizers. Orthopyroxene (low birefringence, with bleb-like exsolution), clinopyroxene (high birefringence), and plagioclase (low birefringence, showing antiperthite exsolution). Cross-polarized light, 2× magnification, field of view = 7 mm.

APPLICATION 3.4

Time and P–T Paths

Careful petrographic analysis can be used to determine the minerals and textures that represent different parts of a P–T path. By combining metamorphic P–T paths with radiometric age data (time, t), we have the ability to document P–T–t paths (i.e. the change in pressure and temperature with time). Different radiometric age-dating methods are associated with distinct "closure" temperatures (i.e. the temperatures at which the radiogenic decay product begins to accumulate), and radiometric dating is key to defining the rates and timing of metamorphic processes.

Time and P–T paths

Mineral	System	Closure T (°C)[a] Slow cooling	Fast cooling
Monazite	U–Th–Pb	~875	>900
Zircon	U–Th–Pb	~850	>900
Rutile	U–Th–Pb	~550	~700
Hornblende	Ar–Ar	~450	~625
Muscovite	Rb–Sr	~300	~650
Biotite	Rb–Sr	~200	~425

[a] After Reiners et al. (2018).

The mineral assemblages and textures of metamorphic rocks are the product of their metamorphic conditions. It is often assumed that at peak metamorphism the mineral assemblage represents equilibrium conditions, with little or no further reaction as the rock cools and decompresses *en route* to the Earth's surface. Samples preserving nonpeak metamorphic conditions usually document disequilibrium – rather than equilibrium – metamorphic conditions, which is also of interest as disequilibrium conditions can be used to reconstruct P–T path segments (prograde/retrograde) and provide additional information for interpreting the metamorphic and tectonic evolution of a sample.

The idealized view of P–T–t paths is that mineral assemblages equilibrate at every stage along the path, from the start of metamorphism to the peak of metamorphism when the final assemblage is formed. This simplistic view neglects kinetic factors related to the energetics of nucleation and growth of minerals which are important for some metamorphic phases and reactions. For example, coexisting andalusite and sillimanite in a rock that equilibrated at P–T conditions corresponding to the stability field of sillimanite documents the sluggish kinetics of the andalusite-to-sillimanite transformation that allows andalusite to persist metastably outside its stability field.

It is also important to recognize the influence of deformation on metamorphic reactions and the extent to which deformation may assist reactions. It is possible that two rocks with the same bulk composition that follow the same P–T–t path, but that have different deformation history – for example, one is pervasively deformed and the other is not (because strain is localized in weaker rocks nearby) – can contain different mineral assemblages. The deformed rock may contain the predicted equilibrium assemblage for the P–T–t conditions attained by the rock, whereas the undeformed or less deformed rock may contain more metastable phases. P–T–t conditions and paths should therefore be considered within their structural context.

P–T–t paths in relation to P–T conditions, path shape, and duration and rate of P–T path segments are used to understand the driving forces of metamorphism within a tectonic framework. The integration of P–T path characteristics, time/rate information, structural data, and other petrologic information can provide significant information about metamorphic and tectonic processes. For example, Manzotti et al. (2018) investigated the Gran Paradiso Massif (Western Alps), a high-pressure micro-continental fragment that was subducted and later exhumed during convergence of the European and Adriatic plates. These researchers were able to

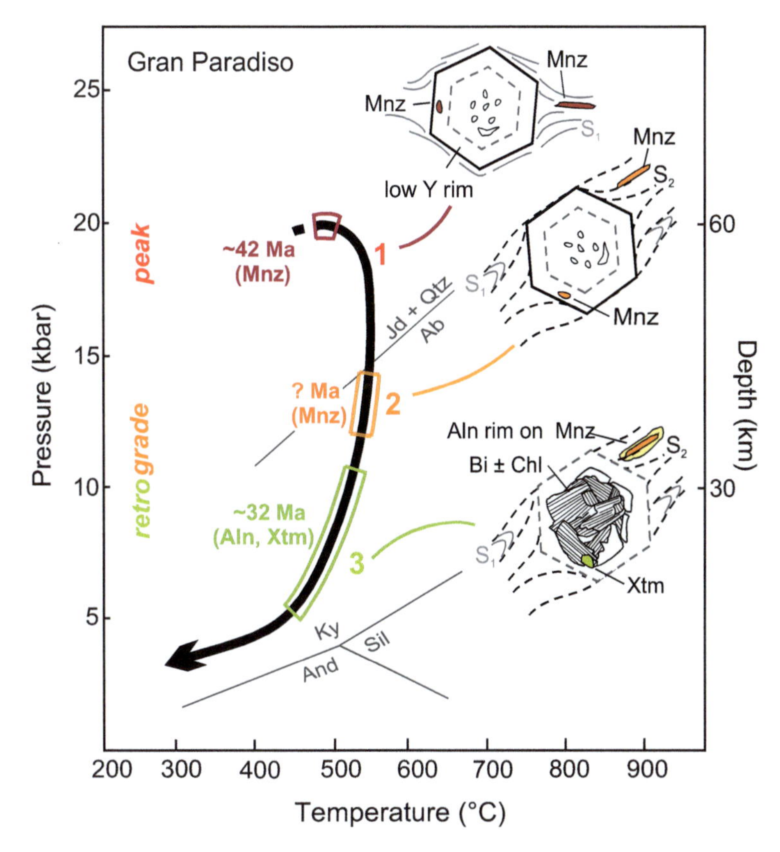

Monazite associated with peak metamorphism (shown in red) in both the rims of garnet and the foliation yielded ages within error of ~42 Ma. Folding of an earlier schistosity by a second deformation event (shown in orange) reoriented some monazite; although the absolute age of the second deformation event is not yet known, it must have post-dated peak conditions. In the final stages of retrograde metamorphism (shown in green), allanite formed coronas around monazite and garnet was statically replaced by aggregates of xenotime, biotite, and chlorite at ~32 Ma.

Image courtesy of P. Manzotti

recognize multiple periods of deformation and date three distinct accessory phases linked to the peak and retrograde metamorphic path, thus generating a P–T–t–d (deformation) path for the Gran Paridiso unit.

From the careful integration of P–T, textural, geochemical, and geochronological data, Manzotti et al. (2018) were able to constrain the time between the peak pressure conditions and the crystallization of xenotime to within ~10 million years; using this information they calculated an exhumation rate for the Western Alps of 2.2–5 mm/year. Their careful documentation of where datable mineral inclusions occur with respect to deformation fabrics developed during metamorphism established the linkages between petrographic textures and pressure, temperature, and time, and provided absolute constraints on the timing and rates associated with exhumation of these eclogite facies rocks.

Slate

Mineralogy: *Qz, Ilt, Sme, Prl, Fsp, Kln, Cc, Dol, Chl, ± Gr*
IUGS classification: *Metasediment (protolith)*
Texture: *Foliated (slaty cleavage)*

PROTOLITH

Slate is an ultrafine- to very fine-grained metamorphic rock with dull luster and a well-developed, continuous cleavage defined by aligned platy phyllosilicates (micas, chlorite) and graphite. This cleavage is also known as *slaty cleavage*. In many slates, however, even at very low metamorphic grade, slaty cleavage can become slightly spaced with the development of anastomosing lenticular phyllosilicate- or quartz-rich domains. Slates originate from low-grade metamorphism of clay-rich sedimentary rocks such as shales, mudrocks, siltstones, and fine-grained sandstones, or from volcanic ash deposits. They are mostly composed of fine-grained (average grain size <0.1 mm) minerals such as quartz, feldspars, micas (illite, smectite, pyrophyllite), chlorite, graphite, kaolinite, carbonates and oxides that cannot be seen without magnification. In many slates it is common to find pyrite or magnetite. Slates have a highly variable coloration, depending on the presence of certain minerals; for example, gray to black varieties have a high concentration of graphite, red varieties are oxidized and hematite-rich, green varieties are chlorite-rich. In the context of contact metamorphism, slate can recrystallize and develop poikiloblasts of cordierite and/or andalusite, forming the so-called "*spotted slates*"; often, secondary alteration removes the poikiloblasts and leaves voids behind. As the degree of metamorphism increases, slates develop coarser mica grains, thus becoming phyllites.

Occurrence

Slates are common rocks in many orogenic belts, where they often form narrow zones that are elongate and parallel to the orogenic belt, called "*slate belts.*" Slates are commonly interspersed with other low-grade metamorphic rocks such as meta-conglomerates, meta-greywackes, meta-tuffs, etc. In slate belts the slaty cleavage is generally steeply inclined and subparallel to the orogenic belt. Classic locations of slate occurrence include Wales and the Scottish Southern Uplands (UK), the Appalachian Mountains and the Pennsylvania and California slate belts (United States). In some cases, slate belts can host important economic deposits (e.g. New South Wales in Australia, the Carolina and California gold deposits).

Graphite-rich slate from Minnesota (United States). Remains of the original bedding (alternating light–dark layers) is visible. Image is ~ 8 cm wide.

Image courtesy of J. St. John

Figure 3.30

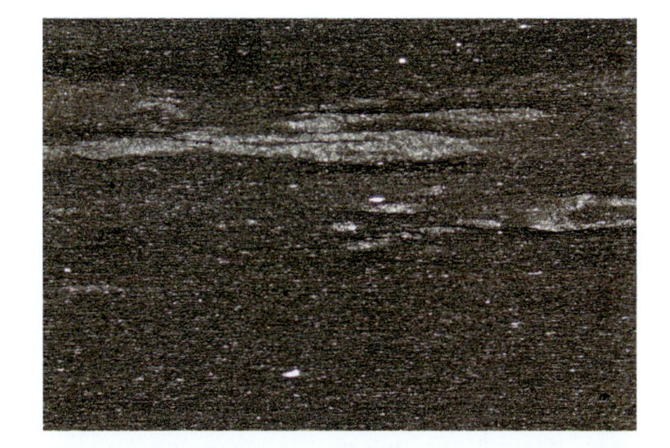

A Slate from Lavagna (Italy). A slaty cleavage (horizontal) is defined by fine-grained phyllosilicates. The lens-shaped lighter zones are quartz-rich and the small white crystals are larger quartz grains. Plane-polarized light, 2× magnification, field of view = 7 mm.

B Slate from Sardinia (Italy). Fine-grained slate with vertical slaty cleavage, cut by several generations of calcite vein. The darker bands (which apparently displace some veins) are pressure-solution surfaces. Plane-polarized light, 2× magnification, field of view = 7 mm.

C Spotted slate from Sardinia (Italy). A fine-grained, continuous cleavage is defined by aligned phyllosilicates and small quartz crystals. The rounded "spots" are incipient cordierite poikiloblasts (the gray birefringence of cordierite can be seen). Cross-polarized light, 10× magnification, field of view = 2 mm.

Phyllite

Mineralogy: *Ms, Qz, Chl/Bt, ± Fsp/Ser, ± Gr, ± Cal*
IUGS classification: *Metasediment (protolith)*
Texture: *Foliated*

PROTOLITH

Phyllite is a fine-grained foliated metamorphic rock characterized by a lustrous sheen which is caused by the presence of oriented phyllosilicates. Phyllites are often fissile and split along planes of weakness. Phyllites are characterized by continuous cleavage (often called *phyllitic cleavage*) in which very fine phyllosilicates (average grain size <0.25 mm and >0.1 mm) are rarely coarse enough to see unaided. Crenulation cleavage is very common. This cleavage develops from slaty cleavage when slates enter the field of greenschist facies metamorphism, determining the growth of new mica minerals with their basal plane more or less perpendicular to tectonic compression. Phyllites are essentially composed of microcrystalline quartz, fine-grained micas (sericite, muscovite), and chlorite. They may contain appreciable quantities of graphite (derived from the original organic material), as well as iron oxides, sulfides, and carbonates (calcite). Most phyllites are black or gray to greenish-gray. Phyllites are relatively coarser-grained than slates, consistent with their slightly higher metamorphic grade, and phyllites are finer-grained than schists.

Occurrence

Phyllite derives from low-grade regional metamorphism of pelitic sediments (shale and mudstone) as well as from finely grained volcanic tuffs. Phyllites are common rocks in the lower part of the greenschist facies. Phyllites are common metamorphic rocks worldwide derived from sedimentary sequences, involved and metamorphosed along convergent plate boundaries, or in accretionary wedges above subduction zones. Together with other metamorphic rocks, they are common rocks in the Appalachians in North America, in the Scottish Highlands, and in the Alps in Europe. As the metamorphic degree increases, phyllites become schists.

Folded phyllite from Fowey, Cornwall (England). The rock has a well-developed foliation, but the individual minerals are indistinguishable. Two veins cross-cut the sample. Image is ~10 cm wide.

Image courtesy of A. Tindle

Figure 3.31

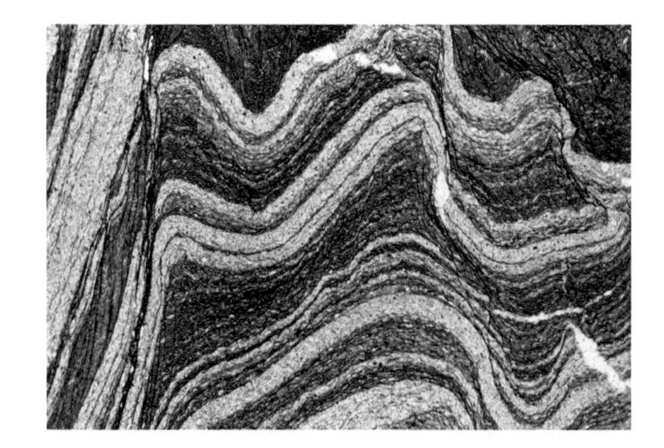

A Folded phyllite from Sardinia (Italy). The rock is characterized by the alternation of light (quartz-rich) and dark (phyllosilicate-rich) layers. Even with the aid of a microscope, it is difficult to distinguish individual minerals. The black, irregular vertical lines are pressure-solution surfaces. Plane-polarized light, 2× magnification, field of view = 7 mm.

B Crenulated phyllite from Calabria (Italy). The phyllitic cleavage (steep layering) is crenulated by a spaced foliation (oriented horizontal). The black material disseminated in the rock is graphite. Plane-polarized light, 2× magnification, field of view = 7 mm.

C Crenulated phyllite from Calabria (Italy). Same image as (B) but with crossed polarizers. The individual minerals are difficult to distinguish, but the birefringence of the phyllosilicates can be seen. Cross-polarized light, 2× magnification, field of view = 7 mm.

Spilite

PROTOLITH

Mineralogy: *Ab, Qz, Chl, Ep, Cc, Prh, Pmp, Tr, Fe oxides + accessory*
IUGS classification: *Metabasalt (protolith)*
Texture: *Massive with common igneous relicts*

Spilite is a metasomatized basaltic to intermediate volcanic or subvolcanic rock in which the original igneous minerals have been partially or completely replaced by a low-grade metamorphism or metasomatism along mid-ocean ridges. Relict magmatic minerals and original igneous textures are commonly preserved. Plagioclase and pyroxene are converted at low-grade to nearly pure albite and chlorite respectively. Plagioclase albitization may occur via:

$$\underset{\text{anorthite}}{CaAl_2Si_2O_8} + \underset{\text{quartz}}{4SiO_2} + \underset{\text{sodium aqueous solution}}{2Na^+} \longrightarrow \underset{\text{albite}}{2NaAlSi_3O_8} + \underset{\text{calcium aqueous solution}}{Ca^{2+}}$$

Primary Fe–Ti oxides are often replaced by a cryptocrystalline titanite aggregate called leucoxene (light gray color in thin-section). Other low-grade metasomatic assemblages include tremolite–actinolite, epidote, calcite, pumpellyite, prehnite, and quartz. Spilite commonly contains amygdales filled with chalcedony, opal, chlorite, calcite, and hydrous Ca–Al silicates.

Occurrence

The interaction between cold seawater and hot mafic–ultramafic rocks at ocean spreading ridges represents the largest active metamorphic/metasomatic system on the planet. Spilites, together with other metasomatic rocks such as epidosite, are widespread in this setting, in oceanic plateaus, in ophiolite complexes (the Semail ophiolite of Oman, the Troodos ophiolite of Cyprus), and in association with continental geothermal fields (Iceland, New Zealand). Unlike fresh mid-ocean ridge basalts, metasomatic processes produce distinctive chemical characteristics in spilites: They are enriched in Na_2O (derived from seawater) and depleted in Ca (due to plagioclase albitization) and MgO (due to pyroxene alteration).

Spilitized greenstone, Catoctin Formation, Pennsylvania (United States). The amygdales in this Neoproterozoic metabasalt have been altered to epidote, and plagioclase phenocrysts altered to albite.

Image courtesy of J. St. John

Figure 3.32

A Spilitized basalt from Liguria (Italy). Some original igneous structure is preserved. The rectangular phenocryst is a former olivine, now pseudomorphed by chlorite (green) and carbonate (high relief). The phenocryst is surrounded by a fine-grained groundmass with albitized plagioclase (pale beige) and oxidized mafic minerals (darker brown crystals). The original interstitial volcanic glass has been replaced by green chlorite. Plane-polarized light, 10× magnification, field of view = 2 mm.

B Spilitized basalt from Liguria (Italy). Same image as (A) but with crossed polarizers. Chlorite (Fe-rich) has deep blue birefringence; plagioclase has low, first-order birefringence; mafic minerals are black due to oxidation. Cross-polarized light, 10× magnification, field of view = 2 mm.

C Spilitized basalt from Michigan (United States). The rock is composed by a fine-grained mixture of chlorite (greenish-brown regions throughout the rock), actinolite, quartz (colorless areas), and epidote (yellowish). Two oval-shaped amygdales are filled with quartz (colorless) and epidote (yellowish). The lower amygdale shows acicular actinolite crystals perpendicular to its margins. Plane-polarized light, 10× magnification, field of view = 2 mm.

Serpentinite

Mineralogy: Lz, Ctl, Atg, Mt, ± Bcr (invariable abundance)
IUGS classification: >75% serpentine
Texture: Granoblastic polygonal

PROTOLITH

Serpentinite is a metamorphic rock made of >75% serpentine group minerals such as antigorite, lizardite, chrysotile, plus magnetite and sometimes brucite. Common secondary minerals include talc, calcite, and magnesite. Serpentinites are easily recognized because serpentine minerals are very distinctive, typically a vibrant green color with a waxy sheen. Serpentinites represent metamorphosed mafic and ultramafic igneous rocks dominated by olivine and pyroxene.

Occurrence

Serpentinization, or the formation of serpentine minerals (ergo serpentinites), results from hydration of magnesium- and iron-rich minerals (i.e. olivine and pyroxene) at low (100 °C) to moderate temperatures. Therefore, serpentinization requires the addition of water, and serpentinites are most commonly associated with tectonic settings such as mid-ocean ridges, the forearc mantle of subduction zones (the Marianas), and ophiolites (e.g. the Bay of Islands ophiolite [Newfoundland], the Lizard Complex [Cornwall, England]). The alteration of nonhydrous olivine and pyroxene into hydrous serpentine [$Mg_3Si_2O_5(OH)_4$] depends on the abundance of Mg + Fe and involves the internal buffering of the pore fluid, a reduction of oxygen fugacity, and the partial oxidation of Fe^{2+} to Fe^{3+}. Mg and Fe diffusion in olivine is slow and often results in the release of hydrogen and the precipitation of magnetite, which imparts a dark color to some serpentinites. The serpentinization reaction of fayalite (Fe end-member of olivine) to magnetite and quartz with water occurs via the reaction:

$$3Fe_2SiO_4 + 2H_2O \rightarrow 2Fe_3O_4 + 3SiO_2 + 3H_2$$

The serpentinization process is exothermic and involves a volume increase (up to 30%) with a concomitant density decrease (fresh peridotite has a density of 3,300 kg m^{-3}, whereas serpentinite has a density of ~2,600 kg m^{-3}). Thus, serpentinization may facilitate brittle deformation (via volume expansion) and uplift (via increased buoyancy), and is considered an important influencing factor on the strength of the lithosphere. In addition, the tectonic setting associated with the serpentinization process controls the abundance of fluid-mobile elements in serpentinites; similar fluid-mobile element enrichment patterns in mantle-wedge serpentinites and arc magmas suggest a linkage between the dehydration of serpentinite and arc magmatism.

Serpentinite from the Coast Range ophiolite of California (United States). Notice its distinctive "waxy" luster. Sample ~10 cm wide.

Image courtesy of J. St. John

Figure 3.33

A Serpentinite from Livorna (Italy). The image is dominated by fibrous antigorite (colorless). Relict olivine + pyroxene (high relief and brownish color) are also present. Plane-polarized light, 2× magnification, field of view = 7 mm.

B Serpentinite from Livorna (Italy). Same image as (A) but with crossed polarizers. The image is dominated by fibrous antigorite (low birefringence). Relict olivine + pyroxene (higher birefringence) are also present. Cross-polarized light, 2× magnification, field of view = 7 mm.

C Serpentinite from unknown location (Oman). The serpentine mineral is chrysotile, which forms a mesh-like pattern of veins. Relict olivine (high birefringence) is still present. Cross-polarized light, 2× magnification, field of view = 7 mm.

Hornfels

Mineralogy: Ab, Ep, Amp, Px, Ksp, Bt, Mu, Qz, Crd, And, Crn, Trd

PROTOLITH

IUGS classification: Pelitic hornfels, calc-silicate hornfels, mafic hornfels

Fabric: Granofels

The hornfels name is used for a group of compact, metamorphosed rocks dominantly composed of silicate and oxide minerals of any grain size with subconchoidal to jagged fracture. They represent contact-metamorphosed pelitic, carbonate, and mafic country rocks that host plutons and define the metamorphic aureoles associated with these intrusions (e.g. the Bergell tonalite in the Italian Alps or the Bugaboo batholith of British Columbia [Canada]).

Pelitic hornfels include contact-metamorphosed clays, sedimentary slates, and shales. With increasing temperatures and appropriate bulk-rock chemistry, metamorphism proceeds from reaction (1) to (3) below.

$$\text{chlorite} + \text{K-feldspar} \longrightarrow \text{biotite} + \text{muscovite} + \text{quartz} + H_2O \quad (1)$$

$$\text{chlorite} + \text{muscovite} + \text{quartz} \longrightarrow \text{cordierite} + \text{andalusite} + \text{biotite} + H_2O \quad (2)$$

$$\text{biotite} + \text{plagioclase} + \text{quartz} \longrightarrow \text{K-feldspar} + \text{cordierite} + \text{orthopyroxene} + \text{melt} \quad (3)$$

In thin-section, biotite is abundant, has a dark reddish-brown color, and displays strong dichroism. Quartz and feldspar may be present, while graphite, tourmaline, and iron oxides may form accessory phases. Faint striping may reflect original bedding, while round or elliptical spots are often made of carbonaceous matter, brown mica, or coarser grains of quartz than occur in the matrix. With increasing temperature, the aluminum silicates andalusite and sillimanite appear ± kyanite. Pink andalusite is pleochroic in thin-section, while white andalusite often has the cruciform inclusion trails characteristic of chiastolite. Sillimanite occurs as small needles in quartz.

Carbonate hornfels include contact-metamorphosed limestone and dolostone. Pure limestone or dolostone recrystallizes to marble, whereas impure rocks form silicate minerals such as diopside, epidote, garnet, titanite, vesuvianite, and scapolite. The rocks are fine-grained and often banded, but much harder than the original rock. Their mineral composition is highly variable and may alternate with biotite hornfels and indurated quartzites.

Mafic/ultramafic hornfels include contact-metamorphosed diabases, basalts, andesites, and their plutonic equivalents of oceanic and/or continental provenance. The original rocks may have contained calcite, zeolite, chlorite, ± other secondary minerals, but they are dominated by feldspar, hornblende, and pyroxene, which are relatively insensitive to temperature variations at low pressures. The original structure of the rock (porphyritic, vesicular, fragmental, etc.) may still be visible after low-temperature hornfelsing, but becomes less evident as temperature increases and alteration progresses.

Massive hornfels in Helena, Montana (United States). Note the typical spotted appearance of cordierite (yellowish) set in a massive groundmass. This close-up view is ~2 cm wide.

Image courtesy of A. Tindle

Figure 3.34

A Pelitic hornfels from Sondrio (Italy). Euhedral (idioblastic) andalusite (high relief, pale, squarish outlines with small inclusions), dark garnet (upper-left) with quartz inclusions along its rim, in a groundmass of biotite (reddish) and quartz (clear) that define a weak foliation. Plane-polarized light, 2× magnification, field of view = 7 mm.

B Pelitic hornfels from Sondrio (Italy). Same image as (A) but with crossed polarizers. Andalusite shows gray birefringence (mostly near extinction), garnet is isotropic, biotite shows high birefringence, and quartz shows gray birefringence. Cross-polarized light, 2× magnification, field of view = 7 mm.

C Carbonate hornfels from Val Martello (Italy). Large vesuvianite (high relief, fractured, sub- to euhedral) dominates the lower part of the image. It is surrounded by calcite (note rhombohedral cleavage) and white mica (colorless). Plane-polarized light, 2× magnification, field of view = 7 mm.

D Carbonate hornfels from Val Martello (Italy). Same image as (C) but with crossed polarizers. Vesuvianite (gray birefringence) shows fractures filled by calcite. Calcite has very high birefringence and white mica a bit lower birefringence. Cross-polarized light, 2× magnification, field of view = 7 mm.

BOX 3.11 Cataclasite, Mylonite, and Pseudotachylite

Rocks in shear zones represent incohesive and cohesive fault rocks. The former is restricted to shallow crustal levels (e.g. breccia and fault gouge). The latter are categorized as foliated or nonfoliated, and classified on the basis of the relative proportion of clast to matrix ± glass in the matrix.

Cohesive fault rocks

% matrix	Nonfoliated	Foliated	Glass in matrix
10–50	Protocataclasite	Protomylonite	Pseudotachylite
50–90	Cataclasite	Mylonite	Pseudotachylite
>90	Ultracataclasite	Ultramylonite	Pseudotachylite

Cataclasitic rocks are unfoliated and characterized by angular porphyroclasts and lithic fragments set in a finer-grained matrix of similar composition. This granitic cataclasite from Baveno (Italy) has typical angular fragments of quartz and feldspar, set in fine-grained (almost isotropic) matrix. Cross-polarized light, 2× magnification, field of view = 7 mm.

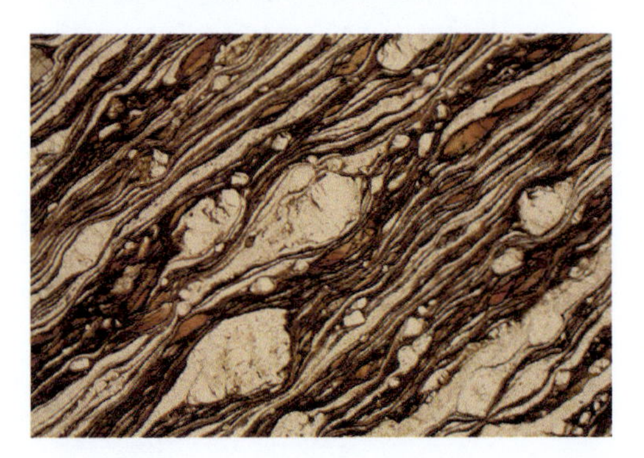

Mylonitic rocks are characterized by rounded porphyroclasts in a foliated and finer-grained matrix. This granitic mylonite from Cesana Torinese (Italy) has feldspar porphyroclasts (colorless) and biotite "mica fish" (brown) wrapped by a mylonitic foliation defined by quartz ribbons alternating with fine-grained biotite-rich layers. Plane-polarized light, 2× magnification, field of view = 7 mm.

Pseudotachylite is a special rock type which reflects local melting of the rock by frictional heating (often associated with earthquakes). In-situ melting and rapid cooling generates a glassy (often devitrified) matrix containing angular rock and mineral fragments. This pseudotachylite vein from Ivrea Verbano (Italy) cuts its gabbro host, with angular, fine-grained rock fragments set in a glassy (black) matrix. Plane-polarized light, 2× magnification, field of view = 7 mm.

BOX 3.12 Marble

Marbles are metamorphosed calcareous rocks dominated by mainly calcite and to a lesser extent dolomite. During regional metamorphism and given the large stability field of calcite, calcite-marbles provide few clues regarding the P–T conditions associated with their formation. At very high temperature, calcite reacts with quartz to form wollastonite ($CaCO_3 + SiO_2 \leftrightarrow CaSiO_3 + CO_2$). On the other hand, dolomitic marbles can develop a series of Ca–Mg silicates sensitive to P–T conditions. During prograde metamorphism, the first mineral to appear is talc according the reaction:

$$3Dol + 4Qz + 1H_2O \rightarrow 1Tlc + 3Cal + 3CO_2$$

As temperature increases, dolomitic marbles enter the tremolite isograd (comparable to the staurolite isograd in pelitic rocks): $5Tlc + 6Cal + 4Qz \rightarrow 3Tr + 2H_2O + 6CO_2$. In this example from Gere Valley (Switzerland), acicular tremolite (orange–violet birefringence) occurs with fibrous talc (green birefringence, upper-left and lower-right); the remaining material is calcite. Cross-polarized light, 10× magnification, field of view = 2 mm.

At higher grade (comparable to the sillimanite isograd in pelitic rocks) diopside appears due to the reaction: $1Tr + 3Cal + 2Qz \rightarrow 5Di + 1H_2O + 3CO_2$. Diopside formation is favored in calcite-rich or silica-poor dolomitic marbles. This example from Mergozzo (Italy) shows bright green diopside (with characteristic cleavage) surrounded by calcite. Plane-polarized light, 2× magnification, field of view = 7 mm.

In silica-free dolomitic marbles, forsterite appears due to the reaction: $1Tr + 11Dol \rightarrow 8Fo + 13Cal + 1H_2O + 9CO_2$. Diopside and forsterite can coexist only if tremolite and calcite are consumed: $3Tr + 5Cal \rightarrow 11Di + 2Fo + 3H_2O + 5CO_2$. This example from Ivrea Verbano (Italy) shows small, anhedral forsterite (high birefringence) in dolomite (larger grains with pastel birefringence). Cross-polarized light, 2× magnification, field of view = 7 mm.

Quartzite

Mineralogy: Qz, ± Mu, ± Ser, ± Grt, ± Opq

PROTOLITH

IUGS classification: >90% quartz

Texture: Granoblastic polygonal

Quartzite is a compact and massive (nonfoliated) metamorphic rock composed predominantly of quartz. Quartzites commonly contain small amounts of other minerals such as muscovite or sericite, as well as chlorite, garnet, feldspar, kyanite, sillimanite, tourmaline, graphite, etc. If accessory minerals are present >10 vol%, they are added as a suffix to the name of the rock (garnet quartzite, muscovite quartzite, etc.). Quartzites are derived from medium- to high-grade, regional and contact metamorphism of quartz-rich sedimentary rocks such as relatively pure quartz sandstones, siltstones, and cherts. Quartzites may sometimes display relict sedimentary structures (e.g. bedding) as indicated by color or grain-size variations. Pure quartzites will never develop a well-defined foliation, although in some cases shape-preferred orientation of quartz can lead to a weak anisotropy. On the other hand, impure quartzites with appreciable quantities of phyllosilicates commonly develop a foliation. Pure quartzites are usually white or pale gray, but they may also be pink or red, reflecting iron oxidation. Other colors are due to impurities associated with minor amounts of other minerals. Quartzites typically have a granoblastic polygonal texture, resulting from the recrystallization of detrital quartz grains under high pressure and temperature conditions. Grain size tends to increase with temperature and with the presence of fluids (aqueous fluids promote larger crystals).

Occurrence

Quartzite is a widespread metamorphic rock. It is common in metasedimentary and volcano-sedimentary sequences, along convergent plate boundaries and in accretionary prisms above subduction zones. In metasedimentary sequences, quartzite is commonly interlayered with marble, schist, phyllite, and slate, representing metamorphosed limestones and shales. In these contexts, quartzite and phyllosilicate-rich rocks metamorphosed under the same conditions can display conspicuously different textures: phyllosilicate-rich rocks develop a well-pronounced foliation seldom seen in pure quartzite. Quartzite derived from deep marine deposits (metachert) or from banded iron formations commonly contain Mn-silicates such as piemontite and spessartine garnet.

Quartzite from Baraboo, Wisconsin (United States). The rock has a massive structure with conchoidal fracture. Its pale pink color is due to hematite. Image is ~5 cm across.

Image courtesy of A. Tindle

Figure 3.35

A Quartzite from Corsica (France). The image is composed entirely of equidimensional quartz with interlobate to polygonal grain boundaries. Cross-polarized light, 10× magnification, field of view = 2 mm.

B Muscovite quartzite from Barge (Italy). A weak foliation is defined by discontinuous layers of muscovite and flattened quartz. Some quartz shows undulose extinction. Cross-polarized light, 2× magnification, field of view = 7 mm.

C Piemontite quartzite from St. Anna (Italy). The protolith of this quartzite is a Mn-rich chert. A coarse foliation is defined by layers of aligned piemontite (red birefringence) and granoblastic quartz. Some muscovite crystals are visible in the lower-right corner. Cross-polarized light, 2× magnification, field of view = 7 mm.

Greenschist

Mineralogy: *Chl, Ep, Act, ±Ab, ± Hbl, ± Qz, ± Cal*
IUGS classification: *Chl schist*
Fabric: *Usually foliated*

PROTOLITH

A greenschist is a schistose metabasite that is typically dark green in color due to the abundance of chlorite, epidote, and actinolite. With metamorphism, the ferromagnesium component of the mafic protolith transforms into chlorite and generates a well-defined schistosity. These green minerals are usually accompanied by albite, calcite, and iron–titanium oxides. Massive (non-schistose) metabasites are known as greenstone. Other minerals that may be present include serpentine, sericite, titanite, and quartz. At higher grade, hornblende or oligoclase may be present. Greenschist is generally a fine- to medium-grained rock because chlorite (and to a lesser extent actinolite) typically forms as small, flat or acicular crystals. Textures and minerals from higher P–T conditions may also be preserved (e.g. primary igneous textures or retrograde greenschist forming after blueschist).

Occurrence

Greenschist is widespread and forms under a relatively narrow temperature range (~300–500 °C) at low to medium pressures (~100–900 MPa). Such metamorphic conditions are attained in a variety of tectonic settings, including subduction zones, orogenic belts, and areas of high heat flow, such as contact metamorphism associated with magmatic intrusions. To the west of the Songliao Basin (China), the Heilongjiang metamorphic belt lies between the Jaimusi and Songliao tectonic blocks. Greenschist of the Heilongjiang complex is thought to represent mafic oceanic crust. It appears as either massive small blocks/thick sheets, or as thin schistose sheets or lenses that surround blueschist. It has a typical greenschist mineral assemblage of epidote + chlorite + plagioclase (albite) + biotite + quartz ± stilpnomelane, and is regarded as the medium-pressure metamorphosed equivalents to the high-pressure blueschists of the Heilongjiang complex. Their chemistry and U–Pb zircon ages indicate a tholeiitic basaltic protolith of ~162 Ma that represents the oceanic crust between the Jaimusi and Songliao blocks.

Foliated greenschist from Ile de Groix, Brittany (France). The typical green color is not very pronounced due to rusty weathering. Numerous albite crystals (whitish, high relief) sit in a dark green matrix of chlorite, calcic-amphibole, and minor epidote. Coin for scale.

Image courtesy of P. Manzotti

Figure 3.36

A Greenschist from Merano (Italy). Chlorite (pale green), epidote (three large pale brown crystals, center), and actinolite (acicular, high relief, pale green – better seen in XPL) define a weak foliation (oriented diagonally). The colorless crystals in the matrix are quartz. Plane-polarized light, 2× magnification, field of view = 7 mm.

B Greenschist from Merano (Italy). Same image as (A) but with crossed polarizers. The large central crystal of epidote has strong concentric compositional zoning and shows variable birefringence. In the matrix, chlorite has low (gray) birefringence and actinolite high (reds, blues) birefringence. Cross-polarized light, 2× magnification, field of view = 7 mm.

C Greenschist from As Sifah (Oman). Large prismatic epidote (brown–pink) surrounded by green chlorite and actinolite (acicular). Other minerals are quartz (colorless) and magnetite (opaque). The sample has a well-developed schistosity. Plane-polarized light, 2× magnification, field of view = 7 mm.

D Greenschist from As Sifah (Oman). Same image as (C) but with crossed polarizers. Epidote has high birefringence, chlorite is near extinction, actinolite has moderate birefringence (lower-left), and quartz has low birefringence. Cross-polarized light, 2× magnification, field of view = 7 mm.

Amphibolite

Mineralogy: Hbl, Pl, ± Ep, ± Qz, ± Bt, ± Grt accessory
IUGS classification: >75% Amp + Pl
Fabric: Weakly foliated to unfoliated (granofelsic)

PROTOLITH

Amphibolite is a medium- to coarse-grained metabasic igneous rock dominated by amphibole (usually hornblende) and plagioclase. Epidote, garnet, and/or biotite are often present. Accessory minerals include titanite, apatite, and ilmenite, while quartz and chlorite are sparse or absent. Amphibolite is typically dark in color and hornblende often forms stubby prisms rather than the elongate blades of actinolite associated with greenschist facies rocks. Most amphibolite is granofelsic, although a weak schistosity can be developed. Metamorphosed ultramafic rocks may produce tremolite. Amphibolite should not be confused with hornblendite – a cumulate igneous rock dominated by hornblende.

Most amphibolite forms by (1) direct hydration of a basic igneous rock, (2) prograde metamorphism after greenschist facies, or (3) retrograde alteration after higher-grade metamorphism (i.e. after granulite or eclogite facies metabasites). Typical mineral reactions may include:

$$\text{Chl} + \text{Czo} + \text{Qz} \longrightarrow \text{Hbl} + \text{An} + \text{fluid}$$
$$\text{Ab} + \text{Act} \longrightarrow \text{Hbl} + \text{Qz}$$
$$\text{Act} + \text{Chl} + \text{Qz} + \text{Czo} \longrightarrow \text{Hbl} + \text{fluid}$$

Some notable parageneses associated with the formation of amphibolite include: chlorite at lower pressures and higher temperatures, cummingtonite (Ca-poor amphibole) at lower pressures, epidote instead of chlorite at higher pressures, garnet at higher pressures, and more calcic-plagioclase with increasing temperature.

Occurrence

Amphibolite is a widespread rock type that is associated with regional metamorphism and crustal thickening during orogenesis, as well as with environments of crustal thinning that have higher heat flow. During burial and prograde metamorphism amphibolite is dehydrated to produce granulite. In contrast, fluids released during dehydration of deeper (hotter) rocks may also hydrate rocks at shallower crustal levels to produce amphibolites during orogenic exhumation and retrograde metamorphism. In regions of higher heat flow, amphibolite generation occurs under lower-pressure conditions.

Metamorphosed Scourie dike from Achmelvich (Scotland). This bimineralic amphibolite has a massive texture defined by amphibole (dark) and plagioclase (white). Note iron staining along some fractures. Image ~12 cm wide.

Image courtesy of D. Waters

Figure 3.37

A Amphibolite from Val Passiria (Italy). Large hornblende (blue–green to very pale) and plagioclase (colorless) define a coarse horizontal foliation. Plane-polarized light, 2× magnification, field of view = 7 mm.

B Amphibolite from Val Passiria (Italy). Same image as (A) but with crossed polarizers. Hornblende shows medium to high birefringence, while plagioclase shows typical low (gray) birefringence. Cross-polarized light, 2× magnification, field of view = 7 mm.

C Garnet amphibolite from Alpe Arami (Switzerland). Hornblende (greenish brown), quartz (colorless), and minor plagioclase (dusty alteration in lower-right) define a coarse foliation. The euhedral to subhedral, high-relief crystals are garnet. Plane-polarized light, 2× magnification, field of view = 7 mm.

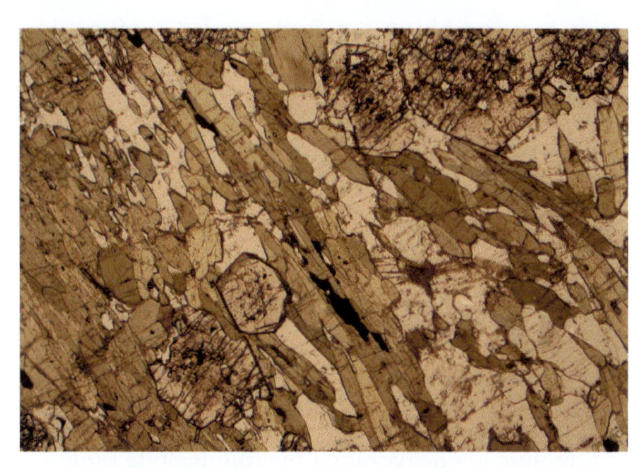

D Garnet amphibolite from Alpe Arami (Switzerland). Same image as (C) but with crossed polarizers. Hornblende shows high birefringence, while garnet is isotropic. Cross-polarized light, 2× magnification, field of view = 7 mm.

Granulite

Mineralogy: *Cpx, Opx, Grt, Pl, Ksp, Qz, ± Ky, ± Sil, ± Rt, ± Il, ±Bt, ± Hbl, ± Crd*

PROTOLITH

IUGS classification: *Mafic granulite (>30% Px), felsic granulite (<30% Px)*

Fabric: *Granofels*

A granulite is a rock metamorphosed at high temperature and low to medium pressure. Though originally defined on the basis of a mafic protolith, it is now accepted that the mineralogy of a granulite reflects its protolith, which can be igneous or sedimentary. Pyroxene (orthopyroxene and clinopyroxene), plagioclase, potassium feldspar, and garnet are typical. When mafic minerals (mostly pyroxene) are abundant (>30%), the rock is called a mafic granulite; if mafic minerals are not abundant (<30%), the rock is called a felsic granulite. Granulite contains plagioclase, whereas eclogite does not. Most granulite has a medium- to coarse-grained granofels texture.

Granulite may form by: (1) dehydration of an amphibolite facies metabasite during prograde metamorphism; (2) retrograde alteration after higher-grade metamorphism (i.e. isothermal decompression from eclogite facies); and (3) dehydration melting. The high-temperature settings associated with granulite genesis often involve partial melting. *Fluid-present* melting occurs once the wet granite solidus has been surpassed and requires the addition of aqueous fluid. *Fluid-absent* dehydration melting uses the structural water in hydrous minerals (biotite and hornblende) and results in melt together with an anhydrous solid residue of pyroxene and garnet. Notable parageneses associated with the formation of granulite via dehydration melting include sapphirine, spinel, sillimanite, and osumilite. Assemblages such as sapphirine + quartz indicate temperatures in excess of 900 °C.

Occurrence

Granulite is a metamorphic rock associated with regional metamorphism and high thermal gradients (i.e. convergence and crustal thickening), contact metamorphism, and regions of crustal thinning and basin subsidence. During burial and prograde metamorphism, amphibolite may be dehydrated to produce granulite. In contrast, during orogenic exhumation and retrograde metamorphism, deeper (hotter) granulite facies rocks may (re)hydrate to produce amphibolite at shallower crustal levels. In regions of higher heat flow, such as continental rift settings, upwelling asthenospheric mantle generates thermal gradients in excess of 30 °C/km and in extreme cases can produce granulite as an anhydrous residue after the extraction of felsic partial melt(s).

Granulite from the Akia terrane (West Greenland). Layers of coarse clinopyroxene (green) and orthopyroxene (brownish) within a finer-grained matrix. The pyroxenes are peritectic crystals, so presumably any melt escaped to pool elsewhere. Image ~0.5 m across.

Image courtesy of D. Waters

Figure 3.38

A Mafic granulite from Hartmannsdorf (Germany). A granoblastic texture is defined by subhedral, equidimensional pyroxene (pink–green pleochroism), plagioclase (colorless), and magnetite (black). Garnet is also present (oval pale crystal, center-right). Plane-polarized light, 2× magnification, field of view = 7 mm.

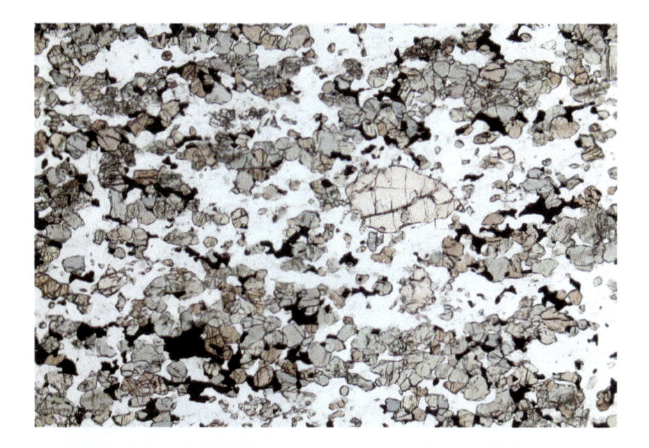

B Mafic granulite from Hartmannsdorf (Germany). Same image as (A) but with crossed polarizers. Pyroxene has high birefringence and plagioclase has first-order gray birefringence with weak polysynthetic twinning. Garnet is isotropic. Cross-polarized light, 2× magnification, field of view = 7 mm.

C Sapphirine granulite from the Val Codera (Italy). Sapphirine (blue–green color, fractured, center) is surrounded by quartz and cordierite (both colorless), but cordierite shows fractures (right) relative to quartz. Pyroxene (pale brown, high relief) and biotite (brown–red) are also present. Plane-polarized light, 2× magnification, field of view = 7 mm.

Blueschist

Mineralogy: *Gln, Ab, ± Rbk, ± Lws, ± Ep, ± Qz, ± Grt, ± Chl, ± Jd, ± Arg, ± Dol, ± Ph, ± Srp, ± Brc, ± Mgs*
IUGS classification: *Gln (≥5%) schist*
Fabric: *Usually foliated*

PROTOLITH

A blueschist is a schistose metabasite with its schistosity defined by lepidoblastic minerals with an elongate or platy shape. Blue alkali-amphibole (glaucophane or riebeckite) is diagnostic and occurs with lawsonite or epidote, albite, and/or quartz and garnet. Lawsonite dehydration frequently results in an epidote + muscovite + albite pseudomorph of lawsonite. Chlorite and albite may be present at lower pressures, while at higher pressures sodic clinopyroxene (jadeite) may occur. In outcrop, blueschist is blue, black, gray, or blue–green. It is rarely coarse-grained because mineral growth is retarded by: (1) the potentially anhydrous basaltic protolith, (2) the low temperature of metamorphism, and (3) rapid exhumation (to avoid thermal re-equilibration). However, porphyritic varieties do occur, such as in the Ile de Groix (France).

Occurrence

Blueschist initially forms during subduction-related metamorphism of the seafloor, hence its basaltic protolith. High-pressure–low-temperature environments associated with subduction zones are dominated by low geothermal gradients (4–14 °C km^{-1}), generating blueschist facies metamorphic conditions. At these conditions of high pressure (>600 MPa) and low temperature (<550 °C), the typical blueschist high-pressure–low-temperature mineral associations are generated at depths of 25–60 km.

Blueschist is often found within orogenic belts in fault contact with greenschist or more rarely eclogite facies rocks. The preservation of blueschist requires rapid exhumation from high- to low-pressure conditions; this occurs via flow or faulting within the accretionary wedge or the upper parts of subducted crust. Exhumation may also be facilitated by buoyancy contrasts between metabasaltic rock and low-density rocks (marble or metapelite) of continental margins.

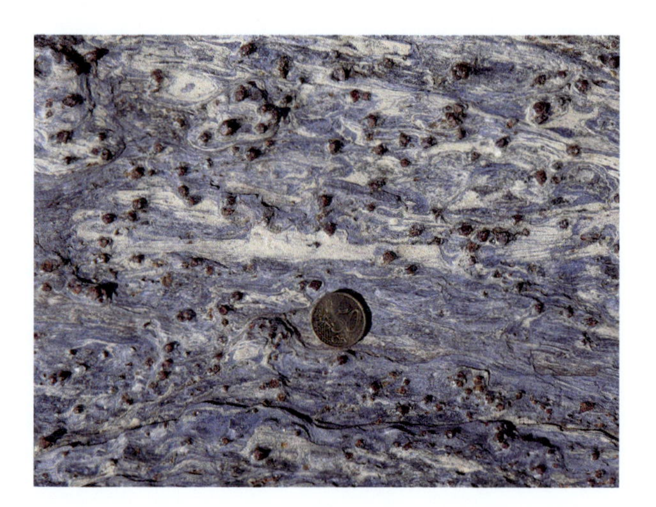

Foliated blueschist from Ile de Groix, Brittany (France) consisting of alternating glaucophane-rich and epidote-rich domains with conspicuous idioblastic garnet. Coin for scale.

Image courtesy of P. Manzotti

Figure 3.39

A Blueschist from Cervinia (Italy). Relict pyroxene with yellow birefringence (center) bordered by glaucophane (light blue) and chlorite (green). Plane-polarized light, 10× magnification, field of view = 2 mm.

B Blueschist from Cervinia (Italy). Same image as (A) but with cross-polarized light. Fine-grained groundmass of albite between glaucophane and chlorite crystals. Pyroxene and groundmass composition consistent with a mafic protolith. Cross-polarized light, 10× magnification, field of view = 2 mm.

C Blueschist from Jenna, California (United States). Two relict garnet porphyroclasts (pink, center) being pseudomorphed by chlorite (light green) and epidote (light yellow) in a groundmass of glaucophane (blue), garnet (light pink), epidote (tan), and muscovite (colorless). Plane-polarized light, 2× magnification, field of view = 7 mm.

Eclogite

PROTOLITH

Mineralogy: Grt, Omp, Ep, Qz, Ky, Pg, Ph, ± accessory Lws, Rt, Py
IUGS classification: >75% Grt + Omp
Fabric: Unfoliated or foliated

An eclogite represents a mafic protolith metamorphosed under very high pressure. It is recognized by the absence of plagioclase and the presence of sodic pyroxene (omphacite) and pyrope (Mg-rich) garnet. Omphacite is responsible for the fantastic green color of many eclogites. The composition of the protolith determines which accessory minerals will be present – for example, kyanite may develop from more aluminous metabasites via the reaction:

$$\underset{\text{anorthorite}}{3CaAl_2Si_2O_8} = \underset{\text{grossular}}{Ca_3Al_2Si_3O_{12}} + \underset{\text{kyanite}}{2Al_2SiO_5} + \underset{\text{quartz}}{SiO_2}$$

The presence of lawsonite is rare and requires special circumstances for its formation and preservation, but if present it defines a particular low-pressure and high-temperature set of conditions.

Occurrence

Eclogite forms at high pressures and lower temperatures; it is associated with xenoliths in mafic and ultramafic host rocks, subducted ocean-floor basalt, and orogenic belts. There is often a close association between eclogite and blueschist developed at higher grades associated with subduction zones, which may reflect differences in protolith composition or availability of fluids. For example, on Syros (Greece), eclogite and blueschist are interlayered and experienced the same metamorphic history, but due to differences in protolith composition developed distinct mineral parageneses. On the Ile de Groix, blocks of eclogite are entrained within blueschist, likely the result of post-metamorphic reactions such as:

$$Zo + Gln = Grt + Omp + Pg + Qz + H_2O$$
$$Ab + Ep + Gln = Omp + Pg + Hbl + H_2O$$

Some eclogites are closely associated with high-pressure granulites, forming either contemporaneously or sequentially. Most commonly, however, these high-pressure rocks are fault-bound, their contacts seldom gradational, and reflect the processes that returned these rocks from great depths to the surface.

Foliated eclogite and blueschist from Syros (Greece). Bright green eclogite pod interfolded with glaucophane-rich blueschist. Image ~40 cm wide.

Image courtesy of J. Walters

Figure 3.40

A Eclogite from Voltri (Italy). The sample consists of omphacite (green with good cleavage), garnet (pink, rounded), and rutile (dark). Retrograde metamorphism generated glaucophane (pale lilac–blue, top half of image). Plane-polarized light, 2× magnification, field of view = 7 mm.

B Eclogite from Voltri (Italy). Same image as (A) but with crossed polarizers. Omphacite has high (second-order) birefringence, garnet is isotropic, and glaucophane has low–medium birefringence. Cross-polarized light, 2× magnification, field of view = 7 mm.

C Eclogite from Junction School, California (United States). The image is dominated by three aligned garnets (pale brown) mantled by chlorite and omphacite (both greenish) that are better distinguished in crossed polars. Minor glaucophane (top-left) is weakly blue. Plane-polarized light, 2× magnification, field of view = 7 mm.

D Eclogite from Junction School, California (United States). Same image as (C) but with crossed polarizers. Garnets are isotropic and have tiny quartz inclusions. Omphacite has low (first-order) birefringence, while chlorite shows deep brown colors. Cross-polarized light, 2× magnification, field of view = 7 mm.

Migmatite

PROTOLITH

Mineralogy: Variable host composition with minimum melt Qtz, Fsp ± Bt, ± Ms
IUGS classification: Composite rock (protolith + melt)
Fabric: Banded to folded with cross-cutting veins or irregular pods

Migmatite is an extremely heterogeneous metamorphic rock of both metamorphic and magmatic parts, and is easily recognized in outcrop or hand sample. Migmatites have complex structures and may appear layered, may form planar to folded or contorted layers, may produce cross-cutting veins, and/or irregular pods. There are many terms for describing migmatites, but the most common and widely accepted include:

paleosome: the unaltered or slightly modified metamorphic parent rock or country rock of a migmatite

neosome: the newly formed partial melt of a migmatite that represents anatexis of the parent rock. Neosome may be further divided into leucosome and/or melanosome.

The leucosome of a migmatite is commonly light-colored (leucocratic) and is the relatively felsic partial melt of its host/country rock. It is often granodioritic to granitic in composition and therefore rich in quartz, K-feldspar, and plagioclase. The leucosome, as a melt, generally lacks any preferred mineral orientation. The paleosome is the dark-colored (melanocratic), more mafic part of a migmatite which represents the more restitic part of the host/country rock – that is, the part that remains after partial melting and is therefore depleted in quartz and feldspar, but enriched in mafic minerals such as biotite, cordierite, garnet, hornblende, or pyroxene, as well as Al-rich minerals. The paleosome often has a well-defined fabric.

Occurrence

Migmatites are widespread rocks occurring in high-grade metamorphic terranes, with granulites, amphibolites, and gneisses in Archaean to Proterozoic cratons, and core complexes of younger Phanerozoic orogenic belts, such as those along the Alpine–Himalayan chain. Their origin is related to partial melting during prograde metamorphism at high temperature, so upper amphibolite to granulite facies conditions. Melting is mainly due to the influx of external hydrous fluids into dry rocks or *fluid-absent* dehydration melting via the reactions:

$$\text{Ms} + \text{Pl} + \text{Qz} = \text{Kfs} + \text{Sil} + \text{Bt} + \text{melt}$$
$$\text{Bt} + \text{Qz} + \text{Sil} = \text{Crd} + \text{Kfs} + \text{melt}$$

Lewisan migmatite from Loch Diabaig (Scotland). Partial melting of these ancient gneissic rocks has resulted in this "mixed" rock of pink granitic melt (neosome) and remaining darker protolith (restite).

Image courtesy of M. Whitehouse

Figure 3.41

A Migmatite from Punta Bados, Sardinia (Italy). Contact between melanosome (top) and leucosome (bottom). Melanosome is dominated by biotite, defining a coarse-grained foliation, while leucosome consists of alkali feldspar (gray birefringence, lower than quartz) and quartz (white–gray birefringence). Cross-polarized light, 2× magnification, field of view = 7 mm.

B Migmatite from Massabesic Gneiss complex, New Hampshire (United States). Leucosome with subhedral to euhedral plagioclase (twinned) and quartz (low birefringence), indicating crystallization from a melt. Cross-polarized light, 10× magnification, field of view = 2 mm.

C Migmatite from Massabesic Gneiss complex, New Hampshire (United States). Leucosome with euhedral microcline (large crystal to the right with straight crystal boundaries, gray birefringence) surrounded by quartz (lower birefringence). Cross-polarized light, 10× magnification, field of view = 2 mm.

BOX 3.13 Skarn Formation

Skarn is a metasomatic rock dominated by Ca–Fe–Mg-rich calc-silicate minerals. It is formed when silica-rich fluids derived from magmatic intrusions infiltrate and circulate in carbonate rocks (limestone or dolostone). Skarns are subdivided into two main groups according to composition: *magnesian skarns* (after dolostone) and *calcic skarns* (after limestone). Skarns are typically zoned and have a variable mineral assemblage (magnetite, scheelite, wolframite, etc.) depending on P–T conditions, fluid(s)/protolith(s) composition, and element mobility. Skarns may contain metal ores of economic interest.

Magnesian skarn forms at about 450–750 °C and 0.5–10 kbar. It is generally characterized by a mineral assemblage of forsterite, Mg-spinel, enstatite, monticellite, åkermanite, merwinite, calcite, dolomite, and/or periclase. In this example from Val di Fassa (Italy), idiomorphic clinopyroxene (high birefringence, some near extinction) is associated with calcite. Cross-polarized light, 2× magnification, field of view = 7 mm.

Calcic skarn forms at about 400–650 °C and 0.5–4 kbar. It typically contains garnet, clinopyroxene, vesuvianite, wollastonite, rhodonite, epidote, scapolite, and/or plagioclase. In this example from Vesuvius (Italy), a large zoned vesuvianite crystal (gray birefringence) is associated with prismatic clinopyroxene (high birefringence). Cross-polarized light, 2× magnification, field of view = 7 mm.

High-temperature skarn, formed at about 700–900 °C and 0.5–1.5 kbar, is characterized by the appearance of melilite (gehlenite). In this example from Brad (Romania), equant melilite crystals with anomalous blue birefringence are associated with calcite (pink birefringence). Cross-polarized light, 2× magnification, field of view = 7 mm.

Bibliography

Textbooks

Fettes, D., Desmons, J. (Eds.), 2007. *Metamorphic Rocks: A Classification and Glossary of Terms*. Cambridge University Press.

Passchier, C., Trouw, R., 2005. *Micro-tectonics* (2nd edn). Springer.

Philpotts, A.R., Ague, J. J., 2021. *Principles of Igneous and Metamorphic Petrology*. Cambridge University Press.

Vernon, R., 2018. *A Practical Guide to Rock Microstructure*. Cambridge University Press.

Yardley, B., Warren, C., 2021. *An Introduction to Metamorphic Petrology* (2nd edn). Cambridge University Press.

Metamorphic Facies

Coombs, D., 1989. Prehnite-pumpellyite facies, in *Petrology: Encyclopedia of Earth Science*. Springer.

Eskola, P., 1920. The mineral facies of rocks. *Norsk Geolologisk Tidsskrift* 6, 143–194.

Liou, J., de Capitani, C., Frey, M., 1991. Zeolite equilibria in the system $CaAl_2Si_2O_8$–$NaAlSi_3O_8$–SiO_2–H_2O. *New Zealand Journal of Geology & Geophysics* 34, 293–301.

Waters, D., 2021. Metamorphism of quartzofeldspathic rocks, in *Encyclopedia of Geology* (2nd edn). Elsevier, pp. 465–477.

Metamorphic Rocks, Textures, and Fabrics

Anenburg, M., Katzir, Y., 2014. Muscovite dehydration melting in Si-rich metapelites: microstructural evidence from trondhjemitic migmatites, Roded, Southern Israel. *Mineralogy and Petrology*, 108(1), 137–152.

Bovay, T., Lanari, P., Rubatto, D., Smit, M., Piccoli, F., 2022. Pressure–temperature–time evolution of subducted crust revealed by complex garnet zoning (Theodul Glacier Unit, Switzerland). *Journal of Metamorphic Geology* 40(2), 175–206.

Copley, A., Weller, O., 2022. The controls on the thermal evolution of continental mountain ranges. *Journal of Metamorphic Geology*, doi: 10.1111/jmg.12664

Daczko, N.R., Piazolo, S., 2022. Recognition of melferite: a rock formed in syn-deformational high-strain melt-transfer zones through sub-solidus rocks: a review and synthesis of microstructural criteria. *Lithos* 430–431, 106850.

Daczko, N.R., Stevenson, J.A., Clarke, G.L., Klepeis, K.A., 2002. Successive hydration and dehydration of a high-P mafic granofels involving clinopyroxene–kyanite symplectites, Mt Daniel, Fiordland, New Zealand. *Journal of Metamorphic Geology* 20, 669–682.

Ferrill, D.A., Morris, A.P., Evans, M.A., et al., 2004. Calcite twin morphology: a low-temperature deformation geothermometer. *Journal of Structural Geology* 26(8), 1521–1529.

Gomez-Rivas, E., Martín-Martín, J.D., Bons, P.D., et al., 2022. Stylolites and stylolite networks as primary controls on the geometry and distribution of carbonate diagenetic alterations. *Marine and Petroleum Geology* 136, 105444.

Gratier, J.P., Dysthe, D.K., Renard, F., 2013. The role of pressure solution creep in the ductility of the Earth's upper crust. *Advances in Geophysics* 54, 47–179.

Lazzarotto, M., Pattison, D., Gagné, S., Starr, P., 2020. Metamorphic and structural evolution of the Flin Flon – Athapapuskow Lake area, west-central Manitoba. *Canadian Journal of Earth Sciences* 57(11), 1269–1288.

Lu, J., Seccombe, P.K., 1993. Fluid evolution in a slate-belt gold deposit. *Mineralium Deposita* 28(5), 310–323.

Manzotti, P., Vosse, V. , Pitra, P., et al. 2018. Exhumation rates in the Gran Paradiso Massif (Western Alps) constrained by in situ U–Th–Pb dating of accessory phases (monazite, allanite and xenotime). *Contributions to Mineralogy and Petrology* 173:24.

Meng, Z. Y., Gao, X. Y., Chen, R. X., et al., 2021. Fluid-present and fluid-absent melting of muscovite in migmatites in the Himalayan orogen: constraints from major and trace element zoning and phase equilibrium relationships. *Lithos* 388, 106071.

Neogi, S., Dasgupta, S., Fukuoka, M., 1998. High P–T polymetamorphism, dehydration melting, and generation of migmatites and granites in the Higher Himalayan Crystalline Complex, Sikkim, India. *Journal of Petrology*, 39(1), 61–99.

Newton, R., Tsunogae, T., 2014. Incipient charnockite: characterization at the type localities. *Precambrian Research* 253, 38–49.

Palin, R., Dyck, B., 2021. Metamorphism of pelitic (Al-rich) rocks, in *Encyclopedia of Geology* (2nd edn). Elsevier, pp. 446–456.

Palin, R. M., White, R. W., 2016. Emergence of blueschists on Earth linked to secular changes in oceanic crust composition. *Nature Geoscience* 9(1), 60–64.

Palmeri, R., Godard, G., Di Vincenzo, G., Sandroni, S., Talarico, F.M., 2018. High-pressure granulite-facies metamorphism in central Dronning Maud Land (East Antarctica): implications for Gondwana assembly. *Lithos* 300, 361–377.

Passchier, C., Simpson, C., 1986. Porphyroclast systems as kinematic indicators. *Journal of Structural Geology* 8 (8), 831–843.

Piazolo, S., Passchier, C.W., 2002. Controls on lineation development in low to medium grade shear zones: a study from the Cap de Creus peninsula, NE Spain. *Journal of Structural Geology* 24(1), 25–44.

Platt, J.P., Vissers, R.L.M., 1980. Extensional structures in anisotropic rocks. *Journal of Structural Geology* 2, 397–410.

Powolny, T., Dumańska-Słowik, M., Anczkiewicz, A.A., Sikorska-Jaworowska, M., 2022. Origin and timing of spilitic alterations in volcanic rocks from Głuszyca Górna in the Intra-Sudetic Basin, Poland. *Scientific Reports* 12(1), 1–26.

Reiners, P.W., Carlson, R.W., Renne, P.R., et al., 2018. *Geochronology and Thermochronology*. Wiley.

Schertl, H.-P., Sobolev, A., 2013. The Kokchetv Massif, Kazakhstan: "type locality" of diamond-bearing UHP-rocks. *Journal of Asian Earth Sciences* 63, 5–38.

Skelton, A., Peillod, A., Glodny, J., et al., 2019. Preservation of high-P rocks coupled to rock composition and the absence of metamorphic fluids. *Journal of Metamorphic Geology* 37, 359–381.

Vernon, R.H., Williams, V.A., D'Arcy, W.F., 1983. Grain size reduction and foliation development in a deformed granite batholith. *Tectonophysics* 92, 123–146.

Weber, S., Diamond, L. W., Alt-Epping, P., Brett-Adams, A. C., 2021. Reaction mechanism and water/rock ratios involved in epidosite alteration of the oceanic crust. *Journal of Geophysical Research: Solid Earth* 126(6), e2020JB021540.

Weisenberger, T., Selbekk, R.S., 2009. Multi-stage zeolite facies mineralization in the Hvalfjördur area, Iceland. *International Journal of Earth Sciences* 98, 985–999.

Willner, A., Rötzler, K., Maresch, W., 1997. Pressure–temperature and fluid evolution of quartzo-feldspathic metamorphic rocks with a relic high-pressure, granulite-facies history from the central Erzgebrige (Saxony, Germany). *Journal of Petrology* 38(3), 307–336.

Zheng, Y.-F., Chen R.-X., 2017. Regional metamorphism at extreme conditions: implications for orogeny at convergent plate margins. *Journal of Asian Earth Sciences* 145, 46–73.

Index

Note: Page numbers in bold indicate a primary location where the mineral is described/defined.

α-quartz **28**, 32
β-quartz **28**, 32

absolute grain size **238**
accretionary prism 330, 374
accretionary wedge 364, 382
acicular **171**, 238, **297**, 376
actinolite 71
aegirine **68**, 234, 244
aegirine–augite **68**, 262
agate 32
age dating *see* dating
agpaitic rock 145
åkermanite **109**, 337, 388
Al_2SiO_5 **84**, 117, 313
albite 10, 17, 28, 38, 59, 73, 77, 83, 96, 108, 333, 347, 366, 376, 382
albitization 366
alkali basalt *see* basalt
alkali feldspar 14, 16, 17, 28, 88, 123, 158, 244
allanite **95**, 151
allotriomorphic 21, 171, 175, 238, *see* anhedral
almandine **101**, 106, 117
alnöites 24, 113
alteration 280, 333, 366, 370
  deuteric 114, 197, 215
  diagenetic 320
  hydrothermal 24, 38, 41, 45, 47, 49, 65, 90, 96, 97, 100, 109, 114, 130, 133, 141, 189, 333, 335
  low-temperature 43
  metasomatic 53, 114, 128, 206
  progressive 47
  retrograde 47, 59
alumina polymorph *see* polymorph
amphibole 53, 97, 105, 151, 230, 276, 297
  alkali 236, 272, 274, 382
  calcic 10, 74, 378
  orthorhombic 53, 66
  sodic 57, 59, 234, 244
  uralitic 69
amphibolite **378**, 380, 386
amphibolite facies 13, 36, 38, 53, 86, 93, 97, 100, 147, 149, 330, 337, **339**, 343, 345, 347, 356, 386
amygdale 34, 41, 125, 366
amygdaloidal texture 199
analcime 19, 21, **34**, 276, 333
anatase 141
anatexis 242, 249, 266, **284**, 330
andalusite 32, 38, 45, **84**, 84, 93, 313, 370
andesine 17, 220, 232, 239, 258, 266
andesite 256, **258**, 262, 268
andradite **103**
anhedral **171**, 238
anhedral granular **175**
ankaramites 254
annite 36
anorthite 10, 17, 71, 96, 97, 100, 104, 337, 366, 384
anorthoclase **12**, 13, 16, 17, 274
anorthosite **228**
  Archean 228
  associated with layered mafic complexes 228
  massif-type Proterozoic 228
  oceanic settings 228
  xenoliths 228
antecryst 63, **257**
anthophyllite **53**
anticlockwise path *see* P/T path
antigorite **46**, 47, 49, 51
antiperthite exsolution *see* exsolution
antitaxial vein *see* veins
apatite **145**, 151, 152, 249
aphanitic 158, 164, **168**, 238
aphyric **169**, 278, 280
aplitic texture 175, *see* anhedral granular
aragonite 30, 33
arfvedsonite **57**, 234, 236, 274
argon dating *see* dating
assimilation 234, 239, 242, 244, 248, 250, 274, 276
atoll 294, **324**
  crystal 324
  structure 324
augen **297**, 356
augite **69**, 74, 123, 256, 258, 264, 272
automorphic 171, *see* euhedral

backscattered electron (BSE) 82, 327
baddeleyite 151
banded texture **187**
barite **150**
Barrovian metamorphism *see* metamorphism
Barrovian zone 332, 354
basalt 218, **254**
  alkali 202, 210, 212, 213, 217, **254**
  enriched mid-ocean ridge tholeiitic basalt (EMORB) 239
  flood 222, 266
  mid-oceanic ridge basalt (MORB) 222, 226, 254, 260, 384
  picritic 254, 270
  tholeiitic **254**, 258
  trachy 260
basaltic andesite **256**
basanite **270**
basanitic tephrite 276
bastite 46, 49
belt
  alpine 26
  collisional 329, 330, 339
  greenstone 220, 228
  metamorphic 68, 74, 115, 331
  mountain 249

belt (cont.)
  ophiolite 217
  orogenic 59, 210, 212, 230, 232, 331, 362, 376, 382, 384
  slate 362
beryl **90**, 249
bimodal volcanic suite 266
biogenic aragonite 33
biotite 14, 36, **36**, 77, 79, 84, 86, 93, 101, 117, 123, 332, 370, 376
bladed 53, 238
bleb-like 189
blebby texture **190**
blueschist 376, **382**, 384
blueschist facies 33, 36, 38, 45, 49, 73, 86, 115, 125, 331, **345**, 347
boninite 217, **256**
boudin 294, 306
boudinage **306**
bound water 198
bowlingite **114**
branching 182
breakdown 10, 19, 36, 53, 77, 84, 86, 88, 101, 147, 295, 324, 326, 343
breccia 372
brewsterite **34**
bronzite 206
brookite 141
brucite **133**
Buchan metamorphism *see* metamorphism
Buchan zone 354
bulk composition *see* composition
burial metamorphism *see* metamorphism
Bushveld igneous complex 143, 208, 215, 222, 224
bytownite 17, 220, 224, 226, 228

calcic skarn *see* skarn
calcite 33, 41, 45, 55, 66, 71, 81, 100, 103, 109, 111, 114, 125, **126**, 128, 151, 276, 373, 388
calcite–aragonite boundary 33
calc-silicate rock 337
carbonate group **125**
carbonate polymorph *see* polymorph
carbonation 341
carbonate hornfels *see* hornfels
carbonatite 126, 252, 276
carpholite 91
cassiterite **134**
cataclasite **372**
cathodoluminescence (CL) 82, 151
cavity texture **199**
chabazite **34**
chadacryst 178
chain silicate **53**
chalcedony 32
chalcopyrite 142
charnockite 17, 77
chemical activity 330
chert 32
chiastolite 84, *see* andalusite
chromite series *see* spinel group
  cleavage 294, 307, 318
  continuous 311, 362, 364
  crenulation **309**, 311, 315, 364
  disjunctive 311
  phyllitic 364
  S–C plane 317
  slaty 309, 362, 364
chemical composition *see* composition
chemical gradient *see* gradient
chlorite 36, **41**, 43, 45, 51, 53, 59, 66, 79, 84, 86, 91, 93, 97, 101, 114, 115, 332, 333, 334, 370, 376
chlorite group **41**
chloritoid 38, 45, 86, 91
chromite 143
chrysotile 47
clinozoisite **96**, 100
clockwise path *see* P/T path
closed system 257
closure temperature 360
coarse-grained **164**, 238
coesite **26**, 32, 74, 105, 123, 347, 348
color index 158
comb structure 322
comb texture 182
composition
  bulk 147, 284, 330, 333
  chemical 74, 123, 158, 174
  crystal 195, 269
  fluid 61, 215, 294, 327, 341
  melt 218, 246
  mineral 74, 82, 124, 187, 240, 246, 303, 307, 341, 354
  protolith 294, 296, 349, 354, 356, 358, 384, 388
  rock 240, 284, 330
compositional layering *see* layering
compressional tectonic setting 284
conchoidal fracture 28, 278, 280
confining pressure 330
consertal texture 189, **191**
contact metamorphism *see* metamorphism
continuous schistosity *see* schistosity
continuous zoning *see* zoning
convection 187, 247
convergent plate boundary 258, 264, 266, 278, 354, 356, 364, 374
cooling rate 170
cordierite 14, 38, 53, 77, 79, 84, 93, 266, 368
corona
  kelyphitic 10, 105
  texture **196**, 197, 294, 324, 325
corundum 79, **137**
crack and seal 322
crenulation cleavage *see* cleavage
cristobalite **27**, 193
  α-cristobalite **27**, 32
  β-cristobalite **27**
crustal thickening 244, 284, 352, 378, 380
cryptocrystalline **169**
cryptocrystalling groundmass *see* groundmass
crystal
  cargo **257**
  composition *see* composition
  faces 170, **171**, 195, 268
  form **171**, 238
  growth 167, 170, 174, 189, 195, 268, 294, 299, 313, 322, 327, 360, 364
  lattice 16, 17, 32
  mush 222, 228, 266, 274
  nucleation 93
  populations 257
  shape **171**, 238, 257, 322
  visibility **238**
crystal size distribution (CSD) 217, **268**
crystal tuff *see* tuff
crystallinity 40, 158, **163**, 238
crystallite **169**, 266, 278, 280
crystallization 21, 101, 120, 123, 151, 182, 187, 199, 214, 217, 268, 278, 322, 349
crystal–melt interface 247
cummingtonite **55**, 378
cumulate 10, 142, 187, 192, 202, 204, 206, 208, 210, 212, 213, 217, 228, 239, 258, 262, 378
cumulate texture 63, 212, 215, 228

dacite 258, **264**, 268
dating
  age 145, 151

argon 152
fission track 152
radiometric 360
Rb–Sr 151
U–Pb 151
decarbonation 337, 341
decompression melting *see* melting
decussate texture **298**
deformation 13, 47, 206, 213, 282, 284, 294, 300, 305, 309, 312, 316, 317, 330, 356, 360, 368
deformation lamellae 312
deformation twins **305**, 312
degassing 247
degree of crystallinity 163
dehydration 46, 53, 133, 322, 337, 339, 341, 343, 345, 354, 368, 378, 380, 382
dehydration melting *see* melting
dendritic 170, 182
depletion 218
devitrification 27, 32, 193, 194, 278, 372
devolatilization 322
diablastic texture 298, *see* decussate
diagenesis 319, 333
diagenetic fluid *see* fluid
differentiation 228, 232, 234, 239, 242, 249, 260, 266, 270, 272
diffusion 16, 167, 170, 278, 294, 326, 327
diopside 10, 69, **71**, 71, 74, 81, 109, 111, 128, 202, 204, 210, 213, 215, 373
diorite **220**
discontinuous compositional zoning *see* zoning
disequilibrium condition 196, 360
disequilibrium texture **324**
disjunctive schistosity *see* schistosity
displacive transformations 32
dissolution 17, 125, 150, 174, 213, 214, 246, 319, 324, 327, 341
dissolution–reprecipitation 13, 149, 197, 226, 303
divergent plate boundary 254, 258
dolerite 178
doleritic texture 178
dolomite 66, 71, 111, **128**, 204, 373, 388
dolomitization 320, 321
dry solidus 284
Duluth Complex 167, 226
dunite **208**, 214
dynamic metamorphism *see* metamorphism

eckermannite **57**
eclogite 326, **384**
eclogite facies 36, 86, 149, 330, 331, 339, 343, 345, **347**, 356, 361, 378
electron microprobe (EMP) 82, 247, 327
element mobility 388
Embayment **174**, 264
EMORB *see* basalt
enclave 220, 242, 264, 274
energy-dispersive spectroscopy (EDS) 82
enstatite 28, 41, 46, 47, 76, **77**, 202, 204, 206, 215, 256, 388
entropy 30
epidosite 366
epidote 59, **97**, 333, 376, 382, 388
epidote group **95**, 97
epitaxy 325
equidimensional **171**, 297, 299
equigranular 164, **164**, 212, 238
equilibrium condition 124, 299, 360
erosion 285, 352
eruption 167, 257, 258, 260, 264, 268, 278, 282
eucrite 83
euhedral **171**, 178, 238, 300
euhedral granular **175**
eutaxitic texture **183**, 282
exchange reaction *see* reaction
exhumation 26, 125, 339, 345, 352, 361, 380, 382
exsolution 266, **329**
antiperthite 17
lamellae 74, 76, 77, 191, **192**, 206, 329
perthite 12, 14, 17, 234
subsolidus 189, 215
texture 21, 24, 69, 199
extensional tectonic setting 266, 278, 284
extensional vein *see* veins

fault
cohesive rock 372
gouge 372
incohesive rock 372
fayalite **111**, 266, 280, 368
feldspar **10**, 17, 83
feldspathoid **18**, 158
feldspathoid rock **248**
felsitic **169**
felsitic texture 193
fenitization 68
ferroactinolite **66**
ferroaugite **69**
ferroeckermannite **57**
ferrohedenbergite 280
ferrokaersutite **63**
ferropargasite **61**
ferrorichterite **65**
ferrosilite **77**, 280
fiamme 183, 282
fibrolite 88, *see* sillimanite
fine-grained 164, **164**, 238
filter-pressing process 266
fission track dating *see* dating
flanking structure 306
flint 32
flood basalt *see* basalt
fluid **341**, 384
circulation 40, 303, 320, 322
composition *see* composition
diagenetic 320
hydrous 57, 129, 333, 335, 374, 380
hypersaline 142
inclusion 120, 131, 145, 322
infiltration 294
intergranular 313, 327, 341
late-stage 34, 121, 199
metamorphic 151
ore-forming 131
overpressure 108
oxidizing 59
fluid-present melting *see* melting
fluorite **131**
foid 158, *see* feldspathoid
foid-bearing alkali-feldspar syenite 236
foid-bearing diorite 220, 248
foid-bearing gabbro 222, 248
foid-bearing monzodiorite 230, 248
foid-bearing monzogabbro 230, 248
foid-bearing monzonite 232, 248
foid-bearing potassium feldspar syenite 248
foid-bearing rock **248**
foid-bearing syenite 234, 248
foid-bearing trachyte 262
foid diorite 250
foid gabbro 250
foidite **276**
foid monzodiorite 250
foid monzogabbro 250
foid monzosyenite 250
foidolite **252**

foid rock 248, **250**
foid syenite 250
foliation 152, 294, 303, **307**, 313, 356, 374
continuous **307**, 311
external 300, 313
internal 300, 313
magmatic 186
matrix 300
mylonitic 372
planar 315
spaced **307**, 309, 311
forsterite 28, 41, 46, 47, 51, 71, **111**, 128, 204, 264, 373, 388
fractional crystallization 212, 215, 220, 222, 224, 226, 228, 230, 232, 234, 236, 242, 244, 248, 250, 252, 254, 258, 262, 264, 270, 272, 274, 280
Franklinite 143
frictional heating 372
fringe structure 322, *see* strain fringe

gabbro 73, **222**
gabbroic rock **220**
gahnite 143
galaxite 143
galena 142
garnet 14, 53, 74, 77, 91, 101, 105, 123, 202, 213, 230, 266, 332, 388
garnet group **101**, 107
gedrite **53**
gehlenite 104, **109**, 388
geobarometers 124
geochronology 122, 149, **151**
geothermal gradient *see* gradient
geothermometers 123, 142
glass 26, 158, 178, 190, 238, 254
fragment 183, 282
inclusion 10, 19, 264
shard 183
volcanic 32, 69, 193, 278, 280, 333
glassy groundmass *see* groundmass
glassy rock **278**
glaucophane 38, **59**, 382
glomeroporphyritic texture **176**, 247, 280
gneiss 297, 316, **356**, 386
augen 297, 356
banded 356
orthogneiss **356**
paragneiss **356**
gneissic banding 356
gneissic texture 354
goethite 114
gold 83, 152, 322, 362
gradient
chemical 309, 326
geothermal 152
thermal 108
grain size and shape **297**
Gran Paradiso massif 360
granite **242**
alkali **244**
monzo 244
peralkaline 244
syeno 244
granitic rock **239**
granitic texture 175
granoblastic texture 294, **298**, 299, 358
granodiorite **240**, 264
granofelsic texture 378, 380
granofels 298, **358**
granophyric texture 189, **190**
granular texture **175**, 238, 242, 244, 358
granularity 158, **164**
granulite 299, 326, **380**, 386
granulite facies 23, 61, 77, 81, 93, 99, 149, 228, 330, 339, **343**, 347, 358, 378, 386
graphic texture 189, **190**
greenschist 50, **376**
greenschist facies 13, 23, 36, 38, 43, 45, 49, 53, 115, 228, 330, 335, **337**, 339, 345, 364, 378
greenstone 376
greisen 120, 134
greisenization 141
grossular 43, **104**, 115, 384
groundmass 158, 167, 186, 238
cryptocrystalline 266, 282
glassy 176, 266, 280, 372
holocrystalline 183
grunerite **55**

harmotome **34**
harzburgite 204, **206**, 214, 218
hastingsite 61
hematite 135, 143
hematite–ilmenite series 135
hercynite 143
holocrystalline **163**, 238
holocrystalline groundmass *see* groundmass
hedenbergite 69, **71**
heulandite **34**, 333, 335
haüyne **24**, 276
holohyaline **163**, 238
hornblende 55, **61**, 66, 71, 349
hornblendite 378
hornfels 298, 330, **370**
carbonate **370**
mafic/ultramafic **370**
pelitic **370**
hornfels facies 343, **349**
hotspot 254, 351
hourglass texture 43, 49, 91
hyalo-ophitic 178
hyalopilitic texture **186**
hydration 46, 47, 95, 278, 333, 339, 341, 368, 378
hydrogrossular **104**
hydrothermal metamorphism *see* metamorphism
hydrous fluid *see* fluid
hypautomorphic 171, *see* subhedral
hypersthene 86, 254
hypidiomorphic 21, 171, 175, 238 *see* subhedral
hypocrystalline **163**, 183
hypohyaline **163**, 238

iddingsite **114**
idiomorphic 171, 238, *see* euhedral
ignimbrite 183, 282
ijolite 252
illite 333, 334, 362
ilmenite 77, 135, 140, 141
imaging **82**
impact metamorphism *see* metamorphism
incongruent melting *see* melting
index mineral 32, 86, 88, 332, 354
inequigranular texture **176**, 238
intergranular fluid *see* fluid
intergranular texture 176, **178**, 189, 254
intergrowth 190, 197, 254
symplectic 10, 74, 79, 189, 325, **326**
texture 18, 21, 38, 86, 117, 135, 142, **189**
vermicular 10, 189, 326
interlobate texture 299
intersertal texture 176, **178**
in-situ melting *see* melting
inverted pigeonite 76
iron–titanium oxide mineral 111, 140, 178

isobaric cooling 351
isothermal decompression 351, 380
IUGS 158, 201, 295

jacobsite 143
jadeite 59, **73**, 74, 108, 382
jasper 32, 99

kaersutite **63**
kalonite 120
kalsilite **18**, 19
katophorite **65**
kelyphitic rim 197
kelyphitic texture **197**
kimberlite 26, 36, 71, 74, 79, 105, 113, 126, 139, 202, 208, 213, 215, 228, 329
kinematic indicator 303, 322
kink band 206, **311**
kirschsteinite 113
knee twin 134
komatiite 170, 213
kyanite 32, 45, 51, **86**, 96, 101, 117, 297, 313, 332, 384
K-feldspar 18, 19, 36, 38, 337, 370

labradorescence 228
labradorite 17, 74, 220, 224, 226, 228, 258
lamellar **171**, 354
lamellar structure 19
lamellar texture 189, **190**
large igneous provinces (LIP) 222, 254
late-stage fluid *see* fluid
latite 258, 260, 262, *see* trachyandesite
laumontite **34**, 41, 333, 335
lawsonite 59, 73, **108**, 382
layered mafic complex 202, 204, 206, 208, 210, 212, 213, 214, 215, 217, 222, 224, 226, 339
layered texture **187**
layering 187, 356
  compositional 354, 356
  magmatic 339
lepidoblastic 382
lepidolite 249
leucite 18, **19**, 28, 262, 270
leucitolite 252
leucocratic 158
leucosome 386
leucoxene 141, 366
lherzolite **202**, 206
lineation 294, 297, **315**
  aggregate **315**
  crenulation **315**
  intersection **315**
  mineral **315**
liquid immiscibility 142
lithic tuff *see* tuff
lithium 249
lizardite 46, 47, **49**

mafic/ultramafic hornfels *see* hornfels
maghemite 143
magma mingling 174, 220, 257
magma mixing 174, 197, 199, 220, 234, 239, 242, 244, 246, 257, 258, 264, 280
magmatic layering *see* layering
magmatic zoning *see* zoning
magmatic underplating 232, 239
magnesioarfvedsonite 57
magnesiohastingsite 61
magnesioriebeckite 59
magnesite 51, **129**
magnesium skarn *see* skarn
magnetite 55, 135, 143
magnetite–ulvöspinel series 135, *see* spinel group
majoritic garnet 26
marble 299, 358, **373**
mechanical twins 305, *see* deformation twins
medium-grained **164**, 238
melanite 103, 252, *see* andradite
melanocratic 158
melanosome 386
melilite 109
melilite group 109
melilitite 109, 388
melilitolite 109
melt composition *see* composition
melteigite 252
melting
  decompressional 285
  dehydration 36, 343, 380, 386
  eclogite 210
  fluid present 380
  incongruent 199
  in-situ 224, 372
  partial 77, 88, 93, 101, 108, 202, 206, 213, 218, 230, 232, 234, 242, 244, 248, 250, 254, 258, 260, 262, 264, 266, 272, 274, 284, 332, 343, 349, 356, 380, 386
  polybaric 218
merwinite 109, 388
mesh structure 114
mesocratic 158
mesosome 386, *see* paleosome
metal ores 131, 322, 337, 388
metamictic 95
metamorphic
  assemblage 91
  aureole 53, 84, 137, 349
  condition 28, 84, 125, 306, 316, 322, 325, 330, 360, 376
  crystallization 152, 361
  environment 97, 108, 324, 329
  evolution 300, 313
  facies 32, 51, **330**
  grade 28, 61, 66, 73, 91, 99, 115, 117, 296, 332, 354, 362
  history 300
  petrography 294
  petrology 32, 284, 294
  process 285, 295, 360
  reaction 36, 294, 322, 327, 345
  rocks, 41, 93, **354**, 356, 358, 360, 362, 364, 368, 374, 386
  series **330**, 345
  terranes 73, 152, 326, 339
  texture **297**
  zone 117, **332**
metamorphic fluid *see* fluid
metamorphism 304, 341
  Barrovian 91, 332
  buchan-type (or buchan) 117
  burial 34, 43, 333, 335, 378, 380
  contact 10, 16, 43, 81, 84, 300, 330, 331, 337, 341, 349, 354, 358, 362, 374, 376, 380
  dynamic 354
  hydrothermal 126, 331
  impact 26
  ocean floor 345
  pneumatolytic 31
  post-deformational 117
  prograde 46, 83, 84, 86, 111, 128, 304, 330, 339, 351, 373, 378, 380, 386
  pyro 93
  regional 38, 53, 55, 59, 73, 84, 104, 117, 337, 339, 343, 354, 356, 364, 373, 374, 378, 380
  retrograde 53, 74, 96, 117, 208, 304, 330, 339, 353, 373, 378, 380
  thermal 27, 333, 335

metapelitic rock 285, 294, 331, 373
metasomatism 43, 53, 68, 107, 121, 122, 145, 189, 202, 204, 206, 333, 337, 366
metastable relict 30
metastable state 14, 32
miarolitic texture **199**
mica fish **318**, 372
micaschist 304
microboudinage **306**
microcline **13**, 14, 16, 234, 242
microcrystalline 167, **168**
micrographic texture 189, **190**
microlite **168**, 247, 257, 268, 280
microlithon 307, 309, 311
mid-ocean ridges 202, 218, 228, 335, 366, 368
migmatite 285, 356, **386**
mineral assemblage 264
mineral composition *see* composition
mineral standard 82
monazite **147**, 151
monticellite **113**, 388
monzodiorite 220, **230**, 232
monzogabbro **230**
monzonite **232**
monzonitic texture 232
MORB basalt *see* basalt
mordenite **34**
Mount Lassen volcano 268
mugearite 12
multivariant equilibria 123
muscovite 14, 36, **38**, 40, 41, 84, 86, 88, 93, 101, 117, 337, 370, 382
mylonite 297, 302, 316, **372**
myrmekite **189**
myrmekitic texture **189**

natrolite 21, 24, **34**
necking 306
nematoblastic 297, *see* acicular
neosome **386**
nepheline **21**, 28, 73, 262, 270
nephelinization 21
nephelinolite 252
nickel–nickel oxide (NNO) 135
non-equidimensional **171**
norite **224**
normal zoning *see* zoning
nosean **24**, 276
nucleation **167**, 198, 278, 294, 360
  density 268
  kinetics 246
  rate 178, 268
  site 167, 170

obsidian 238, 266
ocean floor metamorphism *see* metamorphism
ocean island 254, 262, 266, 270
ocellar texture **199**
ocelli 199
oikocryst 178
oligoclase 17, 220, 232, 239, 240, 242, 258, 262, 266
olivine 10, 74, 105, **111**, 114, 123, 129, 204, 213, 276
olivine group **111**, 113
olivine websterite **214**
olivine-rich anorthosite 226, *see* troctolite
omphacite 10, 38, **74**, 210, 347, 384
opacite rim 36, **198**, 258
opal 32
open system 217, 257, 327
ophimottled 178, 181
ophiolite 202, 204, 206, 208, 210, 212, 213, 214, 215, 222, 224, 226, 256, 335, 366, 368
ophitic texture 176, **178**
orbicular texture **188**
orbicule 188
ore-forming fluid *see* fluid
orogenesis 242, 339, 351, 356, 378
orogenic suite 264
oriented texture **182**
orthoclase 13, **14**, 232, 234, 242
orthopyroxene gabbro 224, *see* norite
oscillatory zoning *see* zoning
overgrowth 257, 327
overgrowth texture **196**
oxidizing fluid *see* fluid
oxygen fugacity (fO_2) **135**, 187, 368

P/T paths **351**
  anticlockwise 351
  clockwise 351
palagonite 334
paleosome **386**
palisade texture 26, 348
panidiomorphic 175, *see* euhedral granular
pantellerite 12, 65
paragonite 40, 96, 108
partial melting *see* melting
patchy zoning *see* zoning
pegmatite **249**
pegmatitic 238
pelitic hornfels *see* hornfels
penetrative fabric 307, 315
periclase 128, 133, 388
peridotite **202**, 274, 368
  alpine 202, 206, 208
  family 204, 206, 208
  fertilized 210
  garnet 105, 197, 329
  mantle 204, 218, 220, 226
  olivine 212
  orogenic 105, 215
  spinel 197, 202
perlitic texture 278
perovskite **137**, 151
perthite exsolution *see* exsolution
pervasive schistosity *see* schistosity
phaneritic **164**, 238
phengite 86, 151
phenocryst 158, 167, 257, 266, 278, 297, 339, 349, 356
phlogopite 36, 276
phyllitic cleavage *see* cleavage
picrite 254
phonolite 262, **274**
phonolitic foidite 276
phonotephrite **272**
phillipsite **34**
phyllite 354, 362, **364**
picritic basalt *see* basalt
piemontite **99**
pigeonite 69, **76**, 123, 254, 258
pilotaxitic texture 183
pitchstone **280**
plagioclase **10**, 14, 59, 73, 88, 158, 202, 370, 376
platy 171, 238, 298, 307, 362, 382, *see* lamellar
pleochroic halo 36, 95, 119, 122, 149
plumasite 137
plumbing system 167, 257
plutonic rock 158, 164, **201**, 222, 224, 226, 228, 230, 232
pneumatolytic metamorphism *see* metamorphism
poikilitic texture 21, 176, **178**, 178, 232, 313

poikiloblast **313**, 362
poikiloblastic texture **313**
poikiloclast **313**
polybaric melting *see* melting
polygonal texture **299**
polymorph 28, 30, **32**, 124, 125, 141, 193, 347
  Al_2SiO_5 **32**, 86, 88
  CO_3 33
  SiO_2 **32**
polymorphism **30**
population density 268
porewater 285
porphyritic texture **158**, 167, 176, 238, 280, 356, 370, 382
porphyroblast 213, **300**, 309, 317, 349
  inter-tectonic **300**
  post-tectonic **300**
  pre-tectonic **300**
  syn-tectonic **300**
porphyroclast 297, **302**, 303, 309, 317, 356, 372
  σ (sigma)-type 302
  δ (delta)-type 302
porphyry copper deposit 232, 260
post-deformational metamorphism *see* metamorphism
prehnite 41, **43**, 330, 333
prehnite–pumpellyite facies 41, 43, 97, **335**, 337
pressure 30, 34, 192, 218, 246, 249, 284, 294, 332, 360
pressure fringe 322, *see* strain fringe
pressure shadow 294, **303**, 322
pressure solution 309, 319, 322
pressure–temperature path (PT path) **360**
pressure–temperature–time path (PTt path) 313
prismatic **171**, 188, 298, 354
prograde metamorphism *see* metamorphism
protocataclasite 372
protolith 295, 354, 356, 376, 382, 384
protomylonite 372
pseudobrookite 135
pseudomorph 26, 43, 46, 49, 77, 88, 129, **304**, 345, 347, 382
pseudotachylite **372**
pumice 183, 266, 282
pumpellyite 41, 43, 97, **115**, 330, 333
pyrite 142
pyrochlore 151
pyroclastic deposit 260, 262, 272
pyroclastic rock 183, 317
pyro metamorphism *see* metamorphism
pyrope 79, 86, **105**, 202, 215, 384
pyrophyllite **45**, 91, 313
pyroxene **68**, 214, 349
  alkali 272
  calcic 69, 276
  clino- 71, 105, 204, 388
  low-Ca 76, 206, 254
  ortho- 10, 14, 55, 77, 79, 105, 370
  sodic 68, 234, 274
pyroxenite **210**, 213, 214, 215, 217
  clinopyroxenite **210**
  olivine clinopyroxenite **212**
  olivine orthopyroxenite **217**
  orthopyroxenite **215**
pyrrhotite 142

QAPF diagram 158
quartz 14, **28**, 36, 38, 41, 43, 45, 55, 59, 66, 73, 77, 81, 84, 86, 88, 91, 96, 97, 108, 115, 117, 126, 151, 158, 264, 313, 333, 348, 366, 370, 373, 376, 380, 384
quartz alkali-feldspar syenite 236
quartz diorite 220
quartz gabbro 222
quartzite 299, 374
quartz monzodiorite 230
quartz monzogabbro 230
quartz monzonite 232
quartz syenite 234

radiate texture **193**
radiogenic decay 360
radiometric age 151, 360
rapakivi texture **197**
Rb–Sr dating *see* dating
reaction
  exchange 19, 123
  rim 196, 264, 325
  texture 324
reconstructive transformations 32
recrystallization 74, 77, 147, 210, 213, 297, 309, 349, 374
refertilization 214, 218
reflecting light microscopy 135, 140, 142, 143
regional metamorphism *see* metamorphism
relative grain size **238**
resorption **174**, 257, 274, 280
retrograde metamorphism *see* metamorphism
reverse zoning *see* zoning
rhyodacite 260
rhyolite 262, **266**
ribbon 316
richterite **65**, 244
riebeckite **59**, 234, 236, 262, 274, 382
rift zone 331, 351
rock composition *see* composition
ruby 137
rutile 123, 135, **141**, 151

saccaroid texture 126
sagenite 141
sanidine 14, **16**, 193, 262, 266, 272, 274
sanidinite 330, 337, 349
sapphire 137
sapphirine **79**, 380
saussurite 100, 222
saussuritization 96, 97
S-C fabrics **317**
scapolite **23**, 388
scheelite 388
schist **354**, 356, 364
schistosity **307**, 309, 354, 358, 376, 378, 382
  continuous 311
  disjunctive 311
  pervasive 296
  spaced 296
secondary twins 305, *see* deformation twins
sector zoning *see* zoning
sedimentary stylolites *see* stylolites
seriate texture 164, **176**, 238
sericite **40**
serpentine 129, 208, **368**
serpentine group **46**, 368
serpentinization 368
shear sense indicator **317**, 318
shear zone 46, 317, 372
shear sense indicator **317**, 318
sheet silicate **36**
siderite **130**
sieve texture **170**, 264, 280
silica oversaturated 27, 158, 248, 250, 254, 262
silica saturated 16, 28, 57, 158, 222, 262, 282
silica undersaturated 16, 21, 24, 63, 109, 139, 141, 222, 248, 250, 252, 262, 270, 276, 282
silica polymorph *see* polymorph
sillimanite 14, 32, 38, 84, **88**, 101, 297, 332
sillimanite isograd 373
skarn 23, 95, 119, 134, 337, **388**

skarn (cont.)
calcic skarn 388
magnesian skarn 388
skeletal texture **170**, 269
slate **362**, 364
slaty cleavage *see* cleavage
smectite 114
sodalite group **24**, 262, 276
solid solution 12, 16, 21, 23, 24, 41, 66, 71, 74, 77, 106, 109, 111, 113, 135, 143, 195, 300, 326, 351
solvus equilibria 123
spaced schistosity *see* schistosity
spessartine **106**
sphalerite 142
spherulite 193, 194
spherulitic texture **193**, 278
spilite 366
spinel group **143**
chromite series **143**, 215
magnetite series **143**
spinel series 41, 79, 105, **143**, 202, 214
spinifex texture 170
spodumene 249
spotted slate 349, 362
staurolite 53, 86, 91, 101, **117**, 332
staurolite isograd 373
stilbite **34**, 333
snowball structure 300
stilpnomelane 50
stishovite 32
strain fringe **322**
strain shadow 300, **303**
stretching 306
stylolites **319**, 320
sedimentary 319
tectonic 319
subduction zone 95, 208, 212, 232, 240, 256, 258, 262, 264, 270, 331, 339, 345, 358, 364, 368, 374, 376, 384
subhedral **171**, 238, 300
subophitic texture 176, **178**, 254
sub-greenschist facies 333, 335, 345, *see* zeolite facies
subhedral granular **175**
sulfide 142
supercooling 167, 182
surface energy 170
swallow-tail texture **170**, 247, 269
syenite 232, **234**, 262
alkali feldspar syenite **236**
peralkaline syenite 234, 236
syenitic rock **230**
symplectic texture **189**
symplectites **189**, 294
synneusis **247**
syntaxial vein *see* veins

tabular **171**, 188, 238, 298, 354
talc **51**, 59, 79, 86, 111, 114, 128
tantalite 249
tectonic denudation 285
tectonic stylolites *see* stylolites
temperature 30, 34, 192, 218, 246, 249, 284, 294, 332, 341, 360
tephriphonolite **272**
tephrite **270**
tephritic foidite 276
thermal conductivity 47
thermal gradient *see* gradient
thermal metamorphism *see* metamorphism
thermobarometry **123**, 351
tholeiitic basalt *see* basalt
thorite 151
titanite **119**, 151
titanohematite series 135
titanomagnetite series 135
Tonalite 220, **239**
topaz **120**, 266
total alkali versus silica (TAS) diagram 158
tourmaline **121**, 249
tourmalinization 121
trachyandesite 232, **260**, 262
trachy basalt *see* basalt
trachyte 234, 260, **262**
trachytic texture **183**
trachytoid texture **183**
tremolite **66**, 71, 111, 128
tremolite isograd 373
trevorite 143
tridymite **31**, 193, 262
α-tridymite 32
β-tridymite 32
tristanite 260
troctolite **226**
tuff 258, **282**, 364
crystal 282
lithic 282
vitric 282
welded 282
tuffaceous rock **282**
tuffaceous sediment 34
tweed structure 14

ultracataclasite 372
ultra-high pressure 26, 329, 330, 339, 343, 347
ultra-high temperature 79, 332
ultramafic complex 178
ultramafic rock 188, **202**, 368
ultramylonite 372
ulvöspinel 135, 143
undercooling 167, 190, 246, 268
undulose extinction 206, 298, 300, 312, 316
univariant equilibria 123
U–Pb dating *see* dating
uralite 61, 66, 68, 71, 77, 222
uraninite 151
urtite 252
uvarovite **107**

variolitic texture 193, **194**
vaterite 33
veins 46, **322**
antitaxial **322**
extensional 306
syntaxial **322**
vermicular *see* intergrowth
vesicular **199**, 370
very coarse grained **164**
vitric tuff *see* tuff
vitrophyre 278, 280
vitrophyric texture 176
volcanic
arc 212, 256, 330
rock 158, 198, 272, 276
sequences 256

wairakite 335
web texture 313
websterite **213**
wehrlite **204**
wehrlitization 204
welded tuff *see* tuff
whiteschist 51, 86
wolframite 388
wollastonite **81**, 126, 128, 373, 388

xenocryst 93, **257**
xenocrystic 101, 105
xenolith 16, 26, 27, 63, 69, 74, 79, 88, 93, 101, 105, 108, 137, 204, 206, 208, 210, 212, 213, 214, 215, 217, 242, 270, 274, 329, 384
xenomorphic 171, *see* anhedral
xenotime **149**, 151
X-ray element mapping 327

zeolite facies 333
zeolite group **34**
zircon **122**, 123, 149, 152, 249
zoisite 43, **100**, 108, 115
zoning **327**
  concentric 10, 77, 83, 97, 121, 174, 327, 351
  continuous **195**
  discontinuous **195**
  magmatic **195**
  normal 77, **195**, 264
  oscillatory **195**, 246, 264, **327**
  patchy 195, 327
  reverse **195**, 264
  sector 195